222 **Automotive lighting technology**
222 Technical demands
223 Legal framework
224 Development of lighting technology
229 Physical principles
242 Front lighting system: Components
262 Rear lighting system: Components
265 Interior lighting system: Components
268 Instrument clusters
270 Display types
272 Special-purpose lamps

274 **Wiper and washer systems**
274 Windshield cleaning
281 Rear-window cleaning
282 Headlamp cleaning systems
283 Components

288 **Automotive microelectronics**
288 Overview
290 Demands on electronic systems
290 History of development

292 **Basic principles of semiconductor technology**
292 Terminology
292 Electrical conductivity

296 **Electronic components**
296 Passive components
297 Semiconductor components

318 **Microcontrollers**
318 Microcontroller developments
320 Microcontroller components
321 Design and operating concept
325 Microcontroller memories
326 Microcontroller peripheral modules
328 Microcontroller busses

330 **Manufacture of semiconductor components and circuits**
330 Semiconductor components
335 Conventional printed-circuit boards
340 Film and hybrid circuits

342 **Glossary and tables for automotive microelectronics**

350 **Automotive sensors**
350 Basics

358 **Position sensors (travel/angle)**
358 Characteristics, Measured variables: Overview
359 Measuring principles
Application examples:
374 Sensor-plate potentiometer
375 Throttle-valve sensor
376 Half-differential short-circuiting-ring sensors
377 Fuel-level sensor
378 Accelerator-pedal sensors
380 Steering-wheel-angle sensors
382 Axle sensors
383 Ultrasonic sensors

386 **Speed and rpm sensors**
386 Measured variables, Measuring principles
Application examples:
389 Relative rpm and speed measurement
395 Absolute rotating-speed measurement
396 Radar sensors
398 Inductive engine-speed sensors
399 Rotational-speed (rpm) sensors and incremental angle-of-rotation sensors
400 Hall-effect phase sensors
402 Wheel-speed sensors
404 Gearbox-rpm sensors
405 Nozzle holder with needle-motion sensor
406 Induction-type sensors for transistorized ignition
407 Hall-effect sensors for transistorized ignition
408 Piezoelectric "tuning-fork" yaw-rate sensor
409 Piezoelectric "oscillating drum" yaw-rate sensors
410 Micromechanical yaw-rate sensors

Contents

412 Acceleration sensors and vibration sensors
412 Measured variables, Measuring principles
Application examples:
418 Hall-effect acceleration sensors
419 Micromechanical bulk silicon acceleration sensors
420 Surface micromechanical acceleration sensors
422 Piezoelectric acceleration sensors
423 Piezoelectric knock sensors

424 Pressure sensors
424 Measured variables, Measuring principles
Application examples:
427 Thick-film pressure sensors
428 Micromechanical pressure sensors
431 High-pressure sensors

432 Force sensors and torque sensors
432 Measured quantities
433 Measuring principles
Application examples:
440 Occupant classification (OC) and detection of child's safety seat

442 Flow meters
442 Measured quantities
444 Measuring principles
Application examples:
448 Sensor-flap (impact-pressure) air-flow sensor LMM
450 Hot-wire air-mass meter HLM
451 Hot-film air-mass meter HFM2
452 Hot-film air-mass meter HFM5

454 Gas sensors, concentration sensors
454 Measured quantities, Measuring principles
Application examples:
457 Air-quality sensors
458 Two-step Lambda oxygen sensors
462 LSU4 planar broad-band Lambda oxygen sensors

464 Temperature sensors
464 Measured quantities
465 Measuring principles
Application examples:
474 Temperature sensors

476 Prospects
476 Development trends
476 Sensor examples

480 Sensor-signal processing
480 Signal conditioning (Evaluation IC)
481 Examples of application

486 Data processing in the vehicle
486 Requirements
486 Microcomputer
486 Electronic control unit (ECU)
489 Complete system

490 Data transfer between automotive electronic systems
490 System overview
490 Serial data transfer (CAN)

492 Index of Technical terms
492 Technical terms
502 Abbreviations

Background information

Starter batteries
109 Battery history

Alternators
163 The history of the generator/alternator

Lighting technology
230 Luminous intensity
233 Luminous flux
242 Luminous intensity and range

Wiper and washer systems
287 A brief history of windshield cleaning

Manufacture of semiconductor components and circuits
334 Micromechanics

Automotive sensors
357 Miniaturization
384 Ranging radar
475 The Bosch Boxberg Test Center

Authors

Vehicle Electrical System
Ing. (grad.) R. Leunig (Vehicle Electrical Systems),
Dipl.-Ing. A. Kerber, and
Dipl.-Ing. M. Gentzsch,
DST Dräxlmaier Systemtechnik GmbH,
Vilsbiburg (Wire Dimensioning),
Dipl.-Ing. W. Gansert (Plug-In Connections)

Electromagnetic Compatibility (EMC) and Interference Suppression
Dr.-Ing. W. Pfaff

Starter Batteries
Dr.-Ing. G. Richter, VB Autobatterie GmbH,
Hannover

Traction Batteries
Dr.-Ing. R. Schenk (Electrical Drives),
Dipl.-Ing. D. Übermeier, VB Autobatterie GmbH,
Hannover;
Dr. rer. nat. U. Köhler, NBT GmbH, Hannover
(Traction Batteries)

Alternators
Dipl.-Ing. R. Meyer

Starter Motors
Dr.-Ing. H. Braun

Workshop Technology for Batteries, Alternators, and Starters
Dr.-Ing. G. Richter, VB Autobatterie GmbH,
Hannover (Battery Maintenance)
Dipl.-Ing. (FH) F. Zauner (Battery Chargers)
H. Weinmann (Alternators and Starter Motors)

Lighting Technology
Dr.-Ing. M. Hamm,
Dipl.-Ing. T. Spingler,
Dipl.-Ing. D. Boebel,
Dipl.-Ing. B. Wörner,
Dipl.-Ing. H.-J. Lipart,
all at Automotive Lighting, Reutlingen

Windshield and Rear-Window Cleaning
Dipl.-Ing. (FH) A. Geis

Microelectronics in the Motor Vehicle
Dr. rer. nat. Ulrich Schaefer
(Basics, Components, Microcontrollers),
Karlheinz Isert
(Manufacturing Printed-Circuit Boards),
Dr. rer. nat. Ulrich Goebel
(Manufacturing Hybrid Circuits)

Sensors in Motor Vehicles
Dr.-Ing. E. Zabler
(Contributions: Sensors in Motor Vehicles,
Features, Measured Quantities, Test Principles,
Examples of Sensors),
Dipl.-Ing. J. Berger
(Various Sensors for Diesel-Engine Management),
Dipl.-Ing. A. Herforth (Fuel-Level Sensor),
Dr. M. Harder (Axle Sensors),
Dipl.-Ing. (FH) H. Arnold (Ultrasonic Receivers),
Prof. Dr. rer. nat. H. Winner
(Distance-Control Radar),
Dipl.-Ing. J.Töpfer
(Transmission Hall-Effect Sensors),
Dipl.-Ing. (FH) R. Giesel (OC Sensor Mat),

Data Processing in Motor Vehicles
Dr. rer. nat. V. Denner

Data Transmission between Electronic Systems
Dr.-Ing. M. Walther

and the Editorial Team in collaboration with the relevant specialist departments at Bosch.

Unless otherwise stated, the above are employees of Robert Bosch GmbH, Stuttgart.

Vehicle electrical systems

This is an integral part of the history of Bosch itself. Over the span of a hundred years it demonstrates the development of the electrical and electronic equipment in the automobile. A development in which Bosch has always played a leading role, starting with its high-voltage magneto in 1902, and proceeding up to the present day with the introduction of the Electronic Stability Program (ESP) in 1995.

The history of vehicle electrical systems

Magneto-ignition system
In 1902 the first electrical system was installed in a vehicle in the form of the magneto-ignition system. Although this consisted of the magneto itself, an ignition distributor, ignition coils, and spark plugs (Fig. 1), there was no question of regarding these few components together with their cables and wires as an on-board electrical system.

Complete automotive systems
It only took another 11 years before and Bosch had the first complete automotive electrical system ready for installation (Fig. 2). This comprised the magneto-ignition system with spark plugs, starter, DC generator, headlamps, battery, and regulator switch. It marked the start of progress towards a genuine on-board system.

The birth of the vehicle electrical system
The already very extensive scope of the electrical installation in the vehicle can be seen from the Bosch wiring diagram from 1958 (Fig. 3). The devices are not represented by symbols but by schematized line drawings. The arrangement of the electrical lines has started to resemble a wiring diagram. With the introduction of the electronically controlled D-Jetronic gasoline-injection system in 1967, electronics entered the automotive world.

Today's vehicle electrical system
In the past few years the increasing complexity and scope of the vehicle's equipment has resulted in a rapid rise in the number of electrical loads. Formerly, the starter, the ignition, and the lighting system were the main points of interest. In the course of the years, though, the vehicle's basic equipment was extended continually by the addition of electronic ignition and fuel-injection systems, together with comfort and convenience systems with their wide range of drive motors, and safety and security systems.

Today the electrical and electronic equipment in the automobile has become so extensive that it would be totally impossible to show this using such a diagram as that in Fig. 3. The diagrammatic presentation showing equipment and wiring in their relative positions to each other was abandoned in favor of a method which showed the on-board electrical system in the form of schematic diagrams. The idea was to present a schematic overview of the installation featuring a high level of clarity, notwithstanding the large number of individual systems and components concerned (refer to the Chapter "Schematic diagram of a gasoline-engine passenger car").

On conventional on-board networks, the interplay of the different subsystems is defined by the allocation of individual lines/wires to individual signals.

Today the immense increases in the amount of data exchanged between the electronic components can no longer be handled with such wiring methods, the wiring harness has become too complicated, and too many pins are needed on the ECUs. This applies particularly on luxury-class vehicles. These problems have been solved by using CAN (Controller Aided Network), a special serial bus system developed by Bosch for automotive applications.

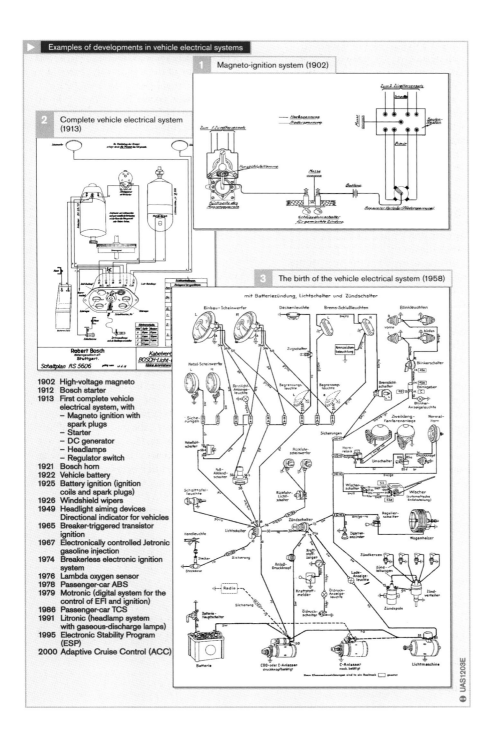

Examples of developments in vehicle electrical systems

1 Magneto-ignition system (1902)

2 Complete vehicle electrical system (1913)

3 The birth of the vehicle electrical system (1958)

1902 High-voltage magneto
1912 Bosch starter
1913 First complete vehicle electrical system, with
 – Magneto ignition with spark plugs
 – Starter
 – DC generator
 – Headlamps
 – Regulator switch
1921 Bosch horn
1922 Vehicle battery
1925 Battery ignition (ignition coils and spark plugs)
1926 Windshield wipers
1949 Headlight aiming devices Directional indicator for vehicles
1965 Breaker-triggered transistor ignition
1967 Electronically controlled Jetronic gasoline injection
1974 Breakerless electronic ignition system
1976 Lambda oxygen sensor
1978 Passenger-car ABS
1979 Motronic (digital system for the control of EFI and ignition)
1986 Passenger-car TCS
1991 Litronic (headlamp system with gaseous-discharge lamps)
1995 Electronic Stability Program (ESP)
2000 Adaptive Cruise Control (ACC)

Electrical power supply in conventional vehicle electrical systems

The electrical system of a motor vehicle consists essentially of a means of storing electricity (battery), a means of generating electricity (alternator) and number of electrical consumers (electric/electronic equipment and devices).

Using the electricity stored in the battery, the vehicle's engine is started by the starter motor (consumer) and, once it is running, its functional parameters adjusted to suit the operating conditions by the engine-related control units (consumers).

When the engine is running, the alternator (electricity generator) supplies electricity which, depending on the voltage level in the electrical system (determined by the alternator speed and the consumers drawing current), is ideally enough to power the consumers and charge the battery as well. If the consumers that are switched on create a larger current draw than the amount of power being supplied by the alternator, the electrical system voltage drops to the battery voltage level and the battery is discharged accordingly. By carefully matching battery, starter motor, alternator and electrical consumers to one another, a balanced electrical system has to be created so that the battery is always sufficiently charged to

- ensure that the engine can reliably be started, and
- power certain electrical consumers for an appropriate period of time while the engine is switched off.

The optimization criteria for the most favorable combination are not only lightweight and small dimensions but also low fuel consumption and, therefore, lower emission arising from the energy conversion in the alternator. The following parameters are important factors in this connection:

Minimum starting temperature
The lowest temperature at which the engine can be started is dependent, among other things, on the battery (capacity, low-temperature test current, charge level, internal resistance, etc.) and the starter motor (type, size and power). If, for example, the engine is to start at –20 °C, the battery must have a minimum charge level, p (Figure 1).

Alternator current output
The current output of the alternator is speed-dependent. At engine idling speed, n_L, the alternator can only generate a certain proportion of its rated current (Figure 2) with the transmission ratios (crankshaft to alternator drive pulley) commonly used of between 1:2 and 1:3. The rated current relates to the current produced at an alternator speed of 6,000 rpm.

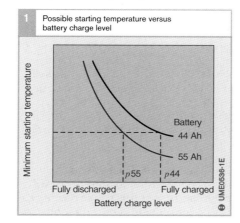

1 Possible starting temperature versus battery charge level

Minimum starting temperature

Battery 44 Ah
55 Ah
p 55 p 44
Fully discharged   Fully charged
Battery charge level

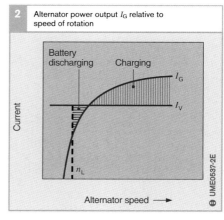

2 Alternator power output I_G relative to speed of rotation

Battery discharging | Charging
I_G
I_V
Current
n_L
Alternator speed →

Fig. 1
p Minimum charge level

Fig. 2
I_V Equipment current draw
n_L Engine idle speed

1 Installed electrical consumers taking account of average duty period (examples)

Electrical consumer	Power consumption	Average power requirement
Motronic, electric fuel pump	250 W	250 W
Radio	20 W	20 W
Side-marker lamps	8 W	7 W
Low-beam headlamps	110 W	90 W
License-plate lamps, tail lamps	30 W	25 W
Indicator lamps, instruments	22 W	20 W
Rear-window heater	200 W	60 W
Interior heater, fan	120 W	50 W
Electric radiator fan	120 W	30 W
Windshield wipers	50 W	10 W
Stop lamps	42 W	11 W
Turn signal lamps	42 W	5 W
Front fog lamps	110 W	20 W
Rear fog lamps	21 W	2 W
Total installed power requirement	1,145 W	
Average power requirement		600 W

Table 1

If the equipment current draw I_v within the vehicle's electrical system is greater than the alternator current output I_G (e.g. at engine idle speed), the battery will be discharged. The electrical-system voltage will drop to the voltage level of the battery under load. If the equipment current draw I_v is less than the alternator current output I_G, a proportion of the excess current flows to the battery and acts as a charge current, I_B.

Vehicle operation
The speed at which the alternator is driven by the engine (and therefore the current output of the alternator) is dependent on the vehicle's operation.

The cumulative frequency diagram for engine speed indicates how frequently a specific engine speed is reached or exceeded for a specific driving pattern (Figure 3).

In urban driving situations in commuter traffic, a car's engine is running at idle speed for a large part of the time due to the frequency of stops at traffic lights and high traffic density. When driving on a freeway, the amount of time that the engine is at idle speed is generally very small. A bus operating on a local passenger service spends additional amounts of time at idle speed because of stopping at the bus stops. Another factor that has a negative effect on the battery's charge balance are electrical consumers that are operated when the engine is switched off (e.g. those that remain switched on at the terminus). Long-distance coaches generally spend only a small proportion of the time with the engine at idle speed, but on the other hand may have long periods at stops when equipment is being operated with the engine switched off.

Consumer power requirements
The various electrical consumers have differing duty periods (lengths of time they are switched on). A distinction is made between

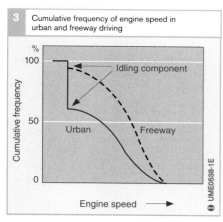

3 Cumulative frequency of engine speed in urban and freeway driving

continuous-duty consumers (ignition system, fuel-injection system, etc.), extended-duty consumers (lights, heaters, etc.) and short-duty consumers (turn-signal indicators, stop lamps, etc.). The use of a number of consumers is dependent on the time of year (air conditioning in the summer, seat heaters in the winter). The frequency with which electric radiator fans switch on depends on the outside temperature and the driving pattern. Commuter journeys in winter are made with the lights on most of the time. The consumer power requirements are not constant over any one journey. They are generally very high in the first few minutes after starting and then diminish (Figure 4):

1. In the future, windshield heaters will require up to 2 kW for 1...3 minutes after the engine is started in order to de-ice the windshield.
2. The secondary-air pump, which pumps additional air into the exhaust ports to burn unused fuel in the exhaust, runs for about 3 minutes after the engine is started.
3. Other electrical consumers such as heaters (rear window, seats, mirrors, etc.), fans and lights are switched on for longer or shorter periods depending on the situation, while the engine management system is in operation all the time.

Engine fuel consumption

Only a small proportion of the fuel used by the engine is required to drive the alternator and transport the weight of the starter motor, battery and alternator (approx. 5 % for a medium-sized car).

Average fuel consumption for a journey of 100 km: for transporting weight of 10 kg approx. 0.1 l, for 100 W of power output to drive alternator approx. 0.1 l.

Alternators with higher mid-range efficiency levels therefore contribute to the engine's fuel efficiency despite being slightly heavier in weight.

Battery charging voltage

Due to the chemical processes that take place in the battery, the charging voltage has to be higher at low temperatures and lower at high temperatures. The gas/voltage graph shows the maximum voltage at which the battery does not produce gas.

Electrical consumers need the supply voltage to be as stable as possible. For light bulbs, the voltage tolerances must be kept tight so that the bulb life and light output are within the specified limits. The voltage regulator prevents the voltage rising above a certain level if the possible alternator current I_G is greater than the sum of the consumer current requirement I_V and the possible battery charging current I_B. The voltage regulator is normally mounted on the alternator.

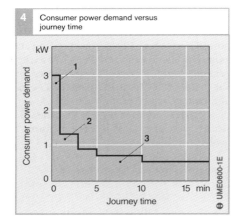

Fig. 4
1 Windshield heater
2 Secondary-air pump
3 Heater, fan, engine management, etc.

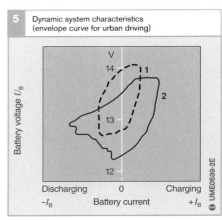

Fig. 5
1 With large alternator and small battery
2 With small alternator and large battery

If there is a wide divergence between regulator and battery-acid temperatures, it is better to mount the regulator-temperature sensor directly on the battery. The voltage drop across the alternator-battery charging cable can be compensated by a regulator which measures the actual voltage at the battery.

Dynamic system characteristics

The interaction between battery, alternator and consumers, and the factors temperature, engine speed and engine-alternator transmission ratio produces the system curve (Figure 5). It is specific to every different combination and every set of operating conditions, and is therefore a dynamic curve. The dynamic system curve can be measured across the battery terminals and recorded using suitable recording equipment.

Battery-charge equation

The battery-charge equation must take account of the factors outlined above. Using a computer program, the battery charge level at the end of a typical driving sequence is determined. A typical sequence for a car is "commuter traffic" (low frequency of high engine speeds) combined with "winter driving" (low battery charging current). For cars with air conditioning, summer conditions may be more demanding (high power consumption). The battery charge level at the end of a driving cycle should be sufficient to restart the engine at ambient temperature.

Electrical system simulation

Contrary to the momentary view using a battery-charge equation, model-based computer simulations can calculate electrical system status at any time in a trip. They can also include electrical-system management systems and assess their effectiveness.

As well as computing the battery-charge equation, the system-voltage curve and the battery-charge cycle can be recorded during a trip. Calculations using computer simulations are always useful in situations where electrical-system topologies need to be compared in order to assess the affect of highly dynamic consumers, or where consumers with a short-duty period are involved.

Electrical system design

The nature of the wiring between alternator, battery and consumers also affects the voltage level and, therefore, the battery charge level. If all consumers are connected close to the battery, the total current ($I_G = I_B + I_V$) flows along the battery-charging cable.

Due to the high voltage drop, the charging voltage is then lower. If, on the other hand, all consumers are connected close to the alternator, the voltage drop is lower and the charging voltage higher. However, this may be undesirable for consumers that are sensitive to voltage peaks or ripple (electronic circuits). It is advisable, therefore, to connect consumers that use more power and are not sensitive to voltage fluctuations close to the alternator, and those that use less power and are sensitive to voltage fluctuations close to the battery (Figure 6).

The use of appropriate cable thicknesses and good connections helps to minimize voltage drops.

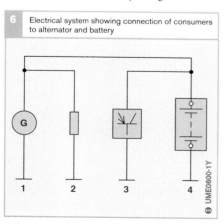

6 Electrical system showing connection of consumers to alternator and battery

Fig. 6
1 Alternator
2 Consumer with high power consumption
3 Consumer with low power consumption
4 Battery

Future electrical systems

Two-battery systems

In 12-volt electrical systems used on standard production vehicles, the battery is a compromise between partially conflicting demands. It has to be dimensioned both for starting the engine and for supplying the electrical system. During the starting sequence, the battery is subjected to high current loads (300...500 A). The associated voltage drop has a negative effect on certain electrical consumers (e.g. devices with microcontrollers) and should be kept as small as possible. Once the vehicle is being driven, however, only relatively low currents are flowing. The capacity of the battery is then decisive. Both properties cannot be optimized simultaneously on the same battery.

In future electrical systems with two batteries (starter battery and general power-supply battery), the battery functions of providing high power for starting and supplying the electrical system when the engine is running will be separated in order to avoid the voltage drop in the electrical system when starting the engine (Figures 7 and 8).

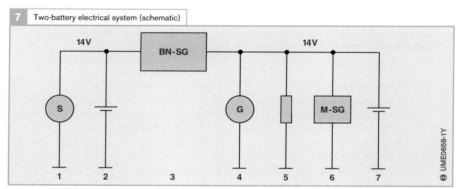

Fig. 7
1 Starter motor
2 Starter battery
3 Electrical-system control unit
4 Alternator
5 Electrical consumer
6 Engine control unit
7 General power-supply battery

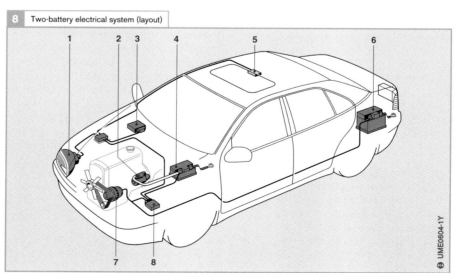

Fig. 8
1 Lighting system (electrical system)
2 Starter motor
3 Engine management (electrical system)
4 Starter battery
5 Other electrical-system consumers (e.g. power sunroof)
6 General power-supply battery
7 Alternator
8 Charging/isolating module

Engine control unit

The control unit (BN-SG) in the two-battery electrical system separates the starter battery and the starter motor from the rest of the electrical system. In this way, it prevents the voltage drop caused by the starting sequence having an effect on the electrical system as a whole.

When the vehicle is not being driven, it also prevents the starter battery from being discharged by consumers that remain switched on when the engine is not running.

By clear separation of the starting system from the remainder of the electrical system, there are theoretically no limits to the voltage level within the starting system. Consequently, the starter battery can be recharged via DC/DC converters in the most effective and most suitable manner, i.e. in the shortest possible time.

If the general power-supply battery has run out of charge, the control unit is capable of temporarily connecting the two sections of the electrical system so as to make the fully charged starter battery available to the system as a whole. In addition, a message to that effect can be sent to the on-board computer via a CAN interface.

In another possible configuration, the control unit would connect only the start-related consumers to whichever battery was fully charged.

Starter battery

The starter battery has to be capable of supplying a high current for a limited period of time (starting sequence). As it is relatively small, it can be fitted close to the starter motor and connected to it by a short lead. Its capacity is similarly lower.

General power-supply battery

The general power-supply battery is intended to supply only that part of the electrical system that does not include the starter motor. It provides relatively low currents (e.g. approx. 20 A for the engine management system) but has a high cyclic capability, i.e. assuming it has a suitable capacity and has not been discharged beyond a certain limit, it can supply and then re-store large amounts of power. Dimensioning is based essentially on the required capacity reserves for consumers that need to be operated without the engine running (e.g. parking lights, hazard warning flashers), the quiescent current consumers and the minimum permissible charge level.

42-V electrical systems

Within a vehicle's electrical system, the complex interaction of electricity generators and consumers has to be coordinated. The engine fuel is used not only to provide motive power but also to generate electrical energy. The aim of development efforts is to make all forms of energy generation and consumption within the vehicle as efficient as possible. To this end, hydraulic energy consumers (such as power steering systems) can be replaced by electrical consumers. The latter are more efficient at using the energy consumed.

Greater safety, economy and comfort also result in the use of more electrical equipment. Examples include heated windshields for greater safety and PTC interior heating for greater comfort with economical diesel engines. The peak electrical power demand will thus increase nearly five times over to more than 10 kW. Equipment power requirements of that order are beyond the capabilities of 14-volt electrical systems.

The VDA (German Automotive Industry Federation) has thus worked together with suppliers, universities and international committees to produce a standard for a 42-volt electrical system for future motor vehicles.

Dual-voltage electrical system with electrical energy management (EMM)

An initial step towards a higher-voltage electrical system is the dual-voltage system which has 14-V and 42-V subsystems. The advantage is that cheap existing components for 14-volt systems can continue to be used. In such a system, the alternator supplies the 42-V high-power consumers directly. The remaining consumers are connected to the 14-V system via a DC converter. Separate batteries for each part of the system can be dimensioned so that combined, they are not significantly heavier than the batteries used at present (Figures 9 and 10).

An electrical energy management (EEM) system coordinates the interaction between alternator, voltage transformer, batteries and electrical consumers when the vehicle is in use. When the vehicle is not in use, the EEM monitors the batteries and switches off consumers and quiescent-current devices as soon as the battery charge reaches a critical level. The EEM coordinator regulates the en-

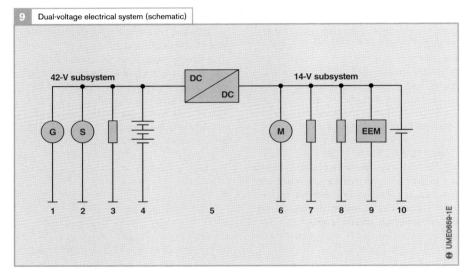

Fig. 9
(Data leads not shown)
42-V subsystem:
1 Alternator
2 Starter motor
3 Electrical consumer
4 Battery
5 Converter (42V/14V)
14-V subsystem:
6 Electric motor
7/8 Consumer groups
9 EEM
10 Battery

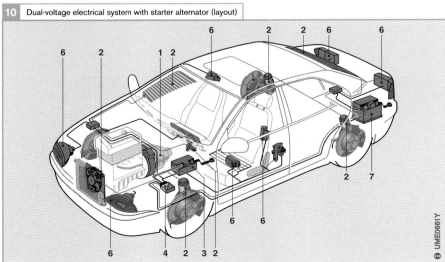

Fig. 10
(Data leads not shown)
42-V subsystem:
1 Starter alternator
2 Consumer groups (42 V)
3 Battery
4 EEM with integral converter (42/14V)
14-V subsystem:
5 Electric motor
6 Consumer groups (14 V)
7 Battery

Table 1. Selected circuit symbols (continued)

Devices for automotive applications

Symbol	Symbol	Symbol
Speaker	Piezoelectric sensor	Velocity sensor (v)
Voltage stabilizer (U const.)	Resistive position indicator	ABS wheel-speed sensor (n)
Inductive sensor, reference-mark controlled	Air-flow sensor (Q_L)	Hall sensor
Flasher, pulse generator, interval relay (G)	Air-mass meter ($\dot{m}/t°$)	Converter, transformer (quantity, voltage) (Q/U)
Lambda oxygen sensor (not heated/heated) (λ, $\lambda/t°$)	Flow sensor, fuel-gauge sensor (Q)	Inductive sensor
	Temperature switch, temperature sensor ($t°$)	

Instrument cluster (dashboard): N1 (U const.), P2 (V), P3 (n), P4 (Q), P5 (t), H1, H2, H3, H4, H5, H6

Table 1 (continued)

Circuit diagrams

Circuit diagrams are idealized representations of electrical devices, rendered in the form of symbols. Such diagrams also include illustrations and simplified design drawings as needed (Figure 7).

The circuit diagram illustrates the functional interrelationships and physical links that connect various devices. It may be supplemented by tables, graphs and descriptions. Circuit diagrams vary according to the intended application (e.g., showing circuit operation) and the selected representation mode.

A "legible" circuit diagram will meet the following requirements:
- The representations must reflect the applicable standards; explanations should be provided for any exceptions
- Electrical current should be portrayed as flowing from left to right and/or from top to bottom

In automotive electrical systems, block diagrams are used to provide a quick overview of circuit and device functions. They are usually unipolar and also dispense with representations of internal circuitry components.

The schematic diagrams in their various permutations (as defined by differences in symbol arrangements) provide a detailed diagram of the circuit. As they illustrate how the circuit operates, they are suitable for use as a reference for repair operations.

The terminal diagram (with equipment connection points) is used by service facilities in replacing defective electrical equipment and when installing supplementary equipment.

Depending upon the type of representation, we distinguish between:
- Unipolar and multipolar representation (according to symbol arrangement)
- Assembled representation, semi-assembled representation, detached representa-

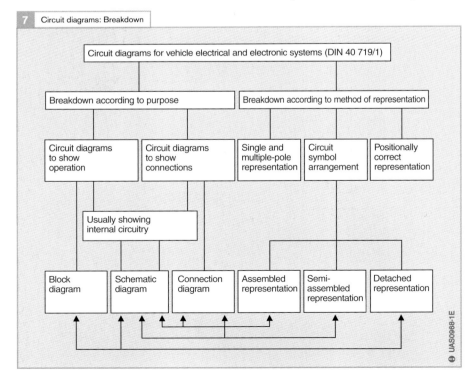

7   Circuit diagrams: Breakdown

Vehicle electrical systems Circuit diagrams and symbols

14 Terminal diagram, detached view

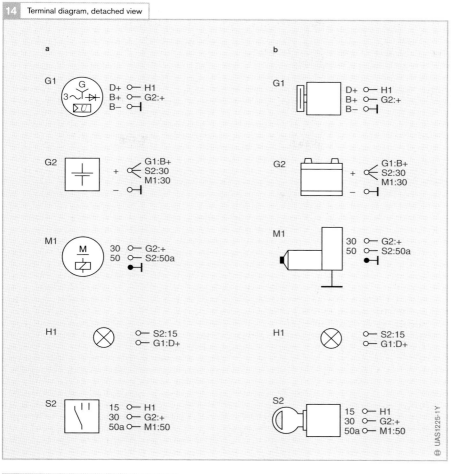

Fig. 14
a With symbols and indications of destination
b With devices and destinations

G1 Alternator with voltage regulator
G2 Battery
H1 Charge indicator lamp
M1 Starter motor
S2 Ignition switch
XX Device ground on vehicle chassis
YY Terminal for ground connection
:15 Conductor potential, e.g., Terminal 15

15 Device designation (Example: Alternator)

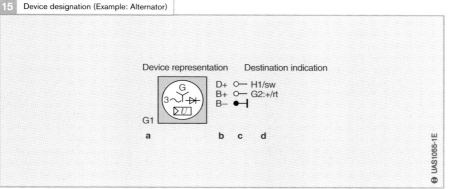

Fig. 15
a Device designation (code letter and sequence number)
b Terminal code on device
c Device to ground
d Destination indication (code letter plus sequence number/terminal designation/wire color code)

Designations for electrical devices

The designations defined in DIN 40719, Section 2 (Table 3) serve as the basis for unambiguous and internationally-recognized labeling conventions for assemblies, components, etc., as represented in circuit diagrams by electrical symbols. The designation as per DIN 40719 is a defined sequence consisting of prefix, letter and numerals. It appears next to the symbol.

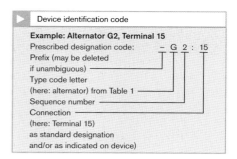

Device identification code

Example: Alternator G2, Terminal 15
Prescribed designation code: – G 2 : 15
Prefix (may be deleted if unambiguous)
Type code letter (here: alternator) from Table 1
Sequence number
Connection (here: Terminal 15) as standard designation and/or as indicated on device)

3 Identification codes for electrical devices

Code letter	Type	Examples
A	System, assembly, component group	ABS control units, car radios, two-way radios, mobile phones, alarm systems, equipment assemblies, triggering units, ECUs, cruise control
B	Transducer for converting non-electrical parameters into electrical values and vice versa	Reference-mark sensors, pressure switches, air horns, horns, Lambda oxygen sensors, loudspeakers, air-flow sensors, microphones, oil pressure switches, various sensors, ignition-triggering devices
C	**Condenser, capacitor**	**All types of condensers/capacitors**
D	Binary element, data storage	On-board computers, digital equipment, integrated circuits, pulse counters, magnetic-tape recorders
E	Various devices and accessories	Heater systems, air-conditioner, lamps, headlights, spark plugs, ignition distributors
F	Protective equipment	Triggers (bimetallic), reverse-polarity guards, fuses, current protection circuitry
G	Current supply, alternator	Batteries, alternators, battery chargers
H	Monitor, signaling, warning, display device	Audio alarms, display lamps, flasher indicators, turn signals, brake-pad wear indicators, stop lamps, high-beam indicators, charge indicator lamps, indicator lamps, signaling devices, oil-pressure warning lamps, optical indicators, signal lamps, warning buzzers
K	Relay, protective device	Battery relays, turn-signal relays, flasher relays, solenoid relays, starter relays, hazard-warning and turn-signal flashers
L	Inductor	Choke coils, windings
M	Motor	Blower motors, fan motors, pump motors for ABS/TCS/ESP hydraulic modulators, windshield washer/wiper motors, starter motors, step motors
N	Regulator, amplifier	Regulators (electronic or electromechanical), voltage stabilizers

Identification codes for electrical devices (continued)

Code letter	Type	Examples
P	Measuring instruments/ Monitoring equipment	Ammeters, diagnosis interfaces, tachometers, manometers, tachographs, test taps/connections, test points, speedometers
R	Resistor	Glow plugs, flame plugs, resistive heater elements, PTC and NTC resistors, potentiometers, regulating resistors, in-line resistors
S	Switch	Switches and contacts of all kinds, ignition points
T	Transformer	Ignition coils, ignition transformers
U	Modulator, converter	DC transformers
V	Semiconductor, electron tube	Darlington transistors, diodes, electron tubes, rectifiers, all semiconductors, varactors, transistors, thyristors, Zener diodes
W	Transmission path, conductor, antenna	Vehicle antennas, shielding, shielded cable, all types of cable, wiring harnesses, (common) ground conductors
X	Terminal, plug, plug-in connection	Terminal studs, all types of electrical connection, spark-plug connectors, terminals, terminal bars, electrical wiring couplings, wiring connectors, plugs, sockets, socket rails, (multiple-pin) plug connections, junction plugs
Y	Electrically operated mechanical devices	Permanent magnets, (solenoid) injection valves, solenoid clutches, electric air valves, electric fuel pumps, solenoids, electric start valves, transmission-shift controls, tractive magnets, kickdown solenoids, headlight leveling controls, ride-height control valves, circuit-control valves, start valves, door locks, central locking systems, auxiliary air devices
Z	Electrical filtering elements	Elements and filters for interference suppression, filter networks, clocks

Table 3 (continued)

DIN 72 552 terminal designations

The system of standard terminal designations prescribed for use in automotive applications has been designed to facilitate correct connection of devices and their wiring, with emphasis on repairs and replacement installations.

The terminal codes (Table 4) are not wire designations, as devices with differing terminal codes can be connected to the opposite ends of a single wire. It is therefore not essential that the terminal codes be provided on the wiring.

The DIN 72 552 codes may be supplemented by the designations defined in DIN-VDE standards for electrical machinery. Multi-pin plug connections large enough to exhaust the range provided by DIN 72 552 are allocated consecutive numbers or letters, avoiding any characters to which the standard has already assigned a specific function.

Table 4. DIN 72 552 terminal designations

Terminal	Definition
1	Ignition coil, distributor Low-tension circuit
	Ignition distributor with two insulated circuits
1 a	to ignition point set I
1 b	to ignition point set II
2	Short-circuit terminal (magneto ignition)
4	Ignition coil, distributor High-tension circuit
	Ignition distributor with two insulated circuits
4 a	Terminal 4, from coil I
4 b	Terminal 4, from coil II
15	Switch-controlled plus downstream from battery (from ignition switch)
15 a	In-line resistor terminal leading to coil and starter
	Glow-plug switch
17	Start
19	Preglow
30	Line from battery positive terminal (direct)
30 a	Series/parallel battery switch 12/24 V Line from battery positive terminal II
31	Return line from battery negative terminal or ground (direct)
31 b	Return line to battery negative terminal or ground via switch or relay (switch-controlled ground)
	Battery changeover relay 12/24 V
31 a	Return line to Battery II negative pole
31 c	Return line to Battery I negative pole
	Electric motors
32	Return line [1]
33	Main connection [1]
33 a	Self-parking switch-off
33 b	Shunt field
33 f	for reduced-rpm operation, speed 2
33 g	for reduced-rpm operation, speed 3
33 h	for reduced-rpm operation, speed 4
33 L	Rotation to left (counterclockwise)
33 R	Rotation to right (clockwise)
	Starter
45	Separate starter relay, output: starter; input: primary current
	Dual starters, parallel activation Relay for pinion-engagement current
45 a	Starter I output Starters I and II input
45 b	Starter II output
48	Terminal on starter and start-repeating relay for monitoring starting process
	Flasher relay (pulse generator)
49	Input
49 a	Output
49 b	Output to second flasher relay
49 c	Output to third flasher relay

[1] Polarity reversal terminal 32/33 possible

Vehicle electrical systems Circuit diagrams and symbols

4   DIN 72 552 terminal designations (continued)

Terminal	Definition
50	**Starter** Starter control (direct)
50 a	**Battery switching relay** Output for starter control
50 b	**Starter control** Dual starters in parallel operation with sequential control
50 c 50 d	**Starting relay for sequential control of engagement current for dual starters in parallel operation** Starter I input at starter relay Starter II input at starter relay
50 e 50 f	**Start-locking relay** Input Output
50 g 50 h	**Start repeating relay** Input Output
51 51 e	**AC generator (alternator)** DC voltage at rectifier DC voltage at rectifier with choke coil for daylight operation
52	**Trailer signaling devices** Supplementary signal transmission from trailer to towing vehicle
53 53 a 53 b 53 c 53 e 53 i	Wiper motor, input (+) Wiper (+), end position Wiper (shunt winding) Electric windshield-washer pump Wiper (brake winding) Wiper motor with permanent magnet and third brush (for higher speed)
55	Front fog lamp
56 56 a 56 b 56 d	Headlights High-beam with indicator lamp Low beam Headlight flasher contact
57	Motorcycle/Moped parking lamps (also for passenger cars, trucks, in some export markets)
57 a 57 L 57 R	Parking lamps Parking lamps, left Parking lamps, right

Terminal	Definition
58	Side-marker lamps, tail lamps, license-plate and instrument illumination
58 b	Tail light mode selection on single-axle tractors
58 c	Trailer gladhand assembly for single-strand tail light with fuse in trailer
58 d	Rheostatic instrument illumination, tail and side-marker lamps
58 L 58 R	left right, license-plate lamps
59	**AC generator (alternator) (magneto generator)** AC voltage output, rectifier input
59 a 59 b 59 c	Charging-armature output Tail-lamp armature, output Stop-lamp armature, output
61	Charge indicator lamp
71 71 a 71 b	**Tone-sequence controller** Input Output to Horns I and II (bass) Output to Horns 1 and 2 (treble)
72	Alarm switch (rotating beacon)
75	Radio, cigarette lighter
76	Speakers
77	Door valve control
54	**Trailer signaling equipment** Trailer gladhand assembly and light combinations Stop lamps
54 g	Pneumatic valve for continuous-duty trailer brake with solenoid control
81 81 a 81 b	**Switches, NC contacts and changeover contacts** Input First output on NC-contact side Second output on NC-contact side
82 82 a 82 b 82 z 82 y	NO contacts Input First output Second output First input Second input Multiple position switch

Table 4 (continued)

Table 4 DIN 72 552 terminal designations (continued)

Terminal	Definition
	Switches, NC contacts and changeover contacts (continued)
83	Input
83 a	Output (Pos. 1)
83 b	Output (Pos. 2)
83 L	Output (Pos. left)
83 R	Output (Pos. right)
	Current relay
84	Input: Actuator and relay contacts
84 a	Output: Actuators
84 b	Output: Relay contacts
	Switching relay
85	Output: Actuator (negative winding end or ground)
86	Input: Actuator Start of winding
86 a	Start of winding or first winding coil
86 b	Winding tap or second winding coil
	Relay contact for NC and changeovers contacts
87	Input
87a	First output (NC-contact side)
87 b	Second output
87 c	Third output
87 z	First input
87 y	Second input
87 x	Third input
	Relay contact for NO contact
88	Input
	Relay contact for NO contact and changeover contacts (NO side)
88 a	First output
88 b	Second output
88 c	Third output
	Relay contact for NO contact
88 z	First input
88 y	Second input
88 x	Third input
	Generator/alternator and voltage regulator
B+	Battery positive terminal
B−	Battery negative terminal
D+	Generator positive terminal
D−	Generator negative terminal
DF	Generator field winding
DF 1	Generator field winding 1
DF 2	Generator field winding 2

Terminal	Definition
	Alternator
U, V, W	Three-phase terminals
	Turn signals (turn-signal flasher)
C	Indicator lamp 1
C 0	Main terminal connection for indicator lamp not connected to turn-signal flasher
C 2	Indicator lamp 2
C 3	Indicator lamp 3 (e.g., for dual-trailer operations)
L	Left-side turn signals
R	Right-side turn signals

Circuit diagram for passenger cars (Examples)

Purpose
Circuit diagrams are the only way to provide an overview of complex automotive electrical systems with their numerous terminals and connections (Fig. 16). Modern systems include more than just the lights, with a large number of electric and electronic devices for open and closed-loop engine management as well as numerous accessories for safety, comfort and convenience.

Design
The schematic diagrams in the following section (Fig. 17...38) portray various vehicle circuits. They are intended to facilitate understanding of the text; they are not intended for use in manufacture or installation.

Sample designation codes
A1 Device code (DIN 40719)
15 Terminal code (DIN 72 552)
1 Section identification code (DIN 40719)

16 Section of an automotive wiring harness

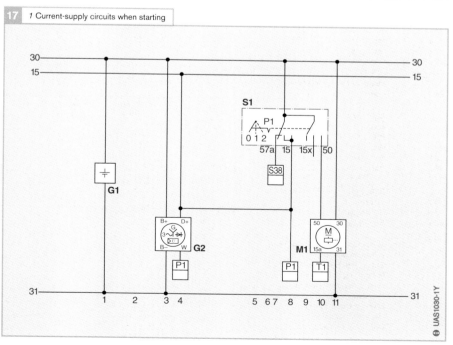

17 *1* Current-supply circuits when starting

Vehicle electrical systems — Circuit diagrams and symbols

18 2 Lighting, signaling circuits

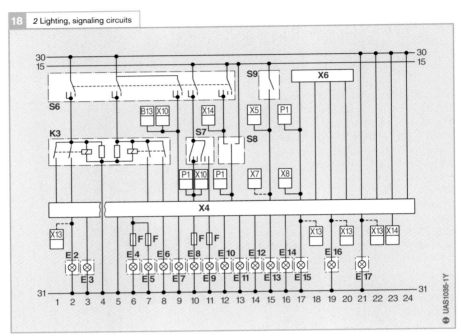

19 2 Lighting, signaling circuits (continued)

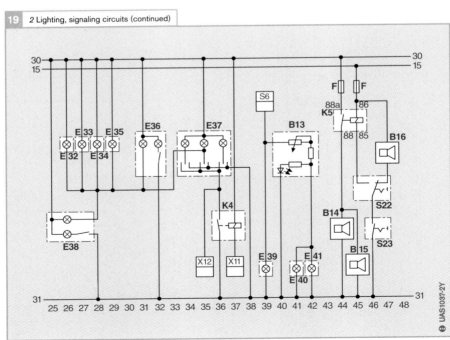

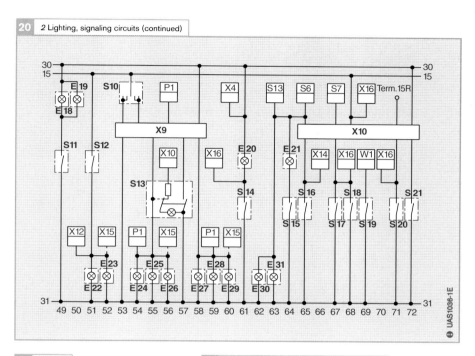
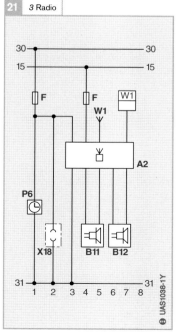
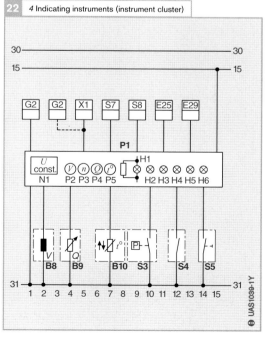

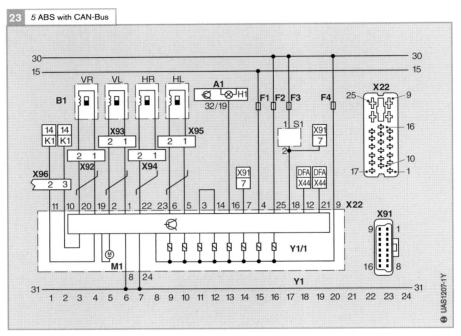

23 5 ABS with CAN-Bus

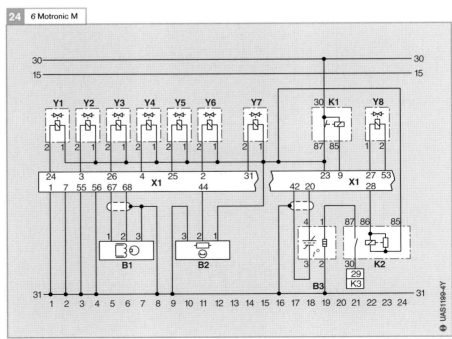

24 6 Motronic M

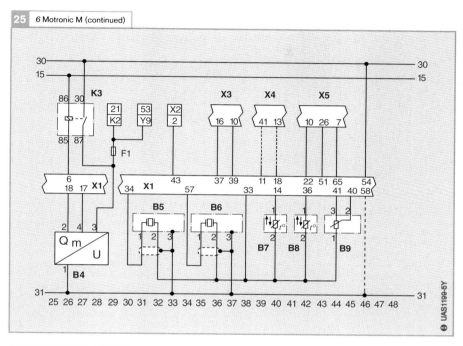

25 6 Motronic M (continued)

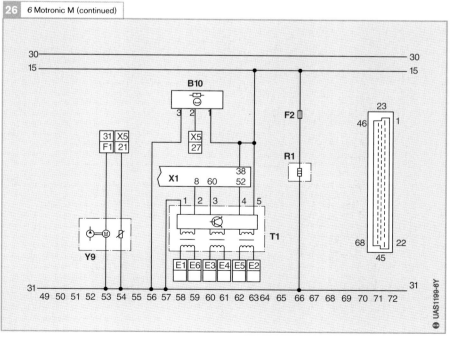

26 6 Motronic M (continued)

27 7 VE/EDC distributor-type injection pump

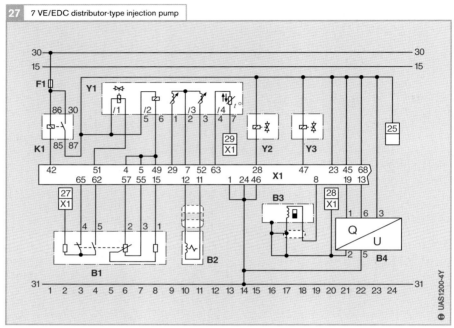

28 7 VE/EDC distributor-type injection pump (continued)

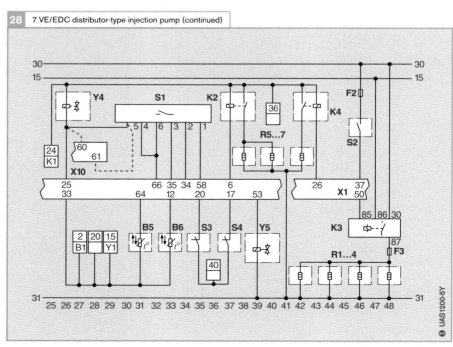

Vehicle electrical systems   Circuit diagrams and symbols

29 7 VE/EDC distributor-type injection pump (continued)

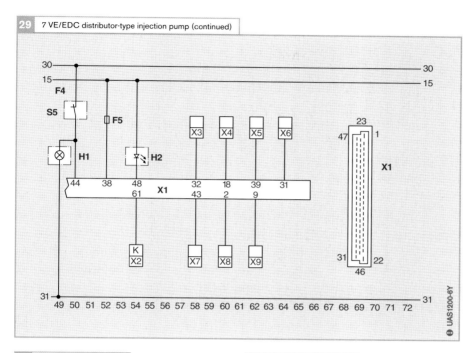

30 8 Diesel preglow circuit

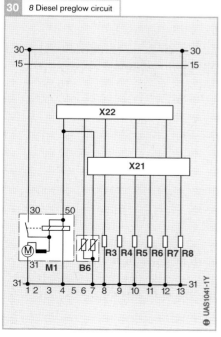

31 9 Vehicle alarm system

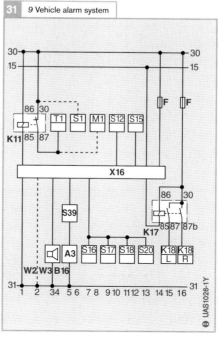

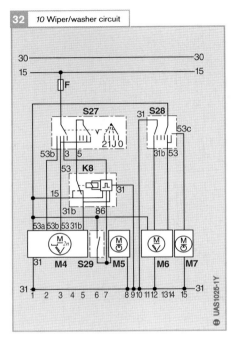

32 *10* Wiper/washer circuit

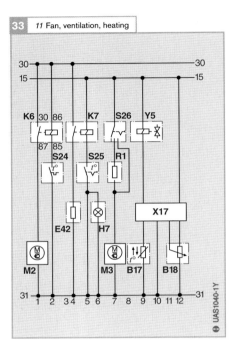

33 *11* Fan, ventilation, heating

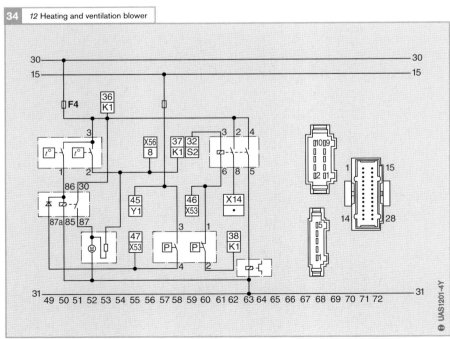

34 *12* Heating and ventilation blower

Vehicle electrical systems   Circuit diagrams and symbols

35 *12* Climatronic automatic climate control (continued)

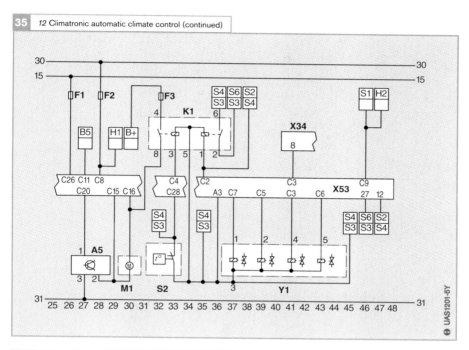

36 *12* Climatronic automatic climate control (continued)

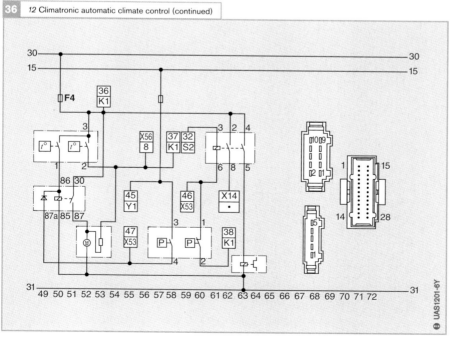

37 *13* AGA Transmission-shift control

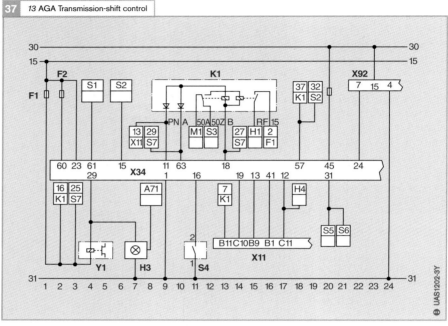

38 *13* AGA Transmission-shift control (continued)

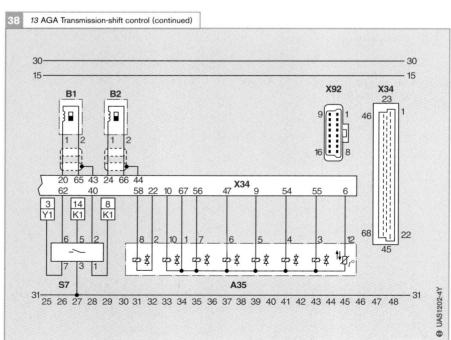

Section designations and device locations

Table 5 contains all of the section designations contained in the section "Circuit diagram for a passenger car with spark-ignition engine". The "Sections in the schematic diagram" (Table 3 below) define specific areas of the schematic diagrams in which a particular system can be found.

Table 6 contains devices and their codes along with section numbers for passenger-car circuit diagrams.

5 Sections

Section	System
1	Starting, current supply
2	Lighting, signaling equipment
3	Clock, radio
4	Display assembly (instrument cluster)
5	ABS with CAN bus
6	Motronic M
7	VE/EDC distributor-type injection pump
8	Diesel preglow system
9	Vehicle alarm
10	Wash/wipe system
11	Heater and ventilation blower
12	Climatronic automatic climate control
13	AG4 transmission-shift control

Table 5

6 Device classification

Code	Device	Section
A1	Warning-lamp display unit	5
A1	Central-flap control motor	12
A2	Radio	3
A2	Dynamic pressure-flap actuator	12
A3	Ignition system with knock control (EZ-K)	9
A3	Temperature-flap actuator motor	12
A4	Dashboard temperature sensor with blower	12
A5	Blower control unit	12
A6	Radiator fan	12
A35	Transmission unit, electric	13
B1	Rotation rate/reference-mark sensor	5, 6
B1	Accelerator-pedal travel sensor	7
B1	Coolant-temperature sensor	12
B1, 2	Vehicle-speed sensor	13, 6
B2	Needle-motion sensor	7
B2	Outside-temperature sensor	12
B2	Transmission-input shaft rpm sensor	13

6 Device classification (continued)

Code	Device	Section
B3	Lambda oxygen sensor	6
B3	Rotation-rate/reference-mark sensor	7
B3	Intake-air temperature sensor	12
B4	Air-mass meter	6, 7
B4	Photosensor	12
B5	Knock sensor 1	6
B5, 7	Coolant-temperature sensor	7, 6
B6	Knock sensor 2	6
B6	Fuel-temperature sensor	7, 8
B8	Speed sensor	4
B8	Intake-air temperature sensor	6
B9	Fuel-level (gauge) sensor	4
B9	Throttle-valve-potentiometer sensor	6
B10	Coolant-temperature sensor	4
B10	Cylinder-identification sensor	6
B11, 12	Speaker	3
B13	Instrument illumination rheostat	2
B14, 15	Supertone horn	2
B16	Horn 2	2, 9
B17	Interior-temperature sensor	11
B18	Setpoint selector	11
E1	Climatronic display unit	12
E2, 3	Fog-warning lamp, L/R	2
E4, 5	Driving lamp, L/R	2
E6, 7	Front fog lamp, L/R	2
E8, 9	Low-beam headlamp, L/R	2
E10, 11	Side-marker lamp, L/R	2
E12, 13	License-plate lamp, L/R	2
E14, 17	Brake light, L/R	2
E15, 16	Tail lamp, L/R	2
E18	Trunk-lid lamp	2
E19	Luggage-compartment lamp	2
E20	Glove-compartment lamp	2
E21	Engine-compartment lamp	2
E22, 23	Backup lamp, L/R	2
E24, 26	Turn signal, LF, LR	2
E25, 28	Hazard warning flasher L/R	2
E27, 29	Turn signal, RF, RR	2
E30, 31	Ashtray lamp, front and rear	2
E32, 33	Footwell lamp, LR, LF	2
E34, 35	Footwell lamp RF, RR	2
E36, 38	Rear map light, R/L	2
E37	Interior map light	2
E39	Vanity-mirror lamp	2
E40	Instrument illumination	2
E41	Control/dashboard illumination	2
E42	Rear-screen defroster	11
F..	Fuses	

Table 6

6 Device classification (continued)

Code	Device	Section
G1	Battery	1
G2	Alternator	1
H1	Charge-indicator lamp	4
H1	ABS warning lamp	5
H1	Preglow indicator lamp	7
H2	Oil-pressure warning lamp	4
H2	Stop lamp	7
H3	Parking-brake indicator lamp	4
H3	Selector-lever illumination	13
H4	Brake-pad wear-indicator lamp	4
H5	High-beam indicator lamp	4
H6	Turn-signal indicator lamp	4
H7	Rear-screen defroster indicator	11
K1	Main relay	6, 7
K1	A/C relay	12
K1	Starter lockout relay	13
K2	Lambda-Oxygen-sensor heater relay	6
K2	Relay for minor heater wire	7
K2	A/C compressor relay	12
K3	Parking-lamp monitor relay	2
K3	Electric-fuel-pump relay	6
K3	Glow-plug relay	7
K3	Radiator-fan starter relay	12
K4	Interior-lamp control relay	2
K4	Heater-line relay	7
K5	Hightone-horn relay	2
K6	Engine-fan relay	11
K7	Rear-screen-defroster relay	11
K8	Intermittent-wiper relay	10
K11	Starter/ignition lockout relay	9
K17	Visual-alarm relay	9
M1	Starter motor	1, 8
M1	Pump motor Hydraulic modulator	5
M1, 3	Fresh-air ventilation fan motor	11, 12
M2	Blower motor	11
M4	Wiper motor	10
M5	Windshield-washer motor	10
M6	Engine-fan relay	10
M7	Rear-screen washer motor	10
N1	Voltage stabilizer	4
P1	Instrument cluster	4
P2	Electric speedometer	4
P3	Tachometer	4
P4	Fuel gauge	4
P5	Engine-temperature gauge	4
P6	Clock	3
R1	Heater-resistance element	6

Code	Device	Section
R1..4	Glow plugs	7
R1	Blower resistor	11
R5..7	Auxiliary heater (with manual transmission)	7
R3..8	Glow plugs	8
S1	Ignition/starter switch	1
S1	Brake-light switch	5
S1	Cruise-control selector unit	7
S1	Light switch	12
S2	A/C switch	7
S2	Evaporator temperature switch	12
S3	Oil-pressure switch	4
S3	Brake-pedal switch	7
S3	Radiator-fan temperature switch	12
S4	Parking-brake switch	4
S4	Clutch-pedal switch	7
S4	A/C system-pressure switch	12
S4	Kickdown switch	13
S5	Brake-pad-wear indicator contact	4
S5	Brake-light switch	7
S6	Light switch	2
S7	Fog-lamp switch	2
S7	Multifunction switch	13
S8	Low-beam switch	2
S9	Brake-light switch	2
S10	Turn-signal switch	2
S11	Trunk-lid lamp switch	2
S12	Backup-lamp switch	2
S13	Hazard-warning-flasher switch	2
S14	Glove-compartment lamp switch	2
S15	Engine-compartment lamp switch	2
S16..18	Door-contact switch, LF, RR, LR	2
S19	Impact switch	2
S20	Door-contact switch, RR	2
S21	Door-handle switch	2
S22	Horn-selector switch	2
S23	Horn-contact switch	2
S24	Thermal switch	11
S25	Rear-screen defroster switch	11
S26	Blower switch	11
S27	Wiper switch	10
S28	Rear-screen wiper washer switch	10
S29	Washer switch	10
S39	Alarm-system code-entry switch	9
T1	Ignition coil	6
W1	Car antenna	3
W1	Socket connector for 16-pin flat cable	12

Table 6 (continued)

6 Device classification (continued)

Code	Device	Section
W2,3	Code entry line	9
X1	Motronic/VE/EDC control-unit plug	6, 7
X3	A/C control-unit plug	6
X4	Lamp-control-module plug	2
X4	Transmission-shift control ECU plug	6
X5	Instrument-cluster connection plug	6
X6	Check Control plug	2
X9	Hazard-warning-flasher relay socket	2
X10	Basic module plug for central bodywork electronics	2, 7
X11	Engine-management- ECU plug	13
X16	Alarm-system-ECU plug	9
X17	A/C & heater-control ECU plug	11
X18	Diagnosis interface	3
X21	Glow-control-unit ECU plug	8
X22	ABS/ABD ECU plug	5
X22	Diagnosis plug	8
X34	Transmission-shift-control ECU plug	12, 13
X44	Navigation-system plug	5
X53	Automatic climate-control plug	12
X91, 92	Diagnosis interface	5, 13
Y1	Hydraulic unit	5
Y1	Injection valve 1	6, 7
Y1	Fuel rail	12
Y1	Shift-lockout solenoid	13
Y2	A/C performance control	7
Y2	A/C solenoid clutch	12
Y2..5	Injection valves	6, 7
Y5	Hot-water valve	11
Y6	Injection valve 6	6
Y7	Canister-purge valve	6
Y8	Idle actuator	6
Y9	Electric fuel pump	6

Table 6 (continued)

Schematic diagram

Bosch has responded to the requirements associated with trouble-shooting on complex, networked systems by developing system-specific circuit diagrams. Bosch makes schematic diagrams for numerous vehicles available on its "P" CD-ROM, which is an integral element within the Bosch ESI Electronic Service Information system. This reference source furnishes vehicle service operations with valuable assistance in localizing defects, while also helping personnel install auxiliary equipment with maximum efficiency. Figure 40 illustrates the schematic diagram for a door-locking system.

The representations in the schematic diagrams diverge from those in the standard circuit diagrams by relying on US symbols with supplementary legends (Figure 39). These legends include component codes (for instance: "A28" for anti-theft system, Table 7) as well as the color codes for wiring (Table 8). Both tables can be accessed with the "P" CD-ROM.

Table 7 — Explanation of component codes

Position	Description
A1865	Electric power-seat system
A28	Theft-deterrent system
A750	Fuse/relay module
F53	Fuse C
F70	Fuse A
M334	Supply pump
S1178	Warning-buzzer switch
Y157	Vacuum actuator
Y360	Actuator, door, right front
Y361	Actuator, door, left front
Y364	Actuator, door, right rear
Y365	Actuator, door, left rear
Y366	Fuel filler-flap actuator
Y367	Actuator, lock, luggage compartment, trunk lid

Table 8 — Wire color code chart

Position	Description
BLK	Black
BLU	Blue
BRN	Brown
CLR	Transparent
DK BLU	Dark blue
DK GRN	Dark green
GRN	Green
GRY	Gray
LT BLU	Light blue
LT GRN	Light green
NCA	No color assignment
ORG	Orange
PNK	Pink
PPL	Purple
RED	Red
TAN	Tan
VIO	Violet
WHT	White
YEL	Yellow

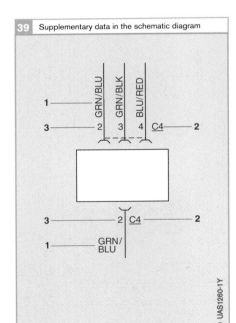

Fig. 39 — Supplementary data in the schematic diagram
1. Wire color
2. Connector number
3. PIN number (dashes between PINs indicate that all PINs are part of the same plug)

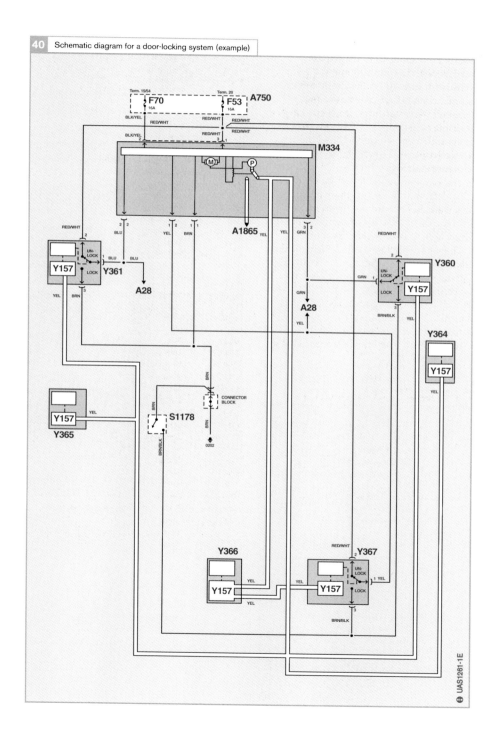

40 Schematic diagram for a door-locking system (example)

Schematic diagrams are classified according to system circuits, with further divisions by subsystem as indicated (Table 9).

Classification of system circuits reflects the standard ESI practice as used for other systems by employing assignments to one of four assembly groups:
- Engine
- Bodywork
- Suspension and
- Drivetrain

As ground-point identification is always important, and absolutely vital for installations of supplementary equipment, CD-ROM "P" supplements the schematic diagrams for specific vehicles with individual diagrams showing ground locations (Figure 41).

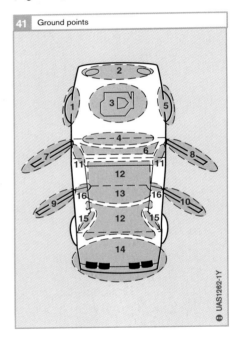

41 Ground points

Fig. 41
1 Left front fender
2 Forward section
3 Engine
4 Firewall
5 Front right fender
6 Footwell panel/ instrument panel
7 Left front door
8 Right front door
9 Left rear door
10 Right rear door
11 A-pillars
12 Passenger compartment
13 Roof
14 Rear section
15 C-pillars
16 B-pillars

9	System sub-circuits
1	Engine management
2	Starter/charging circuit
3	Heating and air conditioning
4	Blower fan
5	ABS
6	Cruise control
7	Power windows
8	Central locking system
9	Instrument panel
10	Wiper/washer system
11	Headlights
12	External lighting
13	Current supply
14	Ground assignments
15	Data cable
16	Shift lockout
17	Theft-deterrent system
18	Passive safety system
19	Electric antenna
20	Alarm system
21	Screen/mirror defroster
22	Supplementary safety systems
23	Interior lighting
24	Power steering
25	Adjustable mirrors
26	Power convertible top
27	Horn
28	Luggage compartment, trunk lid
29	Power seats
30	Electronic damping
31	Cigarette lighter, socket
32	Navigation
33	Transmission
34	Active bodywork components
35	Vibration damping
36	Mobile phone
37	Radio/sound system
38	Vehicle immobilizer

Table 9

The schematic diagrams rely on US symbols that differ from the DIN and IEC standards. Samples of these American symbols are provided in Figure 42.

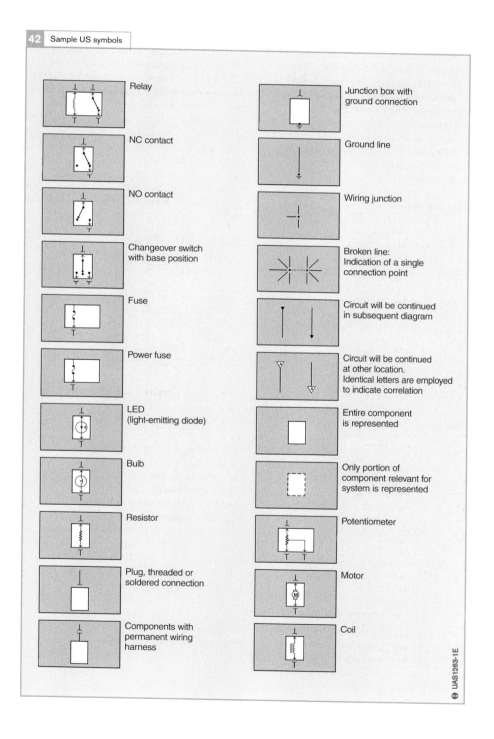

42 Sample US symbols

Electromagnetic compatibility (EMC) and interference suppression

Electromagnetic compatibility consists of two elements. One is understood as the ability of a device to continue providing reliable service when exposed to electromagnetism from external sources. The second aspect focuses on electromagnetic fields generated by the same device; these should remain minimal in order to avoid creating interference that would impinge upon the quality of radio reception, etc. in the vicinity.

Why EMC?

Modern-day vehicles are equipped with a wide range of systems that rely on electrical or electronic componentry to perform an array of functions that were either nonexistent or purely mechanical in earlier automotive applications. The continuing proliferation of electrical and electronic devices within the modern automotive environment has been marked by a proportionate increase in the relative significance of electromagnetic compatibility.

With the exception of two-way transceivers used in special-purpose applications, a car's radio was once the only device in which signal reception was of any importance. In contrast, today's vehicles feature a host of devices that rely on the reception of electromagnetic radio waves. Mobile phones, navigation systems, theft-deterrent systems with remote radio control, and integrated fax and PC units are now being installed and used in vehicles. This trend has led to a commensurate increase in the importance of suppressing interference and ensuring reception of the operationally vital radio waves.

EMC ranges

The design of electrical and electronic systems for automotive applications must focus on three main priority areas:
- Transmitter and receiver
- Electrical and electronic components and
- On-board electronic systems

Transmitter and receiver

All vehicle systems must remain impervious to electromagnetic radiation emitted from such external sources as extremely powerful radio transmitters. In other words, there must be no threat to the vehicle's operational integrity, and functional irregularities representing a potential source of driver irritation are inacceptable. Another consideration is that stationary receivers should remain unaffected by passing traffic. Both considerations are governed by national and international codes (EC ordinances, German StVZO).

Electrical and electronic components

Vehicles contain an extensive array of electrical and electronic components including servo and fan motors, solenoid valves, electronic sensors and ECU's with microprocessors. These devices must rely for their power supply on a single on-board network. They must all function simultaneously in an environment characterized by lack of space and close proximity between units. It is thus vital to avoid mutual interference and feedback phenomena generated by one or several systems so that these do not cause malfunction.

On-board electronic systems

Mobile communications equipment – such as the radio – also exists within an interlinked environment including all of the vehicle's electronic systems. Every device is powered via the same on-board electrical system, and its reception antenna is located in the immediate vicinity of potential interference sources. These considerations make it imperative that strict limits are imposed on the levels of interference emitted by on-board electronic systems. Compliance with official regulations is essential, and it is important to maintain interference-free onboard reception even when conditions are well below optimal.

EMC between various vehicular systems

Shared on-board power supply

All of the motor vehicle's electrical systems rely on a single shared on-board power-supply network. Because the wires and cables leading to the individual systems are frequently combined within a single wiring harness, feedback pulses can easily travel from one system to the I/O ports of its neighbor (Figure 1).

This transfer of interference can be in the form of signal pulses (abrupt, steep jumps in current and voltage) generated during the switching on and off of various electrical components such as motors and solenoid valves. Yet another source is the ignition system's high-tension circuit. Similar to other interference signals (for instance the ripple on the power supply), these signal pulses can propagate through the wiring harness. These interference pulses then proceed to the I/O ports of adjacent systems either directly, through shared conductors such as the power supply (galvanic coupling), or indirectly, through capacitive and inductive coupling stemming from electromagnetic emissions.

Galvanic coupling

Currents for two different circuits (such as a solenoid valve's trigger loop and the circuit for assessing sensor data) flowing through a single conductive path (common ground through the vehicle chassis, etc.) will both generate a voltage owing to the consistent resistance in the shared conductor (Figure 2a). Continuing with this sample scenario, the voltage produced by interference source u_1 has the effect of a supplementary signal voltage in signal circuit 2, and could lead to erroneous interpretation of the sensor signal. One remedy is to use separate return lines for each circuit (Figure 2b).

1 Mutual interference between two systems as transmitted through the shared vehicle power supply (A) and wiring harnesses (B and C)

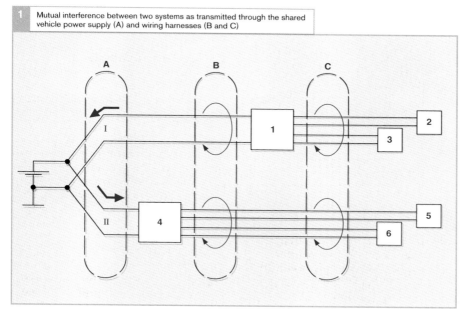

Fig. 1
System I:
1   ECU
2   Actuator
3   Sensor

System II:
4   ECU
5   Actuator
6   Sensor

Capacitive coupling

Capacitive coupling allows variable-periodicity signals such as pulse voltages and sinusoidal DC voltage to produce interference and crosstalk in adjacent circuits, even without the existence of a direct physical link (Figure 3). The potential level of capacitive (interference) voltage is proportional to such factors as the closeness of the neighboring conductor paths and the rise rate of the pulse-shaped voltage shifts (or the frequency of the AC voltage).

The first step is to separate the conductive paths while at the same time extending the ramp periods during which signals rise and fall (or to limit the frequency of the AC voltages to the absolute minimum required for the function).

Inductive coupling

Currents recurring with variable periodicity in one conductor can induce voltage pulses in adjoining circuits. These voltage pulses then generate current in the secondary circuit (Figure 4). This is the inductive principle exploited in transformer design. One prime factor defining susceptibility to overcoupling is the signal's rise and fall time (or AC voltage frequency), reflecting the situation encountered with capacitive coupling. Also significant is the effective mutual inductance, as determined by such factors as the size of the wires and their relative routing. Strategies for avoiding inductive coupling include minimizing the dimensions of circuit wires, keeping critical circuits as far apart as possible from each other, and the avoidance of parallel conductive paths. The tendency toward inductive interference is especially pronounced in circuits carrying low-frequency signals (e.g., coupling into loudspeaker wiring).

Fig. 2
a With shared return conductor path
b With separate return lines

u_1, u_2 Voltage source
Z Internal resistor
Z_a Terminal resistor

Fig. 3
1 Circuit 1
2 Circuit 2

u_1 Voltage source
Z Internal resistor
R_E Input resistance
C_E Input capacitance
$C_{1,2}$ Capacitance between two conductor paths
u_s Interference voltage

Fig. 4
1 Circuit 1
2 Circuit 2

u_1 Voltage source
u_2 Voltage source
Z Internal resistance
Z_a Terminal resistance
L_1, L_2 Inductance of conductors
$M_{1,2}$ Inductive coupling
u_s Interference voltage

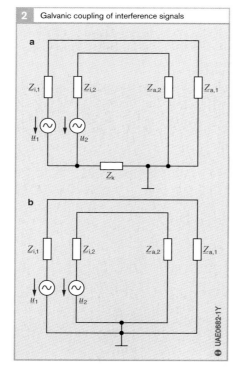

2 Galvanic coupling of interference signals

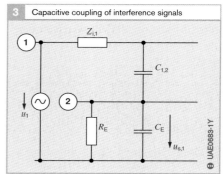

3 Capacitive coupling of interference signals

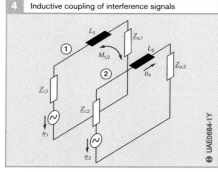

4 Inductive coupling of interference signals

Pulsation in the vehicle electrical system

One strategy for dealing with interference pulses in the vehicle's electrical system entails limiting the amplitude of the interference emanating from the source. On the other hand, the affected electronic components are designed for insensitivity to pulses of specific shapes and amplitudes. The initial step was to list and classify the pulses encountered within vehicular electrical systems (Table 1). Special-purpose pulse generators can be used to generate the test patterns defined in Table 1 as a basis for assessing resistance to these interference waves in exposed devices. Both the test pulses and the examination procedures are codified in standards (DIN 40 839, Section 1; ISO 7637, Section 1) including definitions of the measurement technology for evaluating emissions of pulse-pattern interference. Classifications based on pulse amplitude levels facilitate effective definition of interference sources and the susceptible devices (interference receptors) within each vehicle. It would thus be possible to specify Class II for all of the interference sources within a vehicle while designing all the susceptible devices interference receptors (such as ECU's) to comply with and exceed – by a certain safety margin – Class III. If suppressing interference at the source proves to be cheaper or to involve less technical complexity than reducing sensitivity at the receptors, the logical response might be to shift the definitions to Classes I/II. If the scenario is inverted, with the shielding of potential receptors as the cheaper and simpler solution, then a move to Classes III/IV is warranted.

Because numerous wires are combined within a single wiring harness, each individual conductor is potentially susceptible to inductive and capacitive interference. Although reduced in intensity, the resulting voltage pulses in adjacent wiring can then appear as spurious signals at the input ports and control outputs in neighboring systems. The test procedure for simulating crosstalk interference within wiring harnesses (as defined in DIN 40 839, Section 3 and ISO 7637, Section 3) uses a standardized substitute wiring layout (capacitive

1 Mutual interference within voltage supply

Test pulses as defined in DIN 40 389, section 1				Max. pulse amplitude classes			
Pulse pattern		Internal resistance	Pulse duration	I	II	III	IV
1	⌄	10 Ω	2 ms	–25 V	–50 V	–75 V	–100 V
2	⌃	10 Ω	50 μs	+25 V	+50 V	+75 V	+100 V
3a	⋀⋀⋀	50 Ω	0,1 μs	–40 V	–75 V	–110 V	–150 V
3b	⋁⋁⋁			+25 V	+50 V	+75 V	+100 V
4	⊔	10 mΩ	bis 20 s	12 V / –3 V	12 V / –5 V	12 V / –6 V	12 V / –7 V
5	⌒	1 Ω	bis 400 ms	+35 V	+50 V	+80 V	+120 V

Table 1

clip) with a defined wiring capacitance. Test pulses are fed into this layout and through the specimen's wiring harness to produce overcoupling in the signal and control lines. The effects of low-frequency oscillations within vehicular electrical systems can be simulated by producing the desired signals with a signal generator and projecting these into the wiring harness through an inductive clamp. This process reflects the procedure described above by serving as the basis for the correct balance between the amplitude of radiated interference pulses and the resistance to interference of the potentially susceptible devices (receptors).

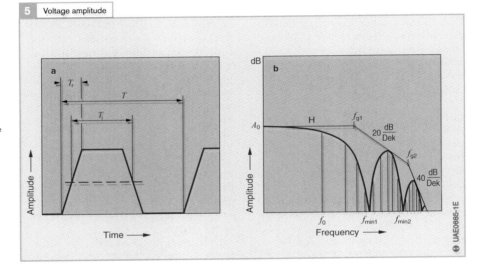

Fig. 5
a   As a function of time
b   As a function of frequency

T Period
T_r Rise time
T_i Pulse duration
$f_0 = T^{-1}$ Fundamental wave
f_g Corner frequencies
f_{min} Periodic minima
H   Envelope curve

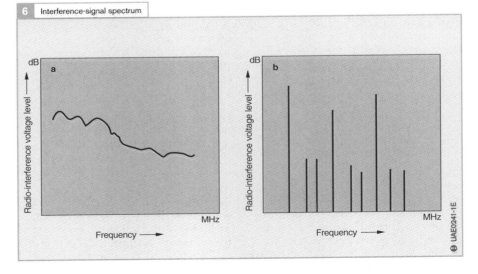

Fig. 6
a   Wide-band interference
b   Narrow-band interference

EMC

DIN 40 839, Section 1
Electromagnetic compatibility (EMC) in road vehicles – conductor-borne interference pulses in supply lines in 12 V and 24 V vehicle electrical systems.

DIN 40 839, Section 3
Electromagnetic compatibility (EMC) in motor vehicles – capacitive and inductive interference in sensor and signal wiring.

DIN 40 839, Section 4
Electromagnetic compatibility (EMC) in motor vehicles – Incident radiation factors.

ISO 7637-0
Road vehicles – Electrical interference from conduction and coupling.
Part 0:   Definitions and general information.

ISO 7637-1
Road vehicles – Electrical interference from conduction and coupling.
Part 1:   Passenger cars and light commercial vehicles with electrical systems rated at 12 V – Electrical transient conduction along supply lines only.

ISO 7637-2
Road vehicles – Electrical interference from conduction and coupling.
Part 2:   Commercial vehicles with electrical systems rated at 24 V – Electrical transient conduction along supply lines only.

ISO 7637-3
Road vehicles – Electrical interference from conduction and coupling.
Part 3:   Vehicles with electrical systems rated at 12 V or 24 V – Electrical transient transmission by capacitive and inductive coupling via lines other than the supply.

ISO 11 451
Road vehicles – Electrical interference from narrow-band electromagnetic radiation – Vehicle test methods.

ISO 11 451-1
Part 1: Definitions and general information.

ISO 11 451-2
Part 2: Remote radiation sources.

ISO 11 451-3
Part 3: On-board transmitter simulation.

ISO 11 451-4
Part 4: Bulk Current Injection (BCI).

ISO 11 452
Road vehicles – Electrical interference from narrow-band electromagnetic radiation – Component test methods.

ISO 11 452-1
Part 1: Definitions and general information.

ISO 11 452-2
Part 2: Absorber-lined chamber.

ISO 11 452-3
Part 3: Transverse Electromagnetic Mode (TEM) cell.

ISO 11 452-4
Part 4: Bulk Current Injection (BCI).

ISO 11 452-5
Part 5: Stripline.

ISO 11 452-6
Part 6: Parallel plate antenna.

ISO 11 452-7
Part 7: Direct radio frequency (RF) power injection.

ISO/TR 10 605
Road vehicles – Electrical interference from electrostatic discharge.

Starter batteries

The battery is a chemical accumulator which stores the electricity generated by the alternator while the engine is running. The stored electricity is needed to start the engine and to supply the electrical equipment in the vehicle's electrical system.

Battery design

Basically, the vehicle electrical system comprises the alternator as energy generator, the battery as energy store, and the starter and other loads as consumers. All these elements must be perfectly matched to each other. The following parameters have an effect upon the special relationship between battery, alternator, and starter:
- Electrical load requirements
- Alternator current output
- Engine speed when the vehicle is actually being driven
- Charging voltage and
- Starting temperature

Electrical loads
In past years, due to the increasingly extensive and complex equipment fitted in the vehicle, the number of electrical loads has increased sharply. Formerly, it was mainly the starter, the ignition, and the lighting system which were the major loads. During the course of the past years though, there has been no letup in the expansion of the vehicle's basic equipment by the addition of electronically controlled ignition and fuel-injection systems, comfort and convenience systems using a variety of different drive motors, and safety and security systems. These loads have a continually increasing share of the power consumed in the vehicle's electrical system.

Starting systems
The starter must crank the IC engine at a given minimum speed and, after the first ignitions, bring the engine up to the minimum self-sustaining speed. In the process, considerable resistances due to compression, piston friction, and bearing friction must be overcome. These are strongly dependent upon engine type and number of cylinders, as well as upon lubricant characteristics and engine temperature. The friction resistances are at a maximum at low temperatures. Even though the starter is only switched on for a brief period, it has the highest current consumption of all the loads (Table 1).

1 Starter power input

Application	Power range (kW)
Passenger car with SI engine	0.7...2.0
Passenger car with diesel engine	1.4...2.6
Buses, trucks, tractors	2.3...9.0

Further electrical loads
The vehicle's electrical loads feature different powers and operating times. A number of the electrical loads are only seasonal (for instance, air-conditioner, heater). A distinction is made between:
- *Permanent loads* (e.g. ignition, fuel injection and/or engine management)
- *Long-time loads* (e.g. lighting, heated rear window) and
- *Short-time loads* (e.g. turn signals, stop lamps)

Operating conditions
In order that serial-production vehicles which finally develop from test vehicles can be equipped with suitable components to cope with everyday energy requirements, the test vehicles are operated under extreme conditions. The demands made on battery, alternator, and vehicle electrical system depend to a great extent on operating conditions and the type of vehicle in which they are installed. In addition to the battery characteristics which are matched to the electrical system of the particular vehicle, such as starting power, Ah capacity, and charge-current input within a temperature range of approx. −30...+70 °C, there are a number of other battery specifications which must also be complied with under certain operating conditions. These include,

for instance, freedom from maintenance, vibration-resistance, and resistance to deep cycling.

Starting temperature
Among other things, the temperature at which the engine can still be started is dependent on battery and starter (size, with/without reduction gear, electrically excited or permanent-magnet excited). If the engine is to be started for instance at temperatures as low as −20 °C, it is imperative that the battery has a minimum charge state. With a larger battery (high Ah capacity), minimum charge state can be less than that with a smaller battery (low Ah capacity). In Europe, for instance, the following minimum start temperatures are stipulated:

Passenger cars	−18...−25 °C,
Trucks and buses	−15...−20 °C,
Tractors	−12...−15 °C.

External loading
Vehicles (cars and light commercial vehicles) which are operated normally on asphalt roads in mixed town and country traffic are not subjected to special external mechanical or cyclical loading.

On commercial and industrial vehicles (passenger cars and trucks), and for special off-road applications such as ski-slope preparation, etc., the short distances involved, together with the effects of vibration and shock, often result in very high external loading (see "Extreme Operating Conditions").

Climatic loading
Generally speaking, batteries are exposed to the effects of damp, dirt, oil, temperature, etc. and must also be able to cope with the specific climatic demands of the application in question.

Installation point
Regarding the installation point, the following critera are important:
- Easy access for battery installation
- Protection against excessive battery heat-up
- Protection against excessive battery cooling
- Protection against damp
- Protection against mechanical damage (e.g. due to excessive vibration). The battery must be securely fastened and not be subjected to vibration
- Protection against oil and fuels, etc.

Regarding battery installation, the engine compartment usually provides better accessibility than other points in the vehicle. Furthermore, in the engine compartment, short lines can be used between battery, alternator, and starter so that voltage losses are kept to a minimum.

To prevent heat accumulation in or around the battery which could accelerate unwanted chemical reactions inside it, the installation position should not be in the vicinity of the engine block or exhaust manifold. The battery should not be exposed permanently to temperatures above 50 °C, or its service life will be reduced (high levels of self-discharge).

When the battery cannot be fitted in the engine compartment, it is installed under the driver's seat or under the rear seat.

Since on batteries with vent plugs the "acid fog" which escapes through the vent holes can attack the metal components in the vicinity, adequate ventilation must be provided around the battery. Even a very slight movement of air eliminates this danger completely.

Formerly, regular servicing, and topping-up with distilled water, was decisive for the battery's service life. The demand for easy accessibility for the workshops or for the driver were therefore more important than they are today. In the case of the completely maintenance-free battery in fact, this form of servicing has become a thing of the past.

Impact of engine speed
The battery charge largely depends on how the vehicle is driven (e.g. in traffic jams, stop-and-go, or without stopping). Driving mode acts on the engine rotational speed which the engine transfers to the alternator.

Power generation in the alternator increases, the higher the engine speed. Long waiting times in traffic jams and at traffic lights, when the engine is only running at idle, result in low alternator rotational speeds and a low charge current (Fig. 1). In addition, an automobile that is not driven on long overland trips will have a poor battery charge balance.

Power demand

Standard version
The power demand, which results from the power consumption of a given vehicle, is determined in accordance with the vehicle operating conditions. It is decisive not only for the dimensioning of the battery but also of the alternator. Under the following test conditions, the original-equipment battery, as specified and installed by the vehicle manufacturer and Bosch, covers the standard power demand of the starter and the loads installed in the vehicle electrical system:
- Winter commuter traffic with day and night driving for 2 weeks, of which 1 week is at 0 °C and the other at −20 °C
- The battery's residual capacity is to be at least 50 % after completing these driving schedules

The following must still function:
- Starting at −20 °C
- The parking lamps for 12 hours
- The hazard-warning and turn-signal system must operate for 3 hours
- All loads which operate with the key removed must run for 2 hours. It must then be possible to start the engine

In addition, the standard values for battery specifications apply

Auxiliary equipment
If auxiliary equipment is chosen for the vehicle, this can lead to considerable increases in power demand. Such equipment includes comfort and convenience systems with additional servomotors for roof and power-window drives, seat and steering-wheel adjustment; as well as seat heating, air-conditioner, cooler unit or similar. The automaker takes this additional power demand into account when dimensioning the electrical components for the vehicle. This means that a vehicle equipped with such auxiliary equipment is delivered with a larger battery, and in some cases with a more powerful alternator. Similarly, specific mechanical, cyclical, or climatic loading may have to be taken into account depending upon the vehicle's application.

1 "Stop-and-go" traffic conditions can be a problem for maintaining battery charge levels if extended-duty consumers (e.g. lights) are switched on

Retrofit equipment

Auxiliary loads
Auxiliary loads which are retrofitted at a later date, or connected from time to time, may necessitate the standard battery being replaced by a more powerful one. Examples are as follows:
- Sophisticated, high-performance car radio and audio systems with high power demands
- Auxiliary lamps and high-mounted stop lamps, fanfare horns, floodlamps and spot lamps, alarm system
- Auxiliary heating system
- Connection of equipment powered by the vehicle's battery through the cigarette lighter, for instance. These include small-power compressors, small lamps, and floodlamps
- Trailers and caravans connected to the vehicle's electrical system

Caravans and mobile homes are often equipped with such electrical appliances as lighting, refrigerator, heating, radio and TV. Here, it is common to fit extra batteries with a separate circuit of their own.

Increased power demand
The above-mentioned auxiliary loads cause an increase in power demand. The existing starter battery though has been carefully matched to the remainder of the vehicle's electrical system, including the alternator, regulator, and starter. Bearing this fact in mind, in order to power the above-mentioned auxiliary loads the original starter battery cannot simply be lifted out and replaced by a battery with a higher Ah capacity. *In such cases, the expert from the Bosch Service Agent should be consulted so that the danger of making the wrong choice is ruled out from the very beginning.* For such replacements, the Bosch aftermarket program contains cross-reference lists with interchangeable batteries which not only have the same physical dimensions, but which also comply with the particular demands.

Extreme operating conditions
It is impossible to completely cover the wide range of different operating conditions which can be encountered in the field with one single standard battery. Such a battery would be far too large for normal operation and far too expensive.

Outside temperatures
Batteries with higher starting power are needed for cold countries with very low temperatures, where starting must often take place at below −20 °C. Such batteries feature an increased number of thinner plates and separators. In temperate and cold zones, there is no need to change the electrolyte's density with its fully-charged freezing limit of −68 °C. In tropical regions, though, the density must be reduced.

Mechanical and cyclical loading of the battery
In the industrial and commercial sectors (bus, taxi, ambulance, delivery van, etc.), the fact that the vehicle is repeatedly driven only short distances means that the current taken from the battery is correspondingly higher. This leads to severe cyclical loading of the battery which is supplemented by further cyclical loading due to high power demands with the vehicle stationary. Such loads include the air-conditioner, the electrohydraulic liftgate, auxiliary heating, and refrigeration unit, etc. In addition to the above cyclical loading, batteries in off-road vehicles, commercial vehicles, construction machines, and tractor vehicles, and those in agriculture and forestry applications, must withstand high vibration and impact stresses during off-road operation and on building and construction sites.

Method of operation

Within the vehicle's electrical system, the battery assumes the role of a chemical storage unit for the electrical energy (which is generated by the alternator when the vehicle is being driven). This energy must be made available to start the engine again after it has been switched off. This is one of the reasons for the battery also being known as the "starter battery". Alternator and starter battery must therefore be correctly matched to each other.

On the one hand, with the engine stopped (and therefore also the alternator), the battery must be able to deliver a high current for a brief period in order to start the engine (this is especially critical at low temperatures). And on the other, when the engine is running at idle or is switched off, it must for limited periods be able to supply some or all of the electrical energy to other important components in the vehicle's electrical system.

The battery also absorbs voltage peaks in the vehicle's electrical system so that these do not damage sensitive electronic components.

Generally speaking, the lead-acid storage battery suffices for meeting these demands, as well as at present still being the most cost-effective energy-storage medium for such assignments.

Typical system voltages are 12 V for passenger cars, and 24 V for commerical vehicles (fitted with two 12-V batteries connected in series).

Electrochemical processes in the lead storage cell
Generation of the cell voltage
If a lead electrode is immersed in dilute sulfuric acid (the electrolyte), positive ions are transferred from the electrode and into the electrolyte due to the effects of so-called "solution pressure". The transfer of the positive lead ions to the electrolyte means that negative charges (electrons) remain on the lead electrode. In other words, the lead electrode is no longer electrically neutral, but has a negative potential referred to the electrolyte.

If 2 electrodes of different materials (for instance, lead [Pb] and lead dioxide [PbO_2]) are immersed in a common electrolyte, different potentials develop at the individual electrodes with respect to the electrolyte (Fig. 1). The difference in potential between the electrodes themselves is the cell voltage.

Since the negative charges remaining on the electrode exert a force of attraction (return force) on the positive ions which have entered the electrolyte, the phenomenon described above, in which charged particles (here the lead ions) are released into the electrolyte, very quickly results in a condition of equilibrium. After a certain period, this return force exactly equals the solution pressure.

If an external voltage is applied across the electrodes, depending on the direction of current, further electrical particles can be released into the electrolyte or they can return from the electrolyte to the lead electrode. It is this fact which makes it possible to recharge the lead-acid storage battery (secondary cell).

In a charged lead cell, the positive electrode consists essentially of lead dioxide, and the negative electrode of pure lead. With the electrolyte (diluted sulfuric acid) used in lead cells, current transportation is via ionic conduction. In the aqueous solution, the sulfuric-acid molecules split into positively charged hydrogen ions and negatively charged residual acid ions. The splitting of the sulfuric-acid molecules is the prerequisite for the electrolyte's conductivity and therefore for the flow of charge or discharge current. When discharge current flows, the positive electrode's lead dioxide and the negative electrode's lead are converted to lead sulfate ($PbSO_4$).

In a discharged lead cell (Fig. 1), both electrodes consist of lead sulfate.

The electrolyte is dilute sulfuric acid (17% pure sulfuric acid [H_2SO_4] and 83% water [H_2O]). The sulfuric-acid component

streamlined body shapes has resulted in an increase in the average engine-compartment temperature. This change in conditions also has an effect on the battery, as a result of which the most recent development in battery design has seen the use of an improved lead alloy for the grids of the positive plates. In addition to having a lower calcium content and a higher tin content, they also contain a proportion of silver.

This alloy combined with a finer grid structure has proved to be extremely durable even at high temperatures – which have the effect of accelerating corrosion. This is true after destructive overcharging with high electrolyte gravity and during (equally undesirable) periods of disuse with low electrolyte gravity. The optimized geometry of the grid structure with additionally optimized electrical conductivity allows better utilization of the active material and amplifies this effect. The center tab for the cell connector provides for balanced fixing of the grid plates inside the battery casing. This technology offers the potential for making the plates even thinner (but stronger) and therefore increasing their number. That will make it possible to further increase cold-starting power without sacrificing quality.

"Robust-design" versions have shorter and thicker positive plates with a stronger frame. The shorter grid plates allow for a greater volume of electrolyte above the plates. Consequently, they are always covered by electrolyte and thus protected against corrosion. Such properties make this type of battery generally "tougher" in practical applications.

Water consumption
When new, starter batteries without antimony as well as those with a reduced antimony content, feature a far lower water consumption than the maximum 6 g/Ah as stipulated by DIN. As a rule the lead-calcium battery has a long-term figure of less than 1 g/Ah.

The fact that the gassing voltage remains at its high initial level throughout the battery's complete service life is responsible for the water-consumption figures for the main- tenance-free batteries being so favorable. High gassing voltage leads to minimum water decomposition.

This has the following advantages:
- In the case of the maintenance-free batteries, the charging voltage only exceeds the gassing voltage at high temperatures. This means that only rarely does gassing (water decomposition) occur at all, so that topping up with distilled water remains unnecessary throughout the battery's service life. This is the reason why the vent plugs are covered by a plate, and can only be opened when necessary by an authorized workshop. Or there are no vent plugs fitted at all.
- It is no longer possible to forget to top-up with distilled water when needed, or to use contaminated water, or even to use so-called "battery improvers".
- Injuries and the danger of damage due to contact with sulfuric acid are a thing of the past.
- Maintenance and servicing costs are reduced.
- It is no longer necessary to install the battery in the vehicle in an easily accessible position.

Battery plate with envelope-type separator

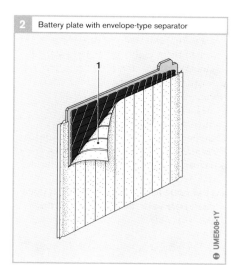

Fig. 2
1   Separator cut open

Characteristics
A vast amount of development work and manufacturing experience are behind the batteries from Bosch. Although this has its price, it certainly provides the user with a high level of reliability and a long service life. These are advantages which are by no means a matter of course for every battery presently being sold on the market.

Bosch starter batteries are completely maintenance-free, and in addition to complying with the performance values stipulated by DIN, they also fulfill the following requirements:
- Freedom from maintenance for the battery's complete usable life.
- No impairment of performance data and charging characteristic as a result of water consumption.
- Performance data and charging characteristic remain as constant as possible throughout the battery's complete usable life.
- Following exhaustive discharge and subsequent extended period of non-use while connected into the vehicle's electrical system, it must still be possible to recharge the battery under normal operating conditions.
- In case of seasonal operation without intermediate charging (but with disconnected ground cable), there is to be no reduction in usable life compared to all-year-round operation.
- It must be possible to store the filled battery for long periods.

External features
The maintenance-free starter battery from Bosch has the following external features:
- Terminal posts protected by caps against inadvertent short-circuit.
- Cover plate over the vent-plug trough prevents the accumulation of dirt and moisture, and serves to cover the vent plugs.
- Grips facilitate easy transportation.
- Labyrinth cover with central gas vent prevents the escape of electrolyte in case the battery is tipped-up inadvertently for a brief period.

Internal features
The internal features of the Bosch maintenance-free batteries are:
- Plate grids of lead-calcium alloy. In some versions, the pasted-on active material is alloyed with silver.
- Battery case without sludge ribs. The plates reach down to the floor of the case (= increased plate surface) with which they are in contact along their complete length (increased stability).
- Microporous envelope-type separators prevent the active material crumbling off of the plates, as well as preventing the formation of short circuits at the bottoms and at the side edges of the plates. The mean pore diameter is smaller than that of conventional separators by the factor of 10, and in addition to the reduced electrical resistance, this is also an effective measure in preventing short-circuits.
- The very low self-discharge level means that all lead-calcium batteries can be filled with sulfuric acid at the factory. This procedure takes place under optimal conditions, and rules out the danger of acid being splashed during the filling and mixing process.

Self-discharge
The self-discharge of the negative and positive plates is inherent in the principle of the lead storage battery. Even without external loads connected, the effects of temperature and other factors lead to the battery becoming electrically "empty" after a given period of time. Considering conventional starter batteries, antimony poisoning leads to an increase of the self-discharge reaction at the negative plates, and the rate of discharge climbs considerably along with increasing battery age. In the field, this means that after 6-month storage at room temperature (20 °C), new, conventional-type starter bat-

teries only have an electrolyte specific gravity of 1.20 kg/l. This corresponds to a state of charge of approx. 65%. Under certain circumstances, old batteries drop to this value within a few weeks.

On the other hand, with the maintenance-free starter batteries from Bosch, the electrolyte specific gravity is still 1.26 kg/l after the same period of time (corresponding to a 90% state of charge), and it takes 18 months for it to drop to 1.20 kg/l (Fig. 3).

The lead-calcium alloy system used in maintenance-free batteries for the plate grids means that the accelerated self-discharge due to antimony poisoning is a thing of the past. The level of negative and positive plate self-discharge stays at a low level throughout the battery's useful life. The self-discharge phenomenon is of particular significance for vehicles used only during specific seasons (agriculture, forestry, and construction sites). It is also important though for second cars and mobile homes which are not driven in winter, or only rarely. It also applies to new vehicles which, since there is no stop in production, are manufactured but have to be parked for long periods due to seasonal fluctuations in sales trends or the transportation time to the salesroom.

In all these cases, the battery must have as high a state of charge as possible before the vehicle is taken out of operation, and it must disconnected from quiescent-current loads the whole time it is not in use. This is done, for instance, by dis-connecting the ground cable.

Starting power
The Bosch maintenance-free starter battery features a higher starting power than a conventional battery. Basically speaking, this is due to the envelope separators with their low specific resistance, and to the increase in plate surface due to the omission of the sediment chamber.

Furthermore, thanks to the lead-calcium alloy used for the plate grids, the maintenance-free battery's starting power remains practically unchanged for years. It doesn't drop to below the DIN figure for new batteries until towards the end of its useful life. Whereas after 75% of its useful life the maintenance-free battery is still above the DIN figure, the conventional battery drops far below this DIN figure considerably sooner (at approx. 40% of useful life). In practice it has already lost about a third of its original starting power after only 75% of its useful life (Fig. 4).

3 Electrolyte specific gravity as a function of storage time (20 °C room temperature)

4 Starting power as a function of useful life

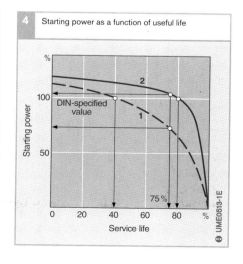

Fig. 3 and 4
1 Conventional starter battery (PbSb)
2 Maintenance-free starter battery (PbCa)

Power input
Low-antimony and antimony-free batteries behave almost the same during the DIN power-input test. In the field, differences are negligible when conventional regulator curves are used, and if the tendency for better power consumption on the part of the maintenance-free lead-calcium-alloy battery at a state of charge below 50 % is ignored. When charging with regulators which take battery temperature into account and which can increase the voltage to above 14.5 V, the maintenance-free battery has clear advantages. This is due to the phenomenon applying to every battery, namely the fact that the lower the temperature, the higher the charging voltage required for a given state of charge. Since its gassing curve is higher during charging, the maintenance-free battery stores the increased charging current without losses due to gassing. In other words it achieves a higher level of charge and therefore provides more starting power.

Useful life
Considering the extremely varied electrical and mechanical loadings placed upon the starter battery in the field, a brief laboratory test on a battery which is immovably clamped down on a test bench is totally inadequate to ascertain its useful life. Most durability tests are aligned to antimony-content batteries and are therefore only of limited use when maintenance-free batteries are concerned. This is why Bosch tests its batteries under field conditions. During such a test, the starting capability of a conventional battery drops to 50 % after 60 months, whereas with lead-calcium batteries it takes 80 months for this level to be reached. Under the cyclical loading conditions encountered on taxis in densely populated areas, on town buses, and on delivery vans, maintenance-free starter batteries have come to the forefront due to the advantages inherent in their separator concept and the reliable protection it affords against premature failure.

Resistance to overcharge
Overcharge is a decisive factor with regard to the battery's usable life. It applies for instance to vehicles which have very high annual mileages, as well as to courier and agricultural vehicles, construction-site machinery, and long-distance haulage trucks. In such cases, the battery is fully charged, the engine turns at high speed, and the alternator only supplies a few loads. Under these circumstances, the charge current leads to overcharging, corrosion, and loosening of the active material. In laboratory tests to simulate these conditions at 40 °C electrolyte temperature and 14 V charge voltage, the maintenance-free starter battery has a considerably longer useful life than the antimony-content battery.

Resistance to exhaustive discharge
To check the battery's resistance to exhaustive discharge, lamps are used to discharge it completely, after which it is left 4 weeks in the short-circuit condition. The battery must then recharge under the conditions of normal vehicle electrical-system connection, it must still be fully operational, and it must only display certain specified reductions in performance.

electrical loads. The vehicle system voltage is 12 V.

When the starting switch is turned, the battery changeover relay automatically switches the two batteries in series so that 24 V is applied across the starting-motor terminals during the cranking process (Fig. 1). All other loads/consumers are still supplied with 12 V.

As soon as the engine has started, that is when the starting switch has been released and the starting motor switched off, the battery changeover relay automatically connects the batteries in parallel again. With the engine turning, the 12 V alternator recharges both batteries.

Components

Battery master switch

Generally, the vehicle's electrical installation is wired such that when the key is pulled from the starting switch the electrical lines leading from the switch are "dead". That is, the lines to the ignition system, the ECU (Motronic, ABS), and to the wipers etc. On the other hand, the lines leading to the ignition/starting switch, to the starting motor, and to the light switch remain "live". In other words there is still voltage on these lines, and if they have frayed or worn-through points these can lead to low resistances which can cause leakage currents or short circuits which result in a discharged battery. And there is even the possibility of fire breaking out. These dangers can be alleviated by fitting a battery master switch (Fig. 2).

The single-pole battery master switch is installed in the battery's ground line (negative terminal) as near to the battery as possible. It should be within convenient reach of the driver. On installations equipped with alternators, due to the danger of voltage peaks (with the attendant destruction of electronic components), it is forbidden to operate the vehicle without the battery connected. On such installations therefore, the battery master switch may only be actuated with the engine at standstill.

Battery relay

Legislation stipulates that in buses, road tankers etc., a battery relay (Fig. 3) must be installed as the master switch to separate the vehicle electrical system from the battery. Not only short circuits are avoided (during repair for instance), but also the decomposition effects due to creepage currents on current-carrying components.

For this type of installation with alternator, in order to prevent excessive voltage peaks it is necessary to fit a 2-pole elec-

2 Battery master switch

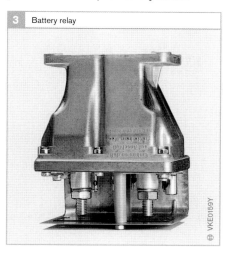

3 Battery relay

tromagnetic battery master switch. This prevents the alternator being separated from the battery when the engine is running.

Battery changeover relay

The battery changeover relay is used for connecting two 12 V batteries in series or in parallel. It is used for instance in commercial vehicles which have a 12 V electrical system but a 24 V starting motor (see "Commercial-vehicle circuits").

Battery cutoff relay

The battery cutoff relay separates the starter battery from a second battery used for ancillary equipment. It protects the starter battery against discharge when the alternator is not delivering charge current. This relay is provided with a diode for protection against false polarity, and a decay diode to suppress the inductive voltage peaks caused by switching (Fig. 4).

Battery charging relay

The battery charging relay is needed when an additional 12 V battery is to be charged in a 24 V vehicle system. It is provided with resistors across which at 10 A a voltage drop takes place which reduces the charging voltage to 12 V. This of course necessitates the 24 V alternator being able to generate the additional 10 A.

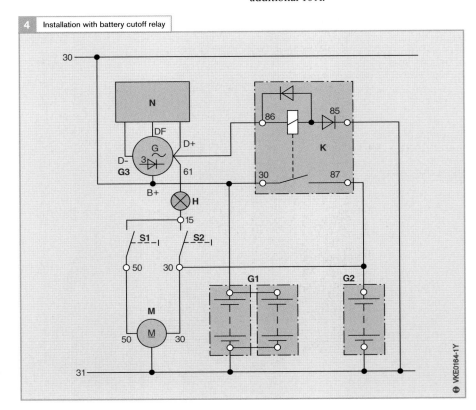

Fig. 4
- G1 Battery for ancillary equipment
- G2 Starter battery
- G3 Alternator
- H Charge-indicator lamp
- K Battery cutoff relay
- M Starting motor
- N Alternator regulator
- S1 Ignition/starting switch
- S2 Driving switch

Battery history

When considering the history of the battery, a number of renowned scientists and inventors deserve particular credit. Above all, such personalities as Luigi Galvani (1789), Alessandro Graf Volta (around 1800), Johan Ritter (around 1800), Gaston Planté (1859) and Camille Faure were decisive in driving on the development of the battery (accumulator). At the end of the 19th century grid plates were already being manufactured. Their basic principles are still found in today's lead batteries.

In other words, basically speaking the lead battery has hardly changed up to the present day: It still contains cells, grid plates, and sulfuric acid.

If we examine the battery more closely, we find that not only has the energy density increased immensely, but also that the previously used materials (separators and case were made of wood) have been superseded to a great extent by plastic, and 100% freedom from maintenance is today standard practice for starter batteries.

In exceptional cases, these can achieve a service life comparable to that of the automobile itself.

The history of batteries in figures
1905: The first batteries were fitted in a motor vehicle (at first only for lighting purposes).
1914: The first starter battery was fitted in a motor vehicle.
1922: The first Bosch motorcycle batteries were installed.
1926: The first Bosch battery charger was introduced.
1927: Starting in 1927, Bosch also developed automotive starter batteries, and in 1936 started to mass-produce them.
After the Second World War, the development of Bosch automotive batteries was marked by:
- The introduction of plastics to battery manufacture (e.g. polystyrene in 1955; polypropylene in 1971)
- The improvement of individual battery components (e.g. the folded-rib separator in 1956; the one-piece battery cover for 6-V batteries in 1964 and for 12-V batteries in 1966; direct cell connectors in 1971; expanded-metal technology for the negative grid in 1985) and
- The production of special-type batteries (e.g. "deep-cycle resistant" in 1969; "low-maintenance" in 1979; "vibration-proof" in 1980; "maintenance-free" in 1982, and "100% maintenance-free" in 1988)

Starter battery: 1951 version

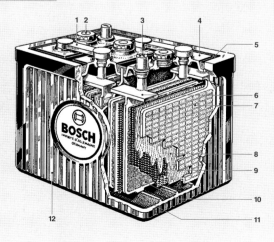

1 Joining bar
2 Vent plug
3 Terminal post
4 Cell cover
5 Sealing compound
6 Plate strap
7 Negative plate
8 Wooden separator
9 Hard-rubber separator
10 Positive plate
11 Cell connector
12 Battery case

Drive/traction batteries

The electric drive is quiet, produces no exhaust emissions and is very efficient. Whereas in purely electrically powered vehicles the electric drive alone powers the wheels, hybrid vehicles have at least two different sources of drive energy, at least one of which is usually an electric drive.

Electrically powered vehicles

Application
In contrast to internal-combustion engine vehicles, on electric-only vehicles the energy accumulator generally determines the vehicle's performance. The capacity of the electric motor is matched to the maximum output of the energy accumulator. The energy accumulator may take the form of an electrochemical battery or a fuel cell and its associated fuel tank.

Depending on the intended application, battery-powered electric vehicles can be classified as either road vehicles (Fig. 1) or industrial trucks. Industrial trucks are used for transporting goods on company premises, and are generally not licensed for use on public roads. Their top speed is below 50 km/h. Due to the low power density of the batteries, the range of battery-powered on-road vehicles is significantly less than that of vehicles powered by internal-combustion engines. The maximum speed of such vehicles is also normally limited to around 130 km/h. Whereas more than half of all new industrial trucks are electrically powered, the percentage of electrically powered on-road vehicles is very low.

Energy supply
There is no shortage of power for electric vehicles that are recharged by plugging them into a wall socket. If electric vehicles in Germany were largely recharged at night, existing power plants could provide enough energy to charge more than 10 million vehicles. That number of electric vehicles would require less than 5 % of Germany's total electricity output.

Any household power outlet can be used to charge the batteries. However, these outlets can provide only 3.7 kW of electrical power, which means that an hour of charging would provide enough power to drive a distance of no more than about 20 km. Shorter recharging periods can be achieved by using a three-phase AC power source (as for industrial trucks). Compared to the refueling times for diesel vehicles in particular, the recharging periods required by comparable electric vehicles in order to cover the same distances are roughly 100 times longer, even in the case of very high charging capacity.

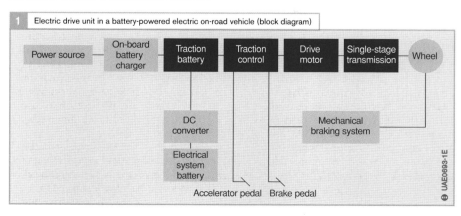

Figure 1: Electric drive unit in a battery-powered electric on-road vehicle (block diagram)

Electric drives

Drivetrains

The drivetrain in an electric vehicle generally consists of the power controller, the motor and the transmission. The power controller translates the position of the accelerator pedal into the appropriate motor current and voltage. In most cases the drive torque is a function of the accelerator pedal position, as in the case of IC engines.

As the cost of the motor is largely determined by the required maximum torque, it is advantageous to use the lowest possible gearing ratio between the motor and the driving wheels. The reduction gearing may offer one or more gear ratios depending on the desired hill-climbing ability and the vehicle's maximum speed within the constraints of the given maximum torque and speed range of the motor. Modern electric cars have single-ratio reduction gears.

A difference between electric drive units and combustion engines is the necessary distinction between short-term and extended-duty performance. Short-term performance is usually limited by the maximum setting of the power controller. The maximum power available over longer periods is defined by the half-hourly output in the case of on-road electric vehicles, which is generally limited by the permissible motor temperature. This distinction also applies to most batteries. Depending on the type of drive, short-term and extended-duty ratings vary by a factor of 1 to 3. Maximum drive power must therefore be monitored and adjusted, if necessary, in accordance with the characteristic thermal limits of the power controller, motor or batteries.

This distinction between short-term and half-hourly operation has among other things also led to the adoption of two maximum-speed ratings for electric on-road vehicles: maximum speed over a distance of 2 x 1 kilometer and maximum speed over a period of 30 minutes.

Series-wound direct-current drive

This type of drive unit has the simplest type of power controller. The motor voltage is set in accordance with the desired current by applying the battery voltage to the motor in a variable on/off ratio and/or chopper frequency by means of a circuit breaker (thyristor or transistor(s)).

For the recovery of braking energy, the power controller must operate as a step-up chopper, which means that additional components are needed. Because the field and armature of the motor are in series, drive power drops in proportion to the square of the motor speed with the full battery voltage applied (Fig. 2a).

Although its efficiency is relatively low, this type of drive is still used in most industrial trucks today because of its simple design and low cost. The low top speeds of these vehicles make it possible to use single-stage reduction gears.

Separately-excited direct-current drive

In this type of drive unit, the motor's magnetic excitation is provided by its own controller (field rheostat). Depending on the size of the motor, the field can be weakened in a ratio of up to approx. 1:4. Field strength starts to diminish at a nominal motor speed obtained with full motor voltage at the armature and maximum field current. During initial acceleration with maximum field current an electronic armature-control device limits the motor current until the motor reaches its nominal speed, with the full motor voltage applied to the armature. In the reduced field-strength range, consistent armature currents produce relatively constant power outputs. Because commutation becomes more difficult as the field current drops, the armature current must usually be reduced before the maximum speed is reached (Fig. 2b and 3a).

As commutating poles are required, this design is somewhat more complex than that of a series-wound motor. The mechanical commutator limits rotational speed to roughly 7,000 rpm. This type of drive unit can be used with a multi-stage transmission to reduce motor cost and weight. Efficient energy recovery during braking is possible without requiring additional components.

However, very few electric cars are being equipped with direct-current drive units. Three-phase AC asynchronous or synchronous drives are now the norm, due in part to their low maintenance requirements. The carbon brushes in DC motors must be regularly replaced, albeit at relatively long intervals.

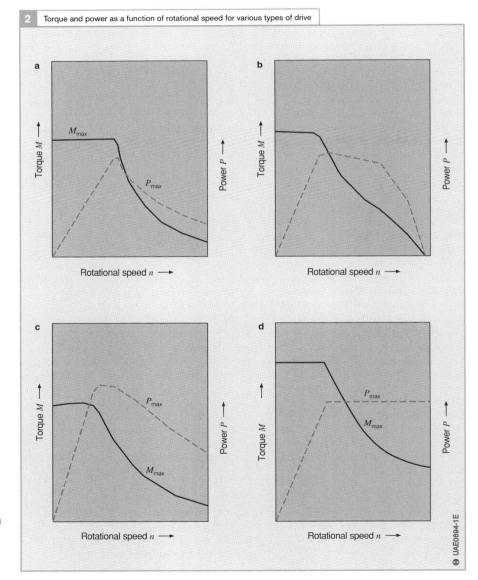

Fig. 2
a Series-wound DC drive
b Separately-excited DC drive
c Asynchronous drive
d Permanently-excited synchronous drive

M_{max} Maximum torque
P_{max} Maximum power

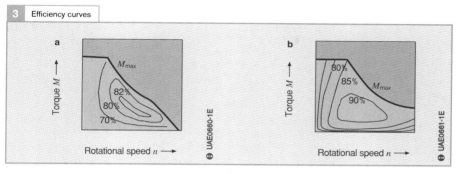

Fig. 3
a Separately-excited DC drive
b Permanently-excited synchronous drive

M_{max} Maximum torque

Asynchronous drive

The motor in an asynchronous drive unit is the simplest and most economical in design, and is also considerably smaller in size and weight than a direct-current motor. In principle, however, the controller in a three-phase drive unit is more complex than that used in DC drives. As with the separately-excited DC motor, operation with reduced field current is possible. Because these motors have no mechanical commutator, they can operate at speeds of up to 20,000 rpm if appropriately designed. This means that single-stage transmissions can be used, even in on-road vehicles. These drives are more efficient than direct-current drives, but not quite as efficient as synchronous drives with permanent magnets. Braking energy can also be recovered with a high degree of efficiency (Fig. 2c).

Permanently-excited synchronous drive

This type of drive is characterized by very-high efficiencies also in the part-load range because it uses permanent magnets to generate the excitation field. Rare-earth magnets with a high energy density allow very compact dimensions combined with high torque. However, rare-earth magnets make the motor more expensive than asynchronous designs, for example. This type of motor is not capable of operation with reduced field current. A quasi weakened-field operating mode can be achieved by increasing the longitudinal component of the stator current by reducing the torque-generating component. At present there are motors with "buried" magnets available which achieve field-weakening ratios of as much as 1:3 due to their pronounced asymmetry. Alternatively, windings with extremely high inductivity are used which permit substantially greater degrees of field reduction. As operation with virtually constant power output is possible, a single-ratio reduction gear is again sufficient in most cases (Figures 2d and 3b).

Battery systems

Whereas costs considerations dictate that the lead-acid battery is the power source used most often in industrial trucks, in modern electric cars it is increasingly being replaced by nickel and lithium batteries.

Lead-acid battery

Although the basic design of the lead-acid battery is the same as that of the starter battery, the combinations of materials and the cell design are specially adapted to the particular requirements of traction operation. The batteries commonly used in industrial trucks are generally combinations of individual cells, whereas a modular design with 3 or 6 cells per module is used in most electric on-road vehicles due to the higher energy density.

Industrial trucks generally use lead-acid batteries with a liquid electrolyte which must be topped up with water on a regular basis. In the case of electric on-road vehicles, this level of maintenance is not acceptable for the vehicle's user. Consequently, maintenance-free batteries with a solid electrolyte (gel) have become standard in these applications. Under real-world conditions, vehicles equipped with lead-acid batteries have a range of 50...70 km per battery charge in city driving. The daily range of an electric vehicle can be increased through intermediate charging of the batteries when the vehicle is parked.

The amount of energy which can be drawn from a lead-acid battery decreases as battery temperature drops. This means that, at temperatures below 0 °C, a battery heater is required in order to enable the battery to be fully charged within a reasonable period of time. The energy required for the heater is usually drawn from the mains supply along with the charging current. The power and energy that can be drawn from the battery similarly diminish as the temperature drops. The result is slower acceleration and shorter range.

Because of the involvement of the electrolyte in the chemical reaction, the usable capacity of lead-acid batteries is dependent on the load current. Driving continuously at full power can therefore reduce vehicle range by 20 % and more.

Batteries in industrial-truck applications can achieve service lives of 7...8 years with 1,200...1,500 cycles. Fleet experience with electric passenger cars indicates that lead-acid batteries can be expected to last for around 5 years and roughly 700 cycles. The shorter service life in electric on-road vehicle applications is primarily a result of the much greater battery load. In these vehicles the battery is discharged in an average of 2 hours or less, whereas discharge times in industrial trucks are generally in the 7...8 hour range.

Nickel-based batteries

Nickel-cadmium batteries and, increasingly, nickel-metal hydride batteries with an alkaline electrolyte are now used in many electrical appliances. Because cadmium is harmful to the environment, it is likely that in the foreseeable future the nickel-cadmium system will be replaced by the nickel-metal hydride system. Whereas appliance batteries are generally of the sealed type, open nickel-cadmium cells – which, like open lead-acid batteries, must be refilled with water at regular intervals – are often more suitable for traction applications. Nickel-metal hydride batteries are always sealed due to the inher-

Table 1 Battery systems

Properties	Lead-acid system open/sealed	Nickel systems Nickel-cadmium (Ni/Cd) Nickel-metal hydride (NiMH)	Lithium systems Lithium-ion Lithium-polymer
Cell voltage	2 V	1,2 V	3...4 V
Energy density	25...30 Wh/kg	35...80 Wh/kg	60...150 Wh/kg
Energy efficiency without heating/cooling	75...80 %	60...85 %	85...90 %
Power density	100...200 W/kg	100...1,000 W/kg	300...1,500 W/kg
Service life in cycles	600...900	> 2,000	> 1,000 projected
Operating temperature	10...55 °C	−20...55 °C	−10...50 bzw. 60 °C
Maintenance-free	Depending on design	Depending on design	yes

Table 2 Commercially available vehicles (examples)

Vehicle type	Type of battery	Engine power	Acceleration 0...50 km/h	Maximum speed	Typical range per charge	Typical line-power consumption
Passenger car	Ni/Cd	21 kW	9 s	90 km/h	80 km	18 kWh/100 km
Passenger car	NiMH	49 kW	7 s	130 km/h	200 km	26 kWh/100 km
Passenger car	Lithium-ion	62 kW	6 s	120 km/h	200 km	23 kWh/100 km
Van	Lead-acid	80 kW	7 s	120 km/h	90 km	35 kWh/100 km

ent characteristics of the system. The low cell voltage of only 1.2 V necessitates a larger number of cells (e.g. for a 6-volt module) than in a lead-acid battery. A battery service life of up to 10 years or 2,000 cycles has been demonstrated in a number of applications. The higher costs resulting from the use of relatively expensive materials and the complex manufacturing process are partially offset by a much longer service life than that of lead-acid batteries.

Nickel-cadmium and nickel-metal hydride batteries are cooled when used in electric on-road vehicles; heating is required only at temperatures below −20 °C. Available capacity is virtually independent of discharge time. The alkaline battery's higher energy density can be exploited both to increase the payload and to extend the vehicle's radius of action. Electric cars typically have a range of approx. 80...100 km using nickel-cadmium batteries.

In the nickel-metal hydride system, the cadmium is replaced by a hydrogen-storing metal alloy. This hydrogen must be stored in a multimetal medium. The nickel-metal hydride battery has a higher energy density and a somewhat longer service life than the nickel-cadmium battery. Nickel-based batteries also have a higher power density, making them particularly interesting for use in hybrid vehicles.

Lithium-based batteries

Lithium systems allow energy densities of over 100 Wh/kg and power densities of over 300 W/kg in vehicle traction batteries. They can be operated at ambient temperature or slightly higher temperatures, and are characterized by high cell voltages of over 4 V. In the demanding electrical-appliance battery market (for products such as laptops and video recorders) the lithium-ion system has already become successfully established.

Lithium systems do not show any memory effect as do nickel-cadmium systems. A disadvantage of the lithium batteries is that they require a relatively complex battery protection system. For example, each individual cell must be monitored because they are not proof against overcharging. In order to prevent endangerment of the environment, these batteries must also be specially protected against short circuits.

Lithium-ion battery

A lithium-ion battery stores lithium ions in electrically reversible form on the negative electrode in a graphite lattice. In the most commonly used appliance batteries of this type, the fundamental component of the positive electrode is cobalt oxide, which makes the system very costly. Attempts are thus being made to use more economical materials such as manganese oxide or nickel oxide. Organic material is used as the electrolyte; water-based electrolytes cannot be used because lithium reacts strongly with water.

Lithium-polymer battery

Another very promising lithium system is the lithium polymer battery. It consists of a thin lithium film, a polymer electrolyte and a positive film electrode made primarily of vanadium oxide. Individual cells are formed by rolling or folding the film, which has an overall thickness of approx. 0.1 mm. The working temperature is approx. 60 °C. Work is also being carried out on developing lithium-ion appliance batteries in which the organic liquid electrolyte is replaced by a polymer electrolyte.

Alternators

In order to supply the power required for the starter motor, for ignition and fuel-injection systems, for the ECUs to control the electronic equipment, for lighting, and for safety and convenience electronics, motor vehicies need an alternator to act as their own efficient and highly reliable source of energy. Energy which must always be available, at any time of day or night.

Generation of electrical energy in the motor vehicle

Onboard electrical energy

Assignments and operating conditions
Whereas, with the engine stopped, the battery is the vehicle's energy store, the alternator becomes the on-board "electricity generating plant" when the engine is running. Its task is to supply energy to all the vehicle's current-consuming loads and systems (Fig. 1). In order that the entire system is reliable and trouble-free in operation, it is necessary that the alternator output, battery capacity, and starter power requirements, together with all other electrical loads, are matched to each other as optimally as possible.

For instance, following a normal driving cycle (e.g. town driving in winter), the battery must always still have sufficient charge so that the vehicle can be started again without any trouble no matter what the temperature. And the ECUs, sensors and actuators for the vehicle's electronic systems (e.g. for fuel management, ignition, Motronic, electronic engine-power control, antilock braking system (ABS), traction control (TCS), etc.) must always be ready for operation.

Apart from this, the vehicle's safety and security systems as well as its signaling systems must operate immediately, the same as the lighting system at night or in fog. Furthermore, the driver-information and convenience systems must always function correctly, and with the vehicle parked, a number of electrical loads should continue to operate for a reasonable period without discharging the battery so far that the vehicle cannot be started again.

As a matter of course, millions of motorists expect their vehicle to always be fully functional, and demand a high level of operational reliability from its electrical system. For many thousands of miles – in both summer and winter.

Electrical loads
The various electrical loads have differing duty cycles (Fig. 2). A distinction is made between permanent loads (ignition, fuel injection, etc.), long-time loads (lighting, car radio, vehicle heater, etc.), and short-time loads (turn signals, stop lamps, etc.).

Some electrical loads are only switched on according to season (air-conditioner in summer, seat heater in winter). And the operation of electrical radiator fans depends on temperaure and driving conditions.

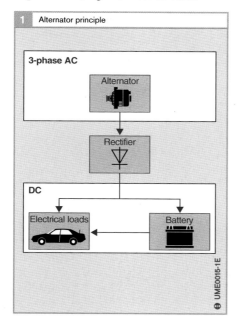

Fig. 1
The 3-phase AC is rectified in the alternator to provide the DC for the vehicle's electrical loads and for charging the battery.

Alternators   Generation of electrical energy in the motor vehicle

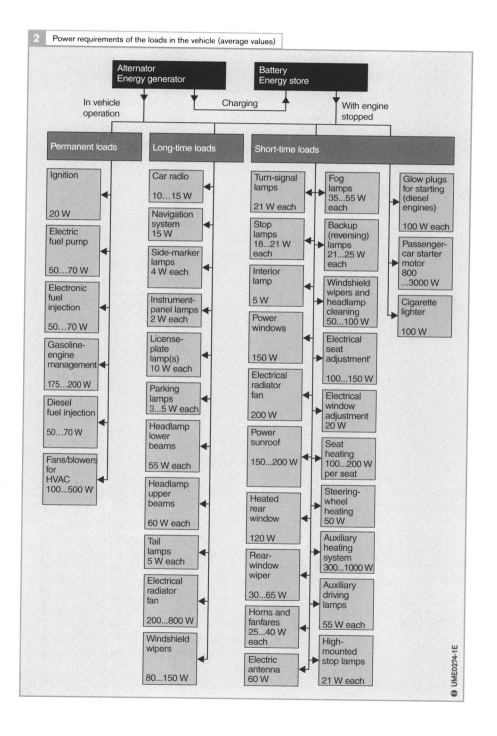

2 Power requirements of the loads in the vehicle (average values)

Charge-balance calculation
Here, a computer program is used to determine the state of battery charge at the end of a typical driving cycle, whereby such influences as battery size, alternator size, and load input powers must be taken into account.

Rush-hour driving (low engine speeds) combined with winter operation (low charging-current input to the battery) is regarded as a normal passenger-car driving cycle.

In the case of vehicles equipped with an air conditioner, summer operation can be even more unfavorable than winter.

Vehicle electrical system
The nature of the wiring between alternator, battery, and electrical equipment also influences the voltage level and, as a result, the state of battery charge.

If all electrical loads are connected at the battery, the total current (sum of battery charging current and load current) flows through the charging line, and the resulting high voltage drop causes a reduction in the charging voltage.

Conversely, if all electrical devices are connected at the alternator side, the voltage drop is less and the charging voltage is higher. This though may have a negative effect upon devices which are sensitive to voltage peaks or high voltage ripple (electronic circuitry).

For this reason, it is advisable to connect voltage-insensitive equipment with high power inputs to the alternator, and voltage-sensitive equipment with low power inputs to the battery.

Appropriate line cross-sections, and good connections whose contact resistances do not increase even after long periods of operation, contribute to keeping the voltage drop to a minimum.

Electrical power generation using alternators

The availability of reasonably priced power diodes as from around 1963, paved the way for Bosch to start with the series production of alternators. Thanks to its design principle, the alternator has far higher electromagnetic efficiency than the DC generator. This fact, together with the alternator's much wider rotational-speed range, enables it to deliver power, and cover the vehicle's increased power requirements, even at engine idle. Since the alternator speed can be matched to that of the engine by means of a suitable transmission, this means that the battery remains at a high charge level even in winter during frequent town driving.

The increased power requirements mentioned above, result from the following factors: The increase in the amount of electrical equipment fitted in the vehicle, the number of ECUs required for the electronic systems (e.g. for engine management and for chassis control), and the safety, security and convenience electronics. The expected power requirements up to the year 2010 are shown in Fig. 3.

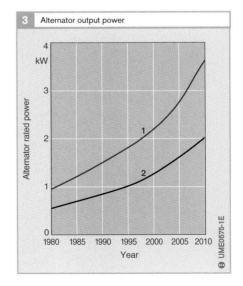

Fig. 3 Alternator output power

Fig. 3
Expected developments for passenger cars up to the year 2010.

1 Luxury car
2 Intermediate-size car

Apart from these factors, typical driving cycles have also changed, whereby the proportion of town driving with extended stops at idle has increased (Fig. 4).

The rise in traffic density leads to frequent traffic jams, and together with long stops at traffic lights this means that the alternator also operates for much of the time at low speeds which correspond to engine idle. Together with the fact that longer journeys at higher speeds have become less common, this has a negative effect on the battery's charge balance. And it is imperative that the battery continues to be charged even when the engine is idling.

At engine idle, an alternator already delivers at least a third of its rated power (Fig. 5).

Alternators are designed to generate charging voltages of 14 V (28 V for commercial vehicles), and 42 V (undergoing development). The three-phase winding is incorporated in the stator, and the excitation winding in the rotor.

The three-phase AC generated by the alternator must be rectified, the rectifiers also preventing battery discharge when the vehicle is stationary.

The additional relay as required for the DC generator can be dispensed with.

Design factors
Rotational speed
An alternator's efficiency (energy generated per kg mass) increases with rotational speed. This factor dictates as high a conversion ratio as possible between engine crankshaft and alternator. For passenger cars, typical values are between 1:2.2 and 1:3, and for commercial vehicles up to 1:5.

Temperature
The losses in the alternator lead to heating up of its components. The input of fresh air to the alternator, or the use of liquid cooling, are suitable measures for reducing component temperature and increasing alternator service life.

Vibration
Depending on installation conditions and the engine's vibration patterns, vibration accelerations of between 500...800 m/s² can occur at the alternator. Critical resonances must be avoided.

Further influences
The alternator is also subjected to such detrimental influences as spray water, dirt, oil, fuel mist, and road salt.

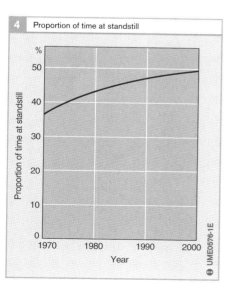

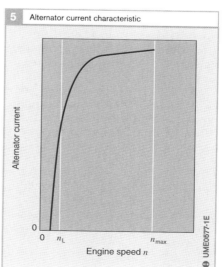

Fig. 4
Developments for urban traffic (large cities) up to the year 2000.

Fig. 5
At constant voltage
n_L Idle speed
n_{max} Maximum speed

Electrical power generation using DC generators

Originally, the conventional lead-acid battery customarily fitted in motor vehicles led to the development of the DC generator, and for a long time this generator system was able to meet the majority of the demands made upon it.

Consequently, until the middle of the seventies, most vehicles were equipped with such DC generators. Today though, these have become virtually insignificant in the automotive sector and will not be dealt with in detail here.

With the DC generator, it proved to be more practical to rotate the magnetic lines of force, while locating the electrically excited magnetic system in the stationary housing. The alternating current generated by the machine is then rectified relatively simply by mechanical means using a commutator, and the resulting direct current supplied to the vehicle electrical system or the battery.

Requirements to be met by automotive generators

The type and construction of an automotive electrical generator are determined by the necessity of providing electrical energy for powering the vehicle's electrical equipment, and for charging its battery.

Initially, the alternator generates alternating current (AC). The vehicle's electrical equipment though requires direct current (DC) for keeping the battery charged and for powering the electronic subassemblies. The electrical system must therefore be supplied with DC.

The demands made upon an automotive generator are highly complex and varied:
- Supplying all connected loads with DC.
- Providing power reserves for rapidly charging the battery and keeping it charged, even when permanent loads are swiched on.
- Maintaining the voltage output as constant as possible across the complete engine speed range independent of the generator's loading.
- Rugged construction to withstand the under-hood stresses (e.g. vibration, high ambient temperatures, temperature changes, dirt, dampness, etc.).
- Low weight.
- Compact dimensions for ease of installation.
- Long service life.
- Low noise level.
- A high level of efficiency.

Characteristics (summary)

The alternator's most important characteristics are:
- It generates power even at engine idle.
- Rectification of the AC uses power diodes in a three-phase bridge circuit.
- The diodes separate alternator and battery from the vehicle electrical system when the alternator voltage drops below the battery voltage.
- The alternator's higher level of electrical efficiency means that for the same power output, they are far lighter than DC generators.
- Alternators feature a long service life. The passenger-car alternator's service life corresponds roughly to that of the engine. It can last for as much as 200,000 km, which means that no servicing is necessary during this period.
- On vehicles designed for high mileages (trucks and commercial vehicles in general), brushless alternator versions are used which permit regreasing. Or bearings with grease-reserve chambers are fitted.
- Alternators are able to withstand such external influences as vibration, high temperatures, dirt, and dampness.
- Normally, operation is possible in either direction of rotation without special measures being necessary, when the fan shape is adapted to the direction of rotation.

Basic physical principles

Electrodynamic principle
Induction
Electromagnetic induction is the basis for the generation of electricity. The principle is as follows:

When an electric conductor (wire or wire loop) cuts through the lines of force of a DC magnetic field, a voltage is generated (induced) in the conductor. It is immaterial whether the magnetic field remains stationary and the conductor rotates, or vice versa.

A wire loop is rotated between the North and South poles of a permanent magnet, and its ends are connected through collector rings and carbon brushes to a voltmeter. The continuously varying relationship of the wire loop to the poles is reflected in the varying voltage shown by the voltmeter. If the wire loop rotates uniformly, a sinusoidal voltage curve is generated whose maximum values occur at intervals of 180°. Alternating current (AC) flows as soon as the circuit is closed (Fig. 1).

How is the magnetic field generated?
The magnetic field can be generated by permanent magnets. Due to their simplicity, these have the advantage of requiring only a minimum of technical outlay. They are used for small generators (e.g. bicycle dynamos).

On the other hand, electromagnets through which DC current flows permit considerably higher voltages and are controllable. This is why they are applied for generation of the (exciter) magnetic field.

Electromagnetism is based on the fact that, when an electric current flows through wires or windings, it generates a magnetic field around them.

The number of turns in the winding and the magnitude of the current flowing through it determine the magnetic field's strength. This excitation field can be further increased by using a magnetizable iron core, which, when it rotates, induces an alternating voltage in the armature coil. In practical generator applications, in order to increase the effects of induction, instead of a single wire loop, a number of wire loops are used to form the "winding" which rotates in the magnetic field.

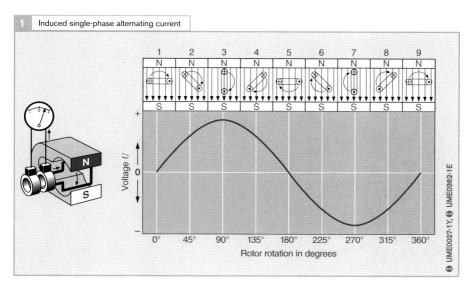

Fig. 1
Voltage curve generated during one full revolution of a winding rotating in a magnetic field. The position of the rotor on the left corresponds to position 3.

When this principle is applied to the generator or alternator, a decisive advantage lies in the fact that the magnetic field, and with it the induced voltage, can be strengthened or weakened by increasing or decreasing the (excitation) current flowing in the (excitation) winding.

Except for slight residual or residual magnetism, the electromagnet in the form of the excitation winding loses its magnetism when the excitation current is switched off. If an external source of energy (e.g. battery) provides the excitation current, this is termed "external excitation". If the excitation current is taken from the machine's own electric circuit this is termed "self-excitation". In electric machines, the complete rotating system comprising winding and iron core is referred to as the rotor.

Principle of operation of the alternator

3-phase current (3-phase AC, Fig. 2) is also generated by rotating the rotor in a magnetic field, the same as with single-phase AC as described above. One of the advantages of 3-phase AC lies in the fact that it makes more efficient use of the electrical generator's potential. The generator for 3-phase AC is designated an "alternator" and its armature comprises three identical windings which are offset from each other by 120°. The start points of the three windings are usually designated u, v, w, and the end points x, y, z. In accordance with the laws of induction, when the rotor rotates in the magnetic field, sinusoidal voltages are generated in each of its three windings. These voltages are of identical magnitude and frequency, the only difference being that their 120° offset results in the induced voltages also being 120° out-of-phase with each other, as well as being out-of-phase by 120° with respect to time.

Therefore, with the rotor turning, the alternator generates a constantly recurring 3-phase alternating voltage.

Normally, with the windings not connected, an alternator would require 6 wires to output the electrical energy that it has generated (Fig. 3a). However, by interconnecting the 3 circuits the number of wires can be reduced from 6 to 3. This joint use of the conductors is achieved by the "star" connection (Fig. 3b) or "delta" connection (Fig. 3c).

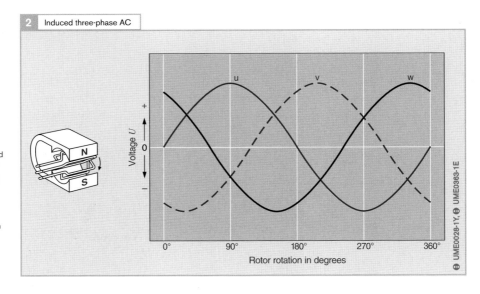

Fig. 2
Voltage curves generated during one revolution of three windings (phases) rotating in a magnetic field. The windings are offset from each other by 120°. The connection of the individual phase voltages results in a 3-phase alternating voltage.

In the case of the "star" connection, the ends of the 3 winding phases are joined to form a "star" point. Without a neutral conductor, the sum of the 3 currents at any instant time is always 0.

Discussions up to this point have centered on the alternator version with stationary excitation field and rotating armature winding in which the load current is induced.

For automotive alternators though, the 3-phase (star or delta connected) winding system is in the stator (the stationary part of the alternator housing) so that the winding is often referred to as the stator winding.

The poles of the magnet together with the excitation winding are situated on the rotor. The rotor's magnetic field builds up as soon as current flows through the excitation winding.

When the rotor rotates, its magnetic field induces a 3-phase alternating voltage in the stator windings which provides the 3-phase current when the alternator is loaded.

Rectification of the AC voltage

The 3-phase AC generated by the alternator cannot be stored in the vehicle's battery nor can it be used to power the electronic components and ECUs. To do so, it must first of all be rectified. One of the essential prerequisites for this rectification is the availability of high-performance power diodes which can operate efficiently throughout a wide temperature range.

Rectifier diodes have a reverse and a forward direction, the latter being indicated by the arrow in the symbol. A diode can be compared to a non-return valve which permits passage of a fluid or a gas in only one direction and stops it in the other.

The rectifier diode suppresses the negative half waves and allows only positive half-waves to pass. The result is a pulsating direct current. So-called full-wave rectification is applied in order to make full use of all the half-waves, including those that have been suppressed.

Bridge circuit for the rectification of the 3-phase AC
The operating principle of the diode in the rectification of an alternating current is shown in Fig. 4 (following page). Half-wave rectification is shown in Fig. 4a, and full-wave rectification in Fig. 4b.

The AC generated in the 3 windings of the alternator is rectified in an AC bridge circuit using 6 diodes (Fig. 5).

Fig. 3
a Windings not connected
b Star connection. Alternator voltage U and phase voltage U_p (partial voltage) differ by the factor $\sqrt{3} = 1.73$
Alternator current I equals phase current I_p
$U = U_p \cdot \sqrt{3}$; $I = I_p$
c Delta connection. Alternator voltage U equals phase voltage U_p. The alternator current I and phase current I_p differ by the factor $\sqrt{3} = 1.73$
$U = U_p$; $I = I_p \cdot \sqrt{3}$

3 Connection of the three windings

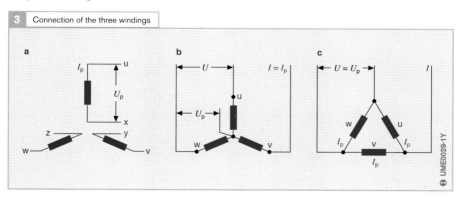

Two power diodes are connected into each phase, one diode to the positive side (Term. B+) and one to the negative side (Term. B−). The six power diodes are connected to form a full-wave rectification circuit. The positive half-waves pass through the positive-side diodes, and the negative half-waves through the negative-side diodes. Rectification takes place.

With full-wave rectification using a bridge circuit, the positive and negative half-wave envelopes are added to form a rectified alternator voltage with a slight ripple (Fig. 5).

This means that the direct current (DC) which is taken from the alternator at Terminals B+ and B− to supply the vehicle electrical system is not ideally "smooth" but has a slight ripple. This ripple is further smoothed by the battery, which is connected in parallel to the alternator, and by any capacitors in the vehicle electrical system. The excitation current which magnetizes the poles of the excitation field is tapped off from the stator winding and rectified by a full-wave bridge rectifier. Older-version alternators have three "exciter diodes". The three "exciter diodes" at Term. D+, and the three power diodes at Term. B− (negative side) form the bridge circuit for the excitation current. With the aim of increasing power output at high speeds (above 3,000 rpm), "auxiliary diodes" can be used with star-connected versions to make full use of the alternator voltage's harmonic component.

Reverse-current block

The rectifier diodes in the alternator not only rectify the alternator and excitation voltage, but also prevent the battery discharging through the 3-phase winding in the stator.

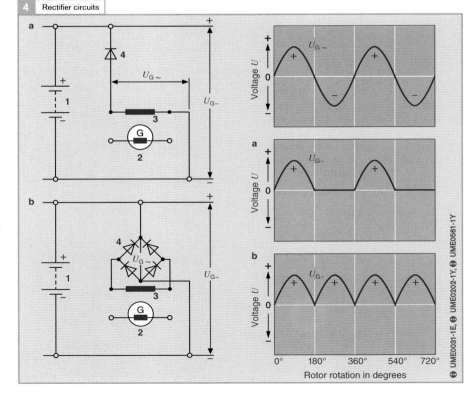

Fig. 4
a Half-wave rectification
b Full-wave rectification.
   $U_G\sim$ AC voltage upstream of the diodes
   U_G- Pulsating DC downstream of the diodes

1 Battery
2 Excitation winding (G)
3 Stator winding
4 Rectifier diodes

With the engine stopped, or with it turning too slowly for self-excitation to take place (e.g. during cranking), without the diodes battery current would flow through the stator winding. With respect to the battery current, the diodes are polarized in the reverse direction so that it is impossible for battery-discharge current to flow. Current flow can only take place from the alternator to the battery.

Rectifier diodes

Regarding their operation, the power diodes on the plus and negative sides are identical. The only difference between them lies in their special design for use as rectifiers in the alternator. They are termed positive and negative diodes, and in one case the diode's knurled metal casing acts as a cathode and in the other as an anode. The metal casing of the positive diode is pressed into the positive plate and functions as a cathode. It is connected to the battery's positive pole and conducts towards B+ (battery positive). The metal casing of the negative diode is pressed into the negative plate and functions as an anode. It is connected to ground (B−). The diode wire terminations are connected to the ends of the stator winding (Fig. 6, overleaf). The positive and negative plates also function as heat sinks for cooling the diodes. The power diodes can be in the form of Zener diodes which also serve to limit the voltage peaks which occur in the alternator due to extreme load changes (load-dump protection).

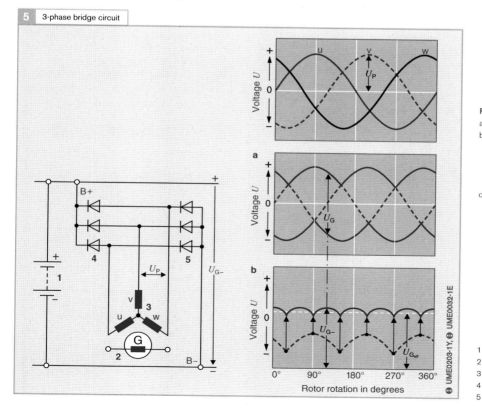

5 3-phase bridge circuit

Fig. 5
a 3-phase AC voltage
b Formation of the alternator voltage by the envelope curves of the positive and negative half-waves
c Rectified alternator voltage.

U_P Phase voltage,
U_G Voltage at the rectifier (negative not to ground),
$U_{G\sim}$ Alternator DC voltage output (negative to ground),
U_{Gms} r.m.s. value of the alternator DC output.

1 Battery
2 Excitation winding
3 Stator winding
4 Positive diodes
5 Negative diodes

The alternator's circuits

Standard-version alternators have the following three circuits:

- Pre-excitation circuit (separate excitation using battery current)
- Excitation circuit (self-excitation)
- Generator or main circuit

Pre-excitation circuit

When the ignition or driving switch (Fig. 7, Item 4) is operated, the battery current IB first of all flows through the charge-indicator lamp (3), through the excitation winding (1d) in the stator, and through the voltage regulator (2) to ground. In the rotor, this battery current serves to pre-excite the alternator.

Why is pre-excitation necessary?
On most alternators, the residual magnetism in the excitation winding's iron core is very weak at the instant of starting and at low speeds, and does not suffice to provide the self-excitation needed for building up the magnetic field.

Self-excitation can only take place when the alternator voltage exceeds the voltage drop across the two diodes (2 x 0.7 = 1.4 V). This serves to support the pre-excitation current which flows through the charge-indicator lamp from the battery. It generates a field in the rotor which in turn induces a voltage in the stator proportional to the rotor speed.

When the engine is started, in order that alternator self-excitation can "get going", the engine must turn at a speed which enables the induced voltage to exceed the voltage drop across the diodes in the excitation circuit. Since the charge-indicator lamp increases the pre-excitation circuit resistance compared to that of the excitation circuit, this speed is above the engine idle speed. It is therefore affected by the charge-indicator lamp's wattage rating.

Charge-indicator lamp
When the ignition or driving switch (Fig. 7, Item 3) is operated, the charge-indicator lamp (3) in the pre-excitation circuit functions as a resistor and determines the magnitude of the pre-excitation current. A suitably dimensioned lamp provides a current which is enough to generate a sufficiently strong magnetic field to initiate self-excitation. If the lamp is too weak, as is the case, for instance, with electronic dis-

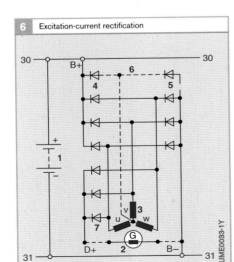

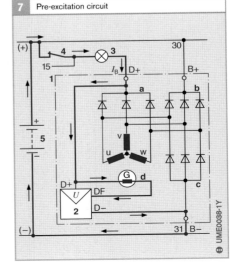

Fig. 6
1 Battery
2 Excitation winding (G)
3 Stator winding
4 Positive-plate diodes
5 Negative-plate diodes
6 Auxiliary diodes
7 Excitation diodes

Fig. 7
1 Alternator
1a Excitation diodes
1b Positive-plate diodes
1c Negative-plate diodes
1d Excitation winding
2 Voltage regulator
3 Charge-indicator lamp
4 Ignition switch
5 Battery

plays, a resistor must be connected in parallel to guarantee adequate alternator self-excitation. The lamp remains on as long as the alternator voltage is below battery voltage. The lamp goes out the first time the speed is reached at which maximum alternator voltage is generated and the alternator starts to feed power into the vehicle electrical system. Typical ratings for charge-indicator lamps are: 2 W for 12 V systems, 3 W for 24 V systems.

Pre-excitation on alternators with multifunctional voltage regulator
Alternators with multifunctional regulators draw their excitation current directly from Term. B+. This means that excitation diodes can be dispensed with (Fig. 8). As from the Series B "Compact" alternator range, the multifunctional regulator has been fitted as standard. When it receives the information "Ignition on" from the L connection, the multifunctional regulator switches on the pre-excitation current. When the rotor starts to turn, the regulator registers a voltage at the phase connection V, whose frequency it uses to calculate the alternator speed. A switch-on speed is set in the regulator, and as soon as this is reached, the regulator switches through the final stage so that the alternator starts to deliver current to the vehicle's electrical system.

Excitation circuit
During alternator operation, it is the task of the excitation current I_{err} to generate a magnetic field in the rotor so that the required alternator voltage can be induced in the stator windings.

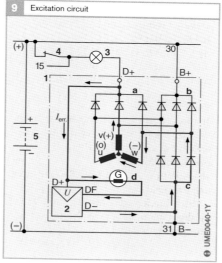

Fig. 9
1 Alternator
1a Excitation diodes
1b Positive-plate diodes
1c Negative-plate diodes
1d Excitation winding
2 Voltage regulator
3 Charge-indicator lamp
4 Ignition switch
5 Battery

8 Excitation circuit without excitation diodes

Alternators Basic physical principles

Since alternators are "self-excited", the excitation current must be tapped off from the current flowing in the 3-phase winding.

Depending on the type of regulator, the excitation current I_{err} takes the following path:
- Either through the excitation diodes (Fig. 9), carbon brushes, collector rings, and excitation winding to Term. DF of the monolithic or hybrid voltage regulator, and from Term. D– of the regulator to ground (B–) or
- Through the positive power diodes (Fig. 8), multifunctional regulator, carbon brushes, collector rings, and excitation winding to ground (B–)

In both cases, the excitation current flows from B– back to the stator winding through the negative power diodes.

Since the alternator provides its own excitation current, one refers to self-excitation.

Generator circuit

The alternating voltage induced in the three phases of the alternator must be rectified by the power diodes in the bridge circuit before it is passed on to the battery and to the loads.

The alternator current I_G, flows from the three windings and through the respective power diodes to the battery and to the loads in the vehicle electrical system. In other words, the alternator current is divided into battery-charging current and load current. In Fig. 11, the curves of the stator-winding voltages are shown as a function of the angle of rotation of the rotor.

Taking a rotor with six pole pairs, for instance, and an angle of rotation of 30°, the voltage referred to the star point at the end of winding v is positive, for winding w it is negative, and for winding u it is zero. The resulting current path is shown in Fig. 10.

Current flows from the end of winding v and through the positive diodes to alternator terminal B+ from where it flows through the battery, or the load, to ground (alternator terminal B–) and via the negative diodes (c) to winding end w. Taking a 45° angle of rotation, current from the v and w winding ends takes the same path to winding end u. In this case, there is voltage present across all of the phases.

Both examples though are momentary values. In reality, the phase voltages and currents continually change their magnitude and direction, whereas the DC supplied for battery charging and for the electrical loads always maintains the same direction.

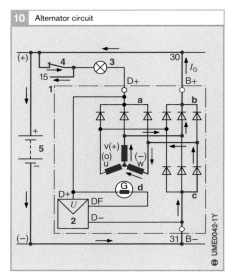

10 Alternator circuit

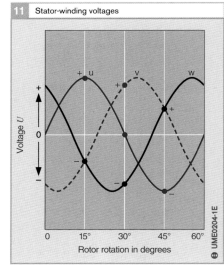

11 Stator-winding voltages

Fig. 10
1 Alternator
1a Excitation diodes
1b Positive-plate diodes
1c Negative-plate diodes
1d Excitation winding
2 Voltage regulator
3 Charge-indicator lamp
4 Ignition switch
5 Battery

Fig. 11
Voltage curves as a function of the angle of rotation of a rotor with 6 pole pairs.

This is due to the fact that, irrespective of the rotor's position, all the diodes are always involved in the rectification process.

For current to flow from the alternator to the battery, the alternator voltage must be slightly higher than that of the battery.

Voltage regulation

Why is it necessary to regulate the alternator voltage?
The regulator has the job of maintaining the alternator voltage, and thus the vehicle system voltage, at a constant level across the engine's complete speed range, independent of load and engine speed.

Presuming constant excitation current, the alternator voltage would be highly dependent upon the alternator's speed and loading. Despite these continually changing operating conditions, steps must be taken to ensure that alternator voltage is regulated to the specified level. This voltage regulation protects the electrical equipment against overvoltage, and prevents battery overcharge.

In addition, the battery's electrochemical properties must be taken into account during battery charging. This means that normally the charging voltage must be slightly higher in cold weather in order to compensate for the fact that the battery is slightly more difficult to charge at low temperatures.

Principle of voltage regulation
The voltage generated by the alternator increases along with alternator speed and excitation current. Considering a fully excited alternator which is not connected to the battery, and which is being driven without load, the voltage without regulation would increase linearly with alternator speed until it reaches about 140 V at a speed of 10,000 rpm. The voltage regulator controls the level of the alternator's excitation current, and along with it the strength of the rotor's magnetic field as a function of the voltage generated by the alternator (Fig. 12).

The voltage-regulation tolerance zone for vehicle electrical systems with 12 V battery voltage is around 14 V, and for systems with 24 V battery voltage around 28 V. The regulator remains out of action as long as the alternator voltage is below the regulator response voltage.

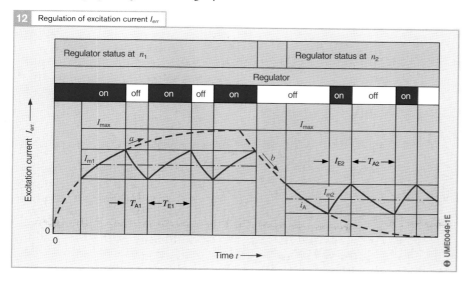

Fig. 12
Regulation of excitation current I_{err}

The relationship between on-time T_E and off-time T_A is decisive for the magnitude of the resulting mean excitation current I_m. The excitation current rises along curve a, and decays along curve b.

Within the tolerance range, if the voltage exceeds the specified upper value, the regulator interrupts the excitation current. Excitation becomes weaker and the alternator voltage drops as a result. As soon as the voltage then drops below the specified lower value, the regulator cuts in the excitation current again, the excitation increases and along with it the alternator voltage. When the voltage exceeds the specified upper value again, the control cycle is repeated. Since these control cycles all take place within a matter of milliseconds, the alternator mean voltage is regulated in accordance with the stipulated characteristic.

The infinitely variable adaptation to the various rotational speeds is automatic, and the relationship between the excitation current "On" and "Off" times is decisive for the level of the mean exciting current. At low rotational speeds, the "On" time is relatively long and the "Off" time short, the excitation current is interrupted only very briefly and has a high average value. On the other hand, at high rotational speeds the "On" time is short and the "Off" time long. Only a low excitation current flows.

Influence of ambient temperature
The alternator's characteristic curves (alternator voltage as a function of temperature) are matched to the battery's chemical characteristics. At low temperatures, therefore, the alternator voltage is increased slightly in order to improve battery charging in the winter, whereby the input voltages to the electronic equipment and the voltage-dependent service life of the light bulbs is taken into account. At higher temperatures, on the other hand, alternator voltage is reduced slightly in order to prevent battery overcharge in summer. Temperature compensation is implemented by the suitable choice of regulator components, e.g. of the Z-diodes. Fig. 13 shows the characteristic curves for 14 V alternator voltage. The voltage level is 14.5 V with an incline of -10 mV/K.

Alternator design
The theoretical principles and interrelationships discussed so far are reflected in the technical design of modern alternators. Individual versions can differ from each other in certain details according to their particular application.

At present, the claw-pole alternator with compact diode assembly is still in use in the majority of older vehicles, but the compact alternator is coming more and more to the forefront.

The major design differences between these two alternator types are the compact alternator's two internally-mounted fans, its smaller collector rings, and the location of the rectifier outside the collector-ring end shield.

The basic construction of a compact alternator is shown in Fig. 14:
- Stator (2) with 3-phase stator winding. The stator consists of mutually insulated, grooved laminations which are pressed together to form a solid laminated core. The turns of the stator winding are embedded in the grooves.

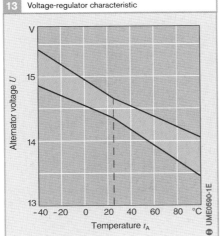

Fig. 13
Permissible tolerance band for the alternator voltage (14 V) as a function of the alternator's intake-air temperature.

Table 2. 14-V standard versions of the compact alternator type range with rated currents

Type range	Size	Rated current at 1,800 rpm A	Rated current at 6,000 rpm A
LIC	GC	37	70
	KC	50	90
	NC (*104 mm)	70	120
	NC (*112 mm)	90	150
LIC-B	GCB1	22	55
	GCB2	37	70
	KCB1	50	90
	KCB2	60	105
	NCB1	70	120
	NCB2	90	160
LI-E	E4	50	110
	E6	65	120
	E8	80	150
	E10	90	200
LI-P	P4	55	95
	P6	70	110
	P8	80	130
	P10	110	180
LI-X	C	95	150
	M	115	180
	H	135	220

* Stator internal diameter

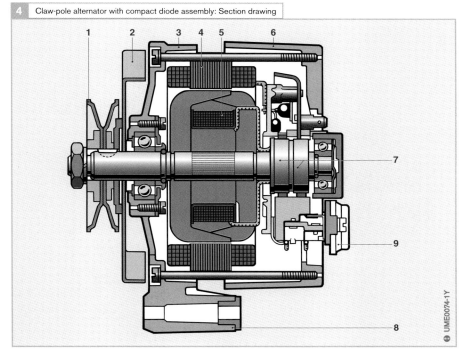

Fig. 4 Claw-pole alternator with compact diode assembly: Section drawing

1. Pulley
2. Fan
3. Drive-end shield
4. Stator core
5. Excitation winding
6. Collector-ring end shield
7. Collector rings
8. Swivel arm
9. Voltage regulator

Alternators — Alternator versions

Type T1 compact-diode-assembly alternators

The compact-diode-assembly alternators in the T1 range are intended for vehicles with high power consumption, above all for buses. The T1 alternators are single-pass ventilated, self-excited 16-pole alternators with integrated rectifier diodes and encapsulated collector rings. A very long maintenance-free service life is ensured by the wide rolling bearings with their large grease pack. Special corrosion-protection measures safeguard the alternators against the effects of splash and salt water. In very demanding applications, cool, dust-free, dry air can be drawn in through an air-intake adapter and hose.

Type DT1 compact-diode-assembly alternators

The Double-T1 alternator (DT1) as shown in Fig. 5 complies with the ever-increasing power demands made on bus alternator systems as a result of rising demands for more comfort. Basically, it consists of two electrically and mechanically coupled T1 alternators in a common housing.

The electronic voltage regulator is installed inside the alternator. Carbon brushes and collector rings are inside a dust-protected collector-ring chamber. A 100 W resistor between D+ and D- causes the charge-indicator lamp to light up in case of an open-circuit field. Fig. 6 shows the circuit diagram of a Double-T1 alternator with two stators and two excitation systems.

Compact alternators

Type LIC compact alternators

Bosch started series production of the Type LIC compact alternators in 1990. These are characterised by two small fans inside the housing which are fitted to the rotor, one on the pulley end and the other on the collector-ring end. Each fan draws in the air from the respective end of the alternator and forces out the (warm) air radially (double-pass ventilation). These two small fans generate considerably less noise than the single large fan in a compact-diode-assembly alternator. Apart from this, they are designed for higher speeds (maximum 18,000...22,000 rpm).

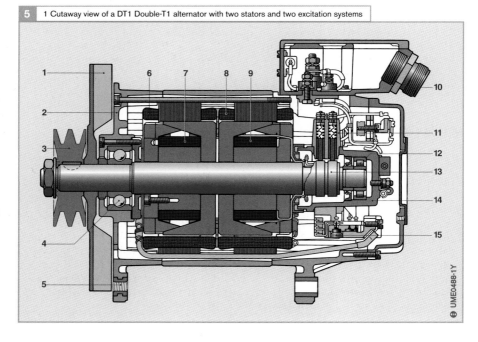

5 1 Cutaway view of a DT1 Double-T1 alternator with two stators and two excitation systems

Fig. 5
1 Fan
2 Drive-end shield
3 Pulley
4 Drive-end ball-bearing assembly
5 Swivel arm
6 Stator winding 1
7 Excitation winding 1
8 Stator winding 2
9 Excitation winding 2 (rotor)
10 Cable entry gland
11 Brush holder
12 Collector-ring roller bearing
13 Collector ring
14 Collector-ring end shield
15 Rectifier assembly

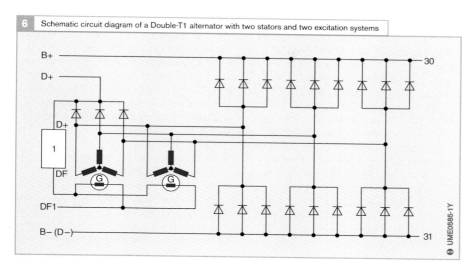

Fig. 6
1 Voltage regulator

These two special features permit a high speed-transforming ratio between crankshaft and alternator which means that for the same speed and size compact alternators can generate up to 25 % more power.

Presuming normal operating conditions, the brush/collector-ring system is designed so that it lasts for the life of the passenger car without replacement being necessary. This applies, even though it operates at higher speeds than in compact-diode-assembly alternators. The small collector rings are located at the outside end of the rotor shaft and with them the carbon brushes have service lives in excess of 250,000 km (155,000 miles).

Type B (LIC-B) Compact alternators
The type B compact alternator (Fig. 7 overleaf) is a further development of the first-generation LIC compact alternator. It has better cooling, as well as being shorter and lighter, while at the same time its power output has been increased. There are 6 sizes of the type B alternator with 14 V rated voltage, and two with 28 V rated voltage. The close spacing of the outputs enables optimal adaptation to the actually required power and the available room in the vehicle's engine compartment.

The basic design of the type B compact alternator is the same as that of a first-generation compact alternator. Further development of the rectifier assembly permits increased air throughput so that cooling improves as a result. Around their complete circumference, the three center laminations of the stator lamination pack are clamped and centered between the end shields. Compared to the first-generation compact alternators, this improves the alternator's resistance to vibration, and the heat transfer from the stator core to the end shields.

The type B compact alternators are equipped with multifunctional voltage regulators (refer to the "Voltage-regulator versions" chapter).

Type E and P (LI-E and LI-P) compact alternators
The E and P type compact alternators are based on the B range. They each comprise four sizes. Referred to the corresponding sizes in the B range, thanks to new stator wire-wrapping techniques efficiency was improved by as much as 10 % and power output by as much as 25 %. The increased efficiency leads to fuel savings of up to 0.5 l per 100 km.

Alternators Alternator versions

The design of the *type E* alternator focuses on "efficiency" and of the *type P* alternator on "power output" in the lower speed ranges.

Type X (LI-X) compact alternators

The type X compact alternators include the three sizes C, M, and H. This flexible, modular system comprises, among other things, windings, diodes, and voltage regulators for a variety of different rated voltages. It permits the construction of all three sizes for the rated voltages 14 V, 28 V, and 42 V. Thanks to basic changes in the stator-manufacturing techniques, and the improvement of component cooling, it was possible to even further increase the power output compared to the E and P types.

Windingless-rotor alternators without collector rings

These alternators are of the self-excitation type. Excitation is by means of the fixed excitation winding mounted on the internal pole. The excitation field magnetizes the alternating pole fingers of the rotating windingless rotor. In turn, the rotating magnetic field of this pole induces a three-phase AC in the stator winding. In the process, the magnetic flux travels from the rotating rotor's pole core, through the stationary internal pole to the windingless rotor, from whose pole finger it then flows to the stationary stator lamination pack. The magnetic circuit closes in the rotor's pole core via the oppositely poled claw half. In contrast to a collector-ring rotor, the magnetic flux must cross two additional air gaps between the rotating polewheel and the stationary internal pole (Fig. 8). Typical for this alternator type is the fact that the housing with

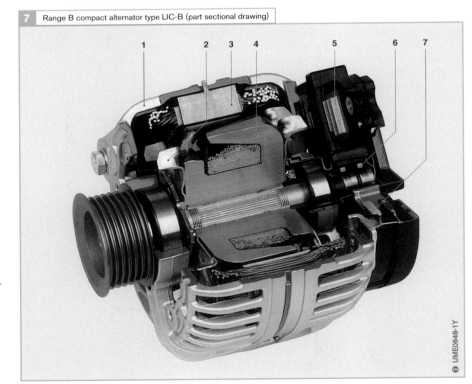

7 Range B compact alternator type LIC-B (part sectional drawing)

Fig. 7
1. Housing with double-pass ventilation
2. Inboard fan
3. Stator
4. Rotor
5. Voltage regulator
6. Outboard collector rings
7. Outboard rectifier

stator lamination pack, the heat sinks with power diodes, the attached transistorized voltage regulator, and the internal pole with excitation winding all belong to the stationary part of the alternator.

The rotating part is comprised merely of the rotor with pole wheel and conductive element (Fig. 9). Six pole fingers of the same polarity form a single north or south pole-finger crown. The two crowns form claw-pole half sections, and are retained by a non-magnetic ring positioned below the mutually-engaged pole fingers.

Type N3 compact-diode-assembly alternators
On the brushless N3 alternators with windingless rotors and high-stability end bearings, the only wear parts are the bearings. These alternators are used in applications where very long service life is of decisive importance (that is, in construction machinery, long-haul trucks, and heavy-duty special-purpose vehicles). Their outstanding feature is that they complete exceptionally high mileages under extremely severe operating conditions. Their design concept is based on minimizing the number of wear components in order to extend the alternator's effective service life. This alternator is practically maintenance-free.

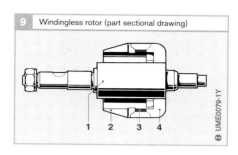

9 Windingless rotor (part sectional drawing)

Fig. 9
1 Rotor shaft with pole core
2 Conductive element
3 Non-magnetic ring
4 Pole-wheel half section

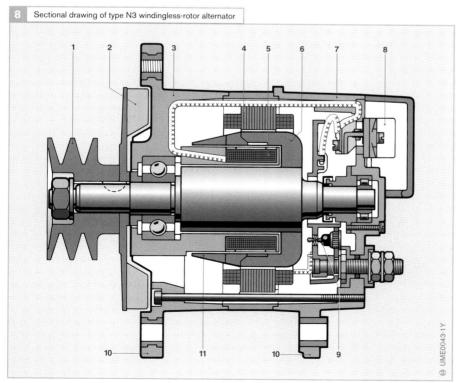

8 Sectional drawing of type N3 windingless-rotor alternator

Fig. 8
1 Double-groove pulley
2 Fan
3 Drive-end shield with stationary internal pole
4 Stator lamination stack
5 Stationary excitation winding
6 Windingless rotor
7 Rear end shield
8 Attached transistor voltage regulator
9 Power diode
10 Swivel arm
11 Conductive element

Liquid-cooled, windingless-rotor compact alternator (LIF)

In the case of air-cooled alternators, it is the cooling fan which is mainly responsible for air-flow noise. At high current outputs, a pronounced reduction of noise can only be achieved by using liquid-cooled alternators (Fig. 10) which utilize the coolant from the engine's cooling circuit.

On modern intermediate-size and luxury cars, the use of liquid-coooled fully-encapsulated alternators is often the only way to achieve a decisive reduction in vehicle noise.

Since a carbon-brush/collector-ring system would not last long enough inside an encapsulated alternator with its high temperatures, the fully encapsulated alternator features a windingless rotor without collector rings. This alternator has a cylindrical aluminum housing provided with a special flange on the drive end to locate it in the special coolant housing. The coolant space

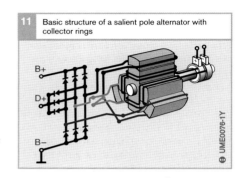

11 Basic structure of a salient pole alternator with collector rings

(jacket) between the alternator and the coolant housing is connected to the engine's cooling circuit. All important sources of heat loss (stator, diodes, voltage regulator, and the stationary excitation winding) are coupled to the alternator housing in such a manner that their heat is transferred efficiently to the coolant. The electrical connections are at the alternator's pulley end.

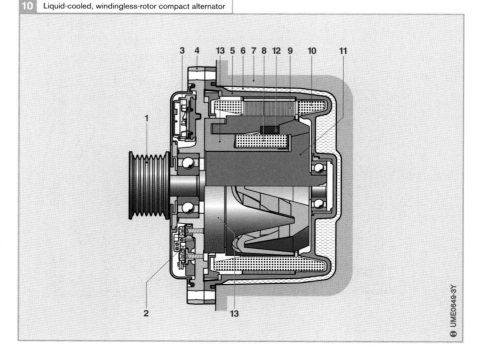

10 Liquid-cooled, windingless-rotor compact alternator

Fig. 10
1 Pulley
2 Rectifier
3 Voltage regulator
4 Drive-end shield
5 Alternator housing (aluminum)
6 Coolant
7 Coolant housing for engine mounting or engine block
8 Stationary excitation winding
9 Stator lamination stack
10 Stator winding
11 Windingless rotor
12 Non-magnetic intermediate ring
13 Conductive element

Type U2 salient-pole collector-ring alternators

Type U2 salient-pole collector-ring alternators Salient-pole collector-ring alternators are mainly used in large vehicles with high power demands (>100 A) and 28-V vehicle electrical systems. These units are therefore ideal for use in rail vehicles, special-purpose vehicles and in marine applications.

Fig. 11 shows a four-pole, self-excited salient-pole alternator. With each rotation, the rotor passes four poles, inducing four half waves in each circuit. In 3-phase operation, this equates to 4 x 3 = 12 half waves for each complete rotation.

The configurations of the 3-phase stator winding, and the current-flow path, correspond to those of the claw-pole alternator. However, the rotor of this basic alternator version (Fig. 12) differs to that of the claw-pole version.

The claw-pole rotor, namely, features a central excitation winding for all poles. In contrast, the salient-pole rotor has four or six individual poles, each carrying its own excitation winding wound directly on the pole.

The salient-pole alternator's slim, elongated cylindrical shape derives from the characteristic shape of the rotor.

The stator with its 3-phase winding is installed in the alternator housing which is closed at the ends by a drive-end shield and a collector-ring end shield. The excitation winding is on the salient-pole rotor which runs in bearings in the end shields. The excitation current is delivered through the collector rings and the carbon brushes.

Rectifier and voltage regulator are installed, remote from the alternator, at a point protected against engine heat, moisture, and dirt. Alternator and regulator are joined electrically by a 6-line wiring harness. This alternator's encapsulated collector rings and its extra-large grease chamber qualify it for high-mileage applications.

12 Cutaway view of a type U2 salient-pole alternator

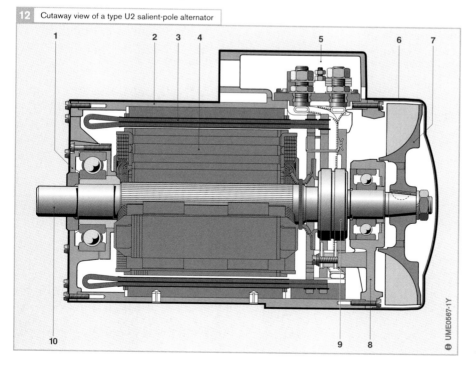

Fig. 12
1 Drive-end shield
2 Housing
3 Stator winding
4 Rotor
5 3-phase connection (remote rectifier and voltage regulator)
6 End cap
7 Fan
8 Collector-ring end shield
9 Collector rings
10 Driveshaft

Voltage-regulator versions

The mechanical electromagnetic contact regulators and the electronic (transistor) versions are the two basic voltage-regulator types.

Whereas the electromagnetic regulator is today practically only used for replacement purposes, the (monolithic or hybrid) transistor regulator is standard equipment on all alternator models.

Electromagnetic voltage regulators

The excitation current is varied by opening and closing a movable contact in the excitation-current circuit. This movable contact is pressed against a fixed contact by a spring, and when the rated voltage is exceeded it is lifted off by an electromagnet.

The contact regulators which are suitable for alternator applications are of the *single-element type*. That is, regulators with a voltage-regulator element comprising an electromagnet, an armature, and a regulating contact. In the single-element, single-contact regulator (Fig. 1), the contact opens and closes as follows: The magnetic force and the spring force of a suspension and adjusting spring are both applied to the regulating armature.

As soon as the alternator voltage exceeds the set value, the electromagnet pulls in the armature and opens the contact (position "b").

This switches a resistor into the excitation circuit which reduces the excitation current and with it the alternator voltage. When the alternator voltage drops *below the set voltage*, the magnetic force is also reduced, so that the spring force predominates and closes the contact again (position "a"). This opening and closing cycle is repeated continually.

The single-element double-contact regulator (Fig. 2) operates with a second pair of contacts which permit 3 switching positions. The regulating resistor is short-circuited in position "a" and a high excitation current flows. In position "b" the resistor and the excitation winding are connected in series and the excitation current is reduced. In position "c", the excitation winding is short-circuited and the excitation current drops to zero (the time constant is independent of the excitation winding's inductance and reactance).

Due to its size and characteristics, this alternator is only suitable for mounting on the vehicle body.

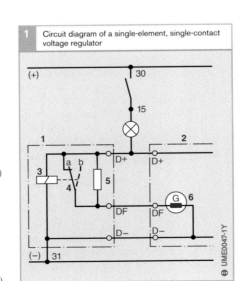

1 Circuit diagram of a single-element, single-contact voltage regulator

Fig. 1
1 Voltage regulator
2 Alternator
3 Electromagnet
4 Regulating contact
5 Regulating resistor
6 Excitation winding (G)

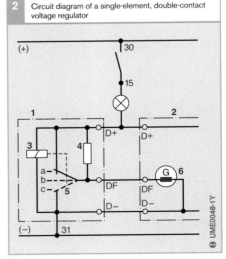

2 Circuit diagram of a single-element, double-contact voltage regulator

Fig. 2
1 Voltage regulator
2 Alternator
3 Electromagnet
4 Regulating resistor
5 Regulating contact
6 Excitation winding (G)

Electronic voltage regulators

Characteristics

Electronic regulators are used solely with alternators. Thanks to its compact dimensions, its low weight, and the fact that it is insensitive to vibration and shock, this regulator can be integrated directly in the alternator.

Whereas the first transistor regulators were built from discrete components, modern-day versions all use hybrid and monolithic circuitry.

The transistor regulator's essential advantages are:
- Shorter switching times which permit closer control tolerances
- No wear (= no servicing)
- High switching currents (less types)
- Spark-free switching prevents radio interference
- Insensitive to shock, vibration, and climatic effects
- Electronic temperature compensation also permits closer control tolerances
- Compact construction allows direct mounting on the alternator, irrespective of alternator size

Operating concept

Basically speaking, the operating concept is the same for all electronic-regulator types. The type EE electronic regulator is used here as an example, and Fig. 3 (overleaf) shows its operation between the "On" and "Off" states.

The operating concept is easier to understand when one considers what happens when the alternator's terminal voltage rises and falls. The actual value of the alternator voltage between terminals D+ and D– is registered by a voltage divider (R1, R2, and R3). A Zener diode in parallel with R3 functions as the alternator's setpoint generator. A partial voltage proportional to the alternator voltage is permanently applied to this diode.

The regulator remains in the "On" state as long as the actual alternator voltage is below the set value (Fig. 3a). The Z-diode's breakdown voltage has not yet been reached at this point, that is, no current flows to the base of transistor Tl through the branch with the Z-diode, Tl is in the blocking state.

With Tl blocked, a current flows from the exciter diodes via terminal D+ and resistor R6 to the base of transistor T2 and switches T2 on. Terminal DF is now connected to the base of T3 by the switched transistor T2. This means that T3 always conducts when T2 is conductive. T2 and T3 are connected as a Darlington circuit and form the regulator's driver stage. The excitation current I_{err} flows through T3 and the excitation winding and increases during the "On" period, causing a rise in the alternator voltage U_G. At the same time, the voltage at the setpoint generator also rises. The regulator assumes the "Off" state as soon as the actual alternator voltage exceeds the setpoint value (Fig. 3b).

The Z-diode becomes conductive when the breakdown voltage is reached, and a current flows from D+ through resistors Rl, R2 into the branch with the Z-diode, and from there to the base of transistor Tl which also becomes conductive. As a result, the voltage at the base of T2 is practically 0 referred to the emitter, and transistors T2 and T3 (driver stage) block. The excitation circuit is open-circuited, the excitation decays, and the alternator voltage falls as a result. As soon as the alternator voltage drops under the set value again, and the Z diode switches to the blocked state, the driver stage switches the excitation current on again.

When the excitation current is open-circuited, a voltage peak would be induced due to the excitation winding's self-induction (stored magnetic energy) which could destroy transistors T2 and T3. A "free-wheeling diode" D3 is connected parallel to the excitation winding, and at the instant of open-circuiting absorbs the excitation current thereby preventing the formation of a dangerous voltage peak.

The control cycle in which the current is switched on and off by connecting the excitation winding alternately to the alternator voltage or short-circuiting it with the free-wheeling diode is repeated periodically.

Alternators Voltage-regulator versions

Essentially, the on/off ratio depends on the alternator speed and the applied load.

The ripple on the alternator DC is smoothed by capacitor C. Resistor R7 ensures the rapid, precise switch-over of transistors T2 and T3, as well as reducing the switching losses.

Hybrid regulators
The transistor regulator using hybrid technology comprises a hermetically encapsulated case, in which are enclosed a ceramic substrate, protective thick-film resistors, and a bonded integrated circuit (IC) incorporating all the control functions.

The power components of the driver stage (Darlington transistors and the free-wheeling diode) are soldered directly onto the metal socket in order to ensure good heat dissipation. The electrical connections are via glass-insulated metal pins.

The regulator is mounted on a special brush holder and directly fastened to the alternator without wiring.

Due to the Darlington circuit in the power stage (two transistors), there is a voltage drop of about 1.5 V in the current-flow direction.

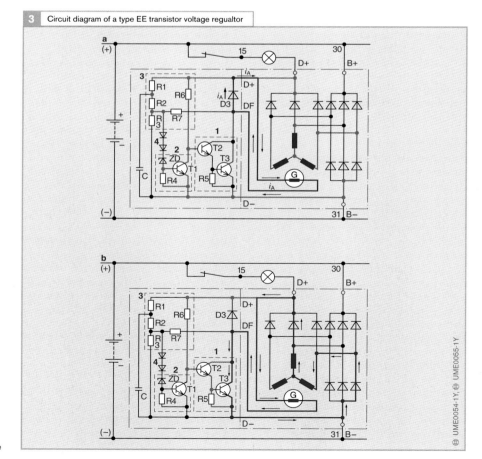

3 Circuit diagram of a type EE transistor voltage regualtor

Fig. 3
a Excitation current switched off by T3
b Excitation current switched on by T3

1 Driver stage
2 Control stage
3 Voltage divider
4 Temperature-compensation diodes

C Voltage-smoothing capacitor

D3 Free-wheeling diode

The circuit diagram (Fig. 4) shows an alternator fitted with an type EL hybrid regulator.

The hybrid regulator's advantages can be summed up as follows: compact construction, low weight, few components, few connections, high reliability in the extreme operating conditions met in automotive applications.

Normally, hybrid regulators using conventional diodes are used with compact-diode-assembly alternators.

Monolithic regulators

The monolithic regulator was developed from the hybrid regulator. The functions of the hybrid regulator's IC, power stage, and free-wheeling diode have been incorporated on a single chip. The monolithic regulator uses bipolar techniques. The compact construction with fewer components and connections enabled reliability to be even further improved. Since the output stage is in the form of a simple power stage, the voltage drop in the current-flow direction is only 0.5 V.

Monolithic regulators in combination with Z-diode rectifiers are used in compact alternators.

Multifunctional voltage regulators

In addition to voltage regulation, the multifunctional regulator can also trigger an LED display instead of the charge-indicator lamp, as well as a fault display to indicate under-voltage, V-belt breakage, or excitation open-circuit. This alternator does without excitation diodes. The signal for "engine running" can be taken from Terminal L. Terminal W provides a signal which is proportional to engine speed. The actual voltage value is taken from Terminal B+ on the alternator.

The standard version of the type B compact alternator has further functions available:

When a load is switched on in the vehicle electrical system, the alternator's excitation follows a ramp. This prevents torque jumps in the alternator's belt drive which, for instance, could otherwise interfere with the smooth running of the engine (LRD: Load-Response Drive; LRS: Load-Response Start).

The regulator's on/off ratio can be picked-off via the DFM terminal. This ratio defines the alternator's loading and can be used for selection circuits (e.g. for switching off low-priority loads when the alternator must deliver full power). Terminal L is designed for relay triggering up to max. 0.5 A.

4 Circuit diagram of an alternator equipped with type EL hybrid electronic voltage regulator

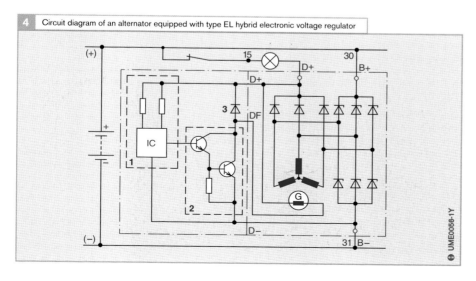

Fig. 4
1 Control stage using thick-film techniques, with resistors and IC
2 Power stage (Darlington stage)
3 Free-wheeling diode

The power loss associated with the charge-indicator lamp in the instrument cluster is often excessive. It can be reduced by using an LED display instead. Multifunctional regulators permit the triggering of lamp bulbs as well as of LED display elements in the instrument cluster.

Overvoltage protection

Usually, with the battery correctly connected and under normal driving conditions, it is unnecessary to provide additional protection for the vehicle's electronic components. The battery's low internal resistance suppresses all the voltage peaks occurring in the vehicle electrical system.

Nevertheless, it is often advisable to install overvoltage protection as a precautionary measure in case of abnormal operating conditions. For instance, on vehicles for transporting hazardous materials, and in case of faults in the vehicle electrical system.

Reasons for overvoltage
Overvoltage may occur in the vehicle electrical system as the result of:
- Regulator failure
- Influences originating from the ignition
- Switching off of devices with a predominantly inductive load
- Loose contacts
- Cable breaks

Such overvoltages take the form of very brief voltage peaks, lasting only a few milliseconds which reach a maximum of 350 V and originate from the coil ignition. Overvoltages are also generated when the line between battery and alternator is open-circuited with the engine running (this happens when an outside battery is used as a starting aid), or when high-power loads are switched off. For this reason, under normal driving conditions, the alternator is not to be run without the battery connected.

Under certain circumstances though, short-term or emergency operation without battery is permissible. This applies to the following situations:
- Driving of new vehicles from the final assembly line to the parking lot
- Loading onto train or ship (the battery is installed shortly before the vehicle is taken over by the customer)
- Service work, etc.

With towing vehicles and agricultural tractors it is also not always possible to avoid operation without the battery connected.

The overvoltage-protection device guarantees that overvoltages have no adverse effects on operation, although it does require extra circuitry.

Types of protection
There are three alternatives for implementing overvoltage protection:

Z-diode protection
Z-diodes can be used in place of the rectifier power diodes. They limit high-energy voltage peaks to such an extent that they are harmless to the alternator and regulator.

Furthermore, Z-diodes function as a central overvoltage protection for the remaining voltage-sensitive loads in the vehicle electrical system.

The limiting voltage of a rectifier equipped with Z-diodes is 25...30 V for an alternator voltage of 14 V, and 50...55 V for an alternator voltage of 28 V.

Compact alternators are always equipped with Z-diodes.

Surge-proof alternators and regulators
The semiconductor components in surge-proof alternators have a higher electric-strength rating. For 14-V alternator voltage, the electric strength of the semiconductors is at least 200 V, and for 28-V alternator voltage 350 V.

In addition, a capacitor is fitted between the alternator's B+ terminal and ground which serves for short-range interference suppression.

The surge-proof characteristics of such alternators and regulators only protect these units, they provide no protection for other electrical equipment in the vehicle.

Overvoltage-protection devices
(only for 28 V alternators)
These are semiconductor devices which are connected to the alternator terminals D+ and D– (ground). In the event of voltage peaks, the alternator is short-circuited through its excitation winding. Primarily, overvoltage-protection devices protect the alternator and the regulator, and to a lesser degree the voltage-sensitive components in the vehicle electrical system.

Generally, alternators are not provided with polarity-reversal protection. If battery polarity is reversed (e.g. when starting with an external battery), this will destroy the alternator diodes as well as endangering the semiconductor components in other equipment.

Overvoltage-protection devices,
non-automatic
This type of overvoltage-protection device is connected directly to the D+ and D– terminals on Tl alternators, e.g. in buses and heavy trucks (Fig. 1). The unit responds to voltage peaks and consistent overvoltage that exceed its response threshold of approx. 31 V. At this point, thyristor Th becomes conductive. The activation voltage is defined by Zener diode ZD, while the necessary response delay is regulated by resistors Rl and R2 along with capacitor C. The unit requires only milliseconds to short circuit the regulator and alternator across D+ and D–. The thyristor assumes responsibility for the short-circuit current. Meanwhile, current from the battery triggers the charge-indicator lamp to alert the driver. The thyristor remains active, reverting to its off-state only after the ignition has been switched off, or the engine and alternator come to rest. The unit will not provide overvoltage protection if the wires at terminals D+ and D– are reversed.

As the charge-indicator lamp also fails to respond, the problem would remain unnoticed if a backup diode DS were not installed between terminals D+ and D– to ensure a signal at the lamp. This diode responds to reversed connections by polarizing to allow current flow, and the indicator lamp remains on continuously.

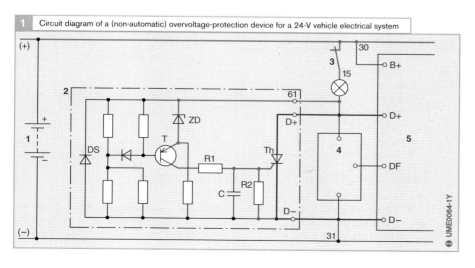

1 Circuit diagram of a (non-automatic) overvoltage-protection device for a 24-V vehicle electrical system

Fig. 1
1 Battery
2 Overvoltage-protection device
3 Driving switch
4 Voltage regulator
5 Alternator

Alternators — Overvoltage protection

Overvoltage-protection devices, automatic

This type of protection device is designed for use with TI alternators (Fig. 2).

The unit incorporates two inputs, D+ and B+ which react to different voltage levels and with varying response times.

Input D+ provides rapid overvoltage protection, as on the device described above.

The second input, B+, responds only to defects at the voltage regulator, while the alternator voltage continues to climb until it reaches the unit's response voltage of approx. 31 V. The alternator then remains shorted until the engine is switched off. Input B+ is thus designed to prevent consequential damage.

This overvoltage-protection device makes it possible for the alternator to operate for limited periods without a battery in the circuit. The alternator voltage collapses briefly when the overvoltage device responds. If the load becomes excessive, renewed alternator excitation is impossible.

Voltage peaks which can be generated by the alternator itself when loads are switched off ("load-dump"), cannot damage other devices in the system because the alternator is immediately short-circuited.

Consequential-damage protection device

This protection device is specially designed for use with the Double-TI alternator with two stators and two excitation systems (Fig. 3).

While the overvoltage-protection device short-circuits the alternator, the consequential-damage protection unit functions as a kind of backup regulator, even with the battery out of circuit. Provided that the alternator's speed and the load factor allow, it maintains a mean alternator voltage of approximately 24 V to furnish emergency capacity.

The consequential-damage protection device responds to operation with battery and a faulty, short-circuited regulator by interrupting the alternator's excitation current approx. 2 seconds after the alternator output passes the response threshold of 30 V.

The unit's relay contact then assumes a backup voltage-control function by operating as a contact regulator.

When the system is operated with the battery out of circuit, the unit reacts to voltage peaks of 60 V or more lasting for more than 1 ms.

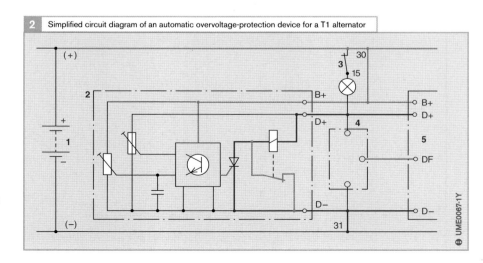

Fig. 2
1 Battery
2 Consequential-damage protection device
3 Ignition switch
4 Voltage regulator
5 Alternator

The charge-indicator lamp flashes to indicate that the system is operating in the backup mode. The system does not charge the battery, as the mean voltages in this mode are very low. Maximum operating times in this backup mode extend to approx. 10 hours, after which the consequential-damage protection device must be replaced.

Free-wheeling diode

The free-wheeling diode (also known as a suppressor diode or anti-surge diode) has already been mentioned in the description of the transistor regulator.

When the regulator switches to the "Off" status, upon interruption of the excitation current a voltage peak is induced in the excitation winding due to self-induction.

Sensitive semiconductor components can be destroyed if precautionary measures are not taken. The free-wheeling diode is connected in the regulator parallel to the alternator's excitation winding. Upon the excitation winding being interrupted, the free-wheeling diode "takes over" the excitation current and permits it to decay, thus preventing the generation of dangerous voltage peaks.

A similar effect can occur on vehicles which are equipped with inductive loads remote from the alternator regulator. Thus, when electromagnetic door valves, solenoid switches, magnetic clutches, motor drives, and relays, etc. are switched off, voltage peaks can be generated in the windings of such equipment due to self-induction, and can endanger the diodes and other semiconductor components.

These induced voltages can be rendered harmless by means of a free-wheeling diode.

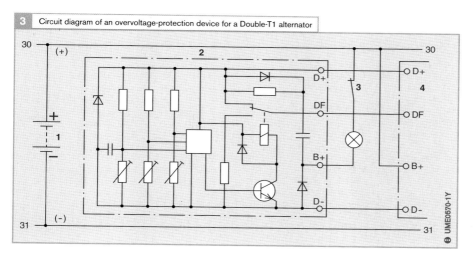

Fig. 3
1 Battery
2 Consequential-damage protection device
3 Ignition switch
4 Alternator with voltage regulator

Cooling and noise

Due above all to the heat developed by the alternator when converting mechanical power into electrical power, and also due to the effects of heat from the engine compartment (engine and exhaust system), considerable increases in the alternator component temperature take place. And when the engine compartment is encapsulated for sound-proofing purposes, temperatures rise even further. In the interests of functional reliability, service life, and efficiency, it is imperative that this heat is dissipated completely. Depending upon alternator version, maximum permissible ambient temperature is limited to 80...120°C, and future temperatures are expected to reach to 135°C. Cooling must guarantee that even under the hostile under-hood conditions encountered in everyday operation, component temperatures remain within the specified limits ("worst-case" consideration).

Cooling without fresh-air intake

For normal operating conditions, through-flow cooling is the most common cooling method applied for automotive alternators. Radial fans for one or both directions of rotation are used. Since both the fan and the alternator shaft must be driven, the cooling-air throughput increases along with the speed. This ensures adequate cooling irrespective of alternator loading.

In order to avoid the whistling noise which can occur at specific speeds, the fan blades on some alternator types are arranged asymmetrically.

Single-flow cooling
Compact-diode-assembly alternators use single-flow cooling. The external fan is attached to the drive end of the alternator shaft. Fig. 1 shows a Gl alternator with a clockwise-rotation fan. Air is drawn in by the fan at the collector-ring or rectifier end, passes through the alternator, and leaves through openings in the drive-end shield.

Double-flow cooling
Due to their higher specific power output, compact alternators are equipped with double-flow cooling (Fig. 2). The compact alternator's two fans are mounted inside the alternator on the driveshaft to the left and right of the rotor's active section.

The two air streams are drawn in axially by the fans through openings in the drive and

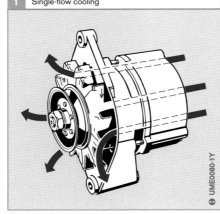

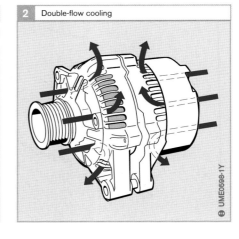

Fig. 1
G1 compact-diode-assembly alternator with clockwise-rotation fan

Fig. 2
Compact alternator

Thus, a correctly dimensioned alternator is decisive for an adequate supply of on-board energy. An under-rated (i.e. overloaded) alternator is not able to keep the battery sufficiently charged, which means that battery capacity cannot be fully utilized.

Consequently, if power demand is increased, for instance as a result of fitting extra equipment, it is advisable to replace the standard fitted alternator by a more powerful version. One of the most important steps to be taken when ascertaining the electrical system's charge balance is the registration of all the installed electrical loads (including retrofitted equipment), together with their power inputs and the average length of time they are switched on (short-time, long-time, or permanent loads). Similarly, the driving cycles as dictated by the traffic situation must also be con- sidered. These include, for instance, low alternator speeds typical for town traffic coupled with repeated standstills, expressway traffic with congestion, and high alternator speeds on clear first-class roads. The time of day (journeys mainly by daylight or during the dark), and the season (winter or summer driving with the related temperatures and weather), also have an affect.

We can sum up as follows:
Even under the most unfavorable operating conditions, in addition to powering all the electrical loads, the alternator power must suffice to keep the battery sufficiently charged so that the vehicle is always ready for operation.

An expert should be consulted before the final selection of alternator size and its matching to the appropriate battery. The following example illustrates the loading of the vehicle's energy household by the electrical loads under a variety of different conditions

Alternator installation and drive
Installation
The motor-vehicle operator usually has little say concerning the alternator or regulator fitted in his/her vehicle. And in every vehicle, the alternator's installation position is dependent upon the conditions prevailing in the engine compartment due to construction and design. However, certain basic factors must always be borne in mind concerning installation:
- Good accessibility for readjusting the V-belt tension and for any maintenance work which may be required.
- Adequate cooling for alternator waste heat as well as for heat conducted and radiated from the engine.
- Protection against dirt, moisture, shock, impact, fuel and lubricants (ingress of gasoline leads to the danger of fire and explosion, and diesel fuel damages the carbon brushes and collector rings).

Almost without exception, alternators which are driven by the engine through normal V-belts are attached by means of a swivel-arm mounting. In addition to the mounting using a swivel bearing, an adjustment facility (to pivot the alternator around a swivel arm) is provided for adjusting the V-belt tension.
If the alternator is driven through a ribbed V-belt (poly-V belt), the alternator is usually rigidly mounted. The belt is adjusted using a belt tensioner (in Fig. 2 overleaf, the belt is tensioned using the alternator).
In special cases, large alternators are cradle-mounted in a recess directly on the engine.
Irrespective of the type of mounting, all alternators must have good electrical connection to the engine block. Furthermore, since current return from the electrical system is in the most cases via ground, a highly conductive ground connection of adequate cross-section must be provided between engine and chassis.

Buses and special-purpose vehicles are often equipped with extra return lines in order to reduce voltage losses and thereby increase safety.

Electric cables and lines only provide efficient connection if they have properly attached terminals or plug connectors.

Alternator drive

Alternators are driven directly from the vehicle engine. As a rule, drive is via V-belts. Less frequently, flexible couplings are used.

The belt drive (using V-belts, ribbed-V belts, etc.) is the most important element in the transmission of power, and as such it is subject to exacting requirements:
- The belt material must have very high flexural strength.
- Belt slip leads to heat-up and wear, and in order to prevent it, longitudinal stretch should remain at a minimum as the belt gets older.

Investigations conducted by ADAC (Germany's largest automobile club) have revealed that V-belt damage is a frequent cause of breakdown. It is therefore important to use V-belts which comply with the above requirements and which are capable of a long service life. Typical for automotive applications are the "open-flank" belt and the ribbed-V belt.

The "open-flank" design (Fig. 3) features high flexibility, coupled with extreme lateral rigidity and resistance to wear. Particularly with small-diameter pulleys, this leads to improved power transmission and longer service life compared to conventional rubber-jacketed V-belts.

V-belt and ribbed V-belt alternator drive
Photo: VW

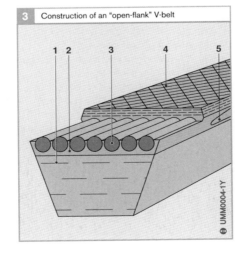

Construction of an "open-flank" V-belt

Fig. 3
1. Short-cut fiber mixture
2. Embedding compound
3. Specially prepared cord
4. Cover fabric
5. Cut flanks

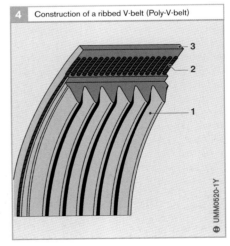

Construction of a ribbed V-belt (Poly-V-belt)

Fig. 4
1. Carcasse
2. Cord
3. Outer cover

The high flexibility of the ribbed-V belt (Poly-V belt, Fig. 4) permits very small bending radiuses. This in turn means that small-diameter pulleys (minimum dia. 45 mm) can be fitted to the alternators thus permitting higher transmission ratios. The back of the belt may also be used to transmit power, thus enabling a number of aggregates (alternator, radiator fan, water pump, power-steering pump, etc.) to all be driven from a single belt with an adequate wrap angle around each pulley wheel.

Usually, a *single V-belt* suffices to drive small-power alternators. With large-power alternators on the other hand, *two V-belts or a single ribbed-V belt* are more common in order to overcome the alternator's resistance to turning which is inherent in its higher power.

Depending upon application, pulley wheels and fan wheels are used which have either been turned, or stamped from sheet metal, and which can be combined with each other as required (Fig. 5).

The pulley wheel's correct diameter depends upon the required transmission ratio between engine and alternator.

Being as the speed ranges covered by the multitude of engines concerned differ considerably from each other, there is a wide variety of pulley wheels available with different diameters.

The transmission ratio must take into account the fact that the alternator's permitted maximum speed must not be exceeded at the engine's maximum speed.

Notes on operation

Battery and regulator must be connected when the alternator is operated. This is the normal operating setup and the installed electronic equipment and semiconductor devices perform efficiently and safely.

Emergency operation without the battery connected results in high voltage peaks which can damage equipment and components. Here, efficient emergency operation is only possible if precautionary measures are taken.

There are three alternatives:
- Zener diodes in the rectifier
- Surge-proof alternator and regulator
- Overvoltage-protection devices

Connecting the battery into the vehicle's electrical system with the wrong polarity immediately destroys the alternator diodes, and can damage the regulator, no matter whether the engine is switched off or running. The same damage can occur if an external voltage source is used as a starting aid and the terminals are reversed.

5 Pulley and fan wheel assembly on a compact-diode-assembly alternator

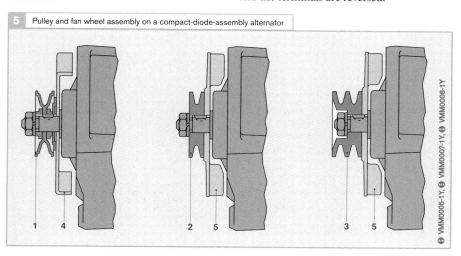

Fig. 5
Fan pulley:
1 Stamped version
2 Turned version (1-groove)
3 Turned version (2-groove)

Fan wheel:
4 Stamped version with support plate
5 Cast aluminum version

Special circuitry is available to safeguard against reverse-polarity damage. When the battery is falsely connected, engine start is blocked in order to protect alternator and regulator.

The charge-indicator lamp acts as a resistor in the alternator circuit. If a correctly rated lamp is fitted which draws enough current, the resulting pre-excitation current provides a magnetic field which is strong enough to initiate alternator self-excitation.

When the charge-indicator lamp lights up, this merely indicates to the driver that the ignition or driving switch is switched on and that the alternator is not yet feeding power into the electrical system. The lamp goes out as soon as the alternator's self-excitation speed is reached and the alternator supplies energy to the electrical system. The lamp therefore provides an indication that alternator and regulator are functioning correctly, that they are correctly connected, and that the alternator is supplying current.

The charge-indicator lamp gives no indication as to whether, and as of what speed, the battery is being charged. When the alternator is heavily loaded, it can happen that even though the lamp has gone out, the battery is not being charged but discharged. The lamp gives no information concerning the state of battery charge even though it is (erroneously) referred to as the "charge-indicator lamp".

If the lamp is defective (broken filament), this means that pre-excitation current cannot flow and self-excitation first sets in at very high speeds. This error is noticeable when the lamp fails to light up with the engine at standstill and the ignition switched on.

If during operation an open-circuit occurs in the lamp line or in the alternator ground line, and the alternator stops supplying current, the driver is not warned of this fact even though the charge-indicator lamp is intact.

Here, in alternators with excitation diodes, it is necessary to connect in an additional resistor (Fig. 6) so that the charge-indicator lamp lights up to inform the driver of open-circuited excitation circuits. If the charge-indicator lamp fails to go out even at high speeds, this indicates a fault in the alternator itself, at the regulator, in the wiring, or at the V-belt.

Mileages and maintenance intervals

Using a variety of statistical methods, and taking typical operating conditions into account, it is possible to calculate specific average service lives, mileages, and driving cycles for different categories of vehicle (passenger cars, commercial vehicles, long-haul trucks, town and long-distance buses, and construction machinery).

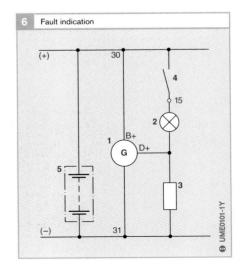

6 Fault indication

Fig. 6
1 Alternator
2 Charge-indicator lamp
3 Resistor
4 Ignition switch
5 Battery

Considering the different fields of application of these vehicle categories, the requirements and criteria for the economic efficiency of their alternators also differ. This leads to there being a range of alternators available for different service lives and maintenance intervals.

Depending upon version and application, passenger-car alternators with encapsulated ball bearings have service lives of 150,000...600,000 km.

Presuming that the engine's service life until it is replaced or has a major overhaul corresponds to that of its alternator, this makes specific maintenance work on the alternator unnecessary. The grease in the bearings suffices for this period.

Due to the use of particularly wear-resistant components, the alternators installed in trucks and buses for instance achieve mileages of 200,000...600,000 km. One prerequisite is that they are equipped with suitable ball bearings which, for instance, feature enlarged grease chambers.

Provided the alternator is installed in a location which is relatively free from dirt, oil, and grease, the carbon-brush wear is negligible due to the low excitation currents involved.

▶ The history of the generator/alternator

At the turn of the century, the introduction of electrical lighting to motor vehicles to take the place of the previously used horse-and-carriage lighting meant that a suitable source of electrical power had to be available in the vehicle. The battery alone was completely unsuitable since, when discharged, it had to be removed from the vehicle for re-charging. Around 1902, Robert Bosch designed a lighting dynamo (which came to be called a "generator") which essentially comprised a stator, an armature with commutator, and a contact breaker for the ignition (see below). The only difficulty here, though, was the fact that the dynamo's voltage was dependent on the engine's speed which varied consideraby.

Endeavours, therefore, concentrated on the development of a DC dynamo with voltage regulation. Finally, electromagnetic control of the field resistor as a function of the machine's output voltage proved to be the answer. Around 1909, using the knowledge available at that time, it thus became possible to build a complete "Lighting and Starting System for Motor Vehicles". This was introduced to the market in 1913 and comprised a dynamo (splashwater-protected, encapsulated 12-V DC dynamo with shunt regulation and a rated power of 100 W), a battery, a voltage-regulator and switching box, a free-wheeling starter with pedal-operated switch, and a variety of different lighting components.

Starter motors

Before an internal-combustion engine can operate independently and generate its own power output, it requires assistance to start it. It needs a certain degree of momentum before the torque produced by the ignition stroke is sufficient to overcome the resistance of the exhaust, induction and compression strokes. In addition, when an engine is first started, the bearings are not properly lubricated so that high levels of friction have to be overcome. In short, the process of starting an internal-combustion engine is one that requires a large amount of force.

Development of starting systems

Manual starting methods

On January 29, 1886, the inventor Carl Benz registered his motor carriage, a new type of road-going vehicle driven by a gasoline engine, with the German Imperial Patent Office.

In August 1888, his wife Berta Benz and their two sons embarked on their legendary and courageous journey from their home in Mannheim in the south of Germany to Pforzheim. It was the first-ever journey across country in the history of the automobile (Figure 1). As this excursion took place without the knowledge of her husband, push-starting was almost certainly the only way that Berta Benz could get the three-wheeler car going. Carl Benz himself would have been able to start it by spinning the flywheel. But it is doubtful whether his wife or the two boys would have had the physical strength required to start the engine in that way.

For many years after Berta Benz' daring journey, muscle power remained the chief source of energy for starting the internal-combustion engine. Whether by push-starting, spinning the flywheel or using a crank handle, starting a motor car required physical exertion and often a good deal of sweat. No wonder, then, that from an early stage, engineers began to look for an easier method of getting the engine going.

Right from the early days of the Twentieth Century, the inventive efforts of numerous engineers were devoted to this problem. An enormous variety of starting devices was thought up, including spring-loaded, compressed-air, hydraulic and inertia starters (Figure 2).

The electric starter motor

The development of the electric starter motor was a major breakthrough. However, its effective use depended on the availability of a sufficiently powerful battery. Lead-acid accumulator batteries with sufficient capacity started to be produced around 1910, which meant that from then on nothing more stood in the way of the spread of the electric starter motor.

1 Berta Benz tests out her husband's invention (source: DaimlerChrysler Classic, Group Archive)

In the beginning, starter motors were linked directly to the crankshaft by a chain or belt-drive system and, in some cases, they also performed the function of a dynamo – a concept that has been revived in some of today's very latest designs. At that point, however, separate devices optimized for the distinct functions of starting and electricity generation became the established norm. As early as 1913 there was an electric starter motor with planetary gearing and an overrunning clutch (Figure 3).

In order to reduce the technical complexity, starter-motor designs were subsequently developed which allowed separation of the starter-motor drive from the driven component. Initially, the starter-motor drive pinion was brought into engagement with the ring gear on the flywheel by manual means involving a lever, pedal or cable linkage – a principle that remained in isolated use right up until the 1950s. However, it wasn't long before electrical actuation by means of the ignition switch and engagement of the pinion with the aid of a solenoid switch became the established method. The basic design of the electric starter motor had thus been determined by roughly the end of the Second World War.

The years that followed were marked by a wide variety of design improvements. In particular,

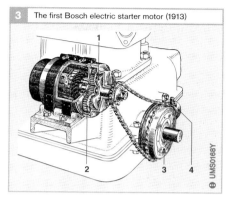

The first Bosch electric starter motor (1913)

Fig. 3
1 Starter motor
2 Planetary gear
3 Freewheel
4 Magnetic brake

these involved increasing the reliability of the pinion-engagement action and reducing size and weight. On commercial vehicles, increasingly large engines were used. Accordingly, the mechanical stresses on the starter motor were much greater. These were overcome by multi-stage pinion-engaging systems and specially designed overrunning clutches.

By the beginning of the 1980s, the development of ceramic permanent magnets had reached a stage where they could be used in starter motors for cars. This made it possible to dispense with the more costly electromagnet.

As a further means of reducing size and weight, starter motors with internal reduction gear were introduced in 1982. They have remained the market-leading design ever since.

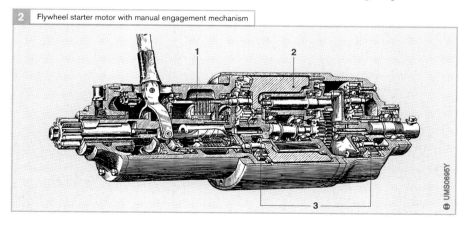

Flywheel starter motor with manual engagement mechanism

Fig. 2
1 Multiplate overrunning clutch
2 Flywheel
3 Planetary gear

Starting the internal-combustion engine

A starter motor starts an internal-combustion engine by engaging its pinion gear in a ring gear that typically has about 130 teeth (on cars). On vehicles with a manually shifted transmission, the ring gear is on the engine flywheel, while on vehicles with automatic transmission, it is on the torque-converter housing. When in its resting position, the starter-motor pinion (which typically has 10 teeth) is disengaged from the ring gear, but remains only a few millimeters away from it (Figure 1, Diagram 1). When the driver turns the ignition key to the starting position, the starter motor first of all establishes the mechanical link between itself and the engine.

Operating sequence of the starter motor
Engagement
When the ignition switch is in the starting position, it completes an electric circuit that energizes the starter-motor solenoid switch. The magnetic field created by the solenoid coil draws in the solenoid armature, thus operating the engagement lever so that the pinion gear is moved outwards and comes into contact with the ring gear.

Under ideal circumstances, the teeth of the pinion will be in line with the gaps in the ring gear so that the two gears mesh perfectly and the mechanical link between the starter motor and the engine is immediately established (Figure 1, Diagram 2). As it reaches the end of its travel, the solenoid armature closes a switch which completes the starter-motor main circuit. The starter motor thus starts to rotate so that its drive shaft and pinion gear drive the ring gear with the result that the engine starts to turn as well (Figure 1, Diagram 4).
This ideal sequence of events, whereby the pinion meshes perfectly with the ring gear first time, actually rarely occurs in practice because the 0.4 mm of play between the teeth of the two gears leaves very leeway for smooth meshing.

The typical scenario (which occurs around 70 % of the time) is a situation in which the teeth of the pinion collide with those of the ring gear as the pinion moves outwards. The pinion can then not move any further outwards and mesh with the ring gear unless it is rotated. However, in this case too, the solenoid armature continues to be drawn further into the coil so that the engagement lever compresses the meshing spring. Consequently, the pinion is pressed with increasing force against the side of the ring gear (Figure 1, Diagram 2).

As it reaches the end of its travel, the solenoid closes the switch in the starter-motor main circuit (as previously explained). As a result, the pinion starts to turn and at some point reaches a position where it will mesh with the ring gear. At that point, the tension in the meshing spring forces the pinion rapidly outwards. In this way, the necessary mechanical link between starter motor and engine is established in this more complex scenario.

On most starter motors, pinion engagement is also assisted by a helical spline. This has two effects:
- As the pinion moved outwards, it turns slightly in the opposite direction to its normal direction of rotation, which assists the meshing action and
- When the starter motor starts to rotate, the effect of the helix throws the pinion outwards as it engages with the ring gear

As starter motors for commercial vehicles have to produce high levels of torque, appropriate meshing methods have to be adopted in order to prevent excessive stress on the teeth of the pinion and ring gear (refer to the section "Starter motors for commercial vehicles"). They ensure that there is sufficient overlap of the teeth before the starter motor delivers its full power.

Turning the engine
When the starter motor begins to rotate, the transmission ratio between the pinion and the ring gear produces a large amount of torque acting on the crankshaft of the engine.
The frictional resistance is overcome and the engine starts to turn over.

The machine constant, c_1, is derived from the number of poles and the characteristics of the armature winding.

Since a voltage is induced in a wire coil that is moving inside a magnetic field, the induced voltage

$$U_{ind} = 2 \cdot \pi \cdot c_2 \cdot B \cdot l_{Fe} \cdot d \cdot n,$$

where n is the speed of rotation of the armature, occurs in the armature winding. Due to the effect of commutation, this voltage appears externally as DC voltage. It is acting in the opposite direction to the supply voltage even though the polarity of the conductors on the rotating armature is continually changing. The machine constant, c_2, is determined as described above.

The electric circuit of a starter motor is shown in Figure 2. The voltage at the battery terminals, U_K, is the product of the open-circuit voltage, U_L, minus the voltage drop due to the battery's internal impedance, R_i. The voltage available at the starter motor, U_S, is further reduced by the voltage drop due to the supply-cable impedance, R_{perm}. The voltage at the commutator is in turn diminished by the voltage drop, U_{br}, due to the brushes. Approximately 1.2 V is generally lost for each pair of brushes regardless of the current. For positive and negative brushes combined, therefore, $U_{br} \approx 2.4$ V. The impedance of the starter motor itself is R_S.

For the electric circuit as a whole, therefore,

$$U_L = (R_i + R_{perm} + R_S) \cdot I + U_{br} + U_{ind}$$

and with the inclusion of the speed-dependent induced voltage,

$$U_L = (R_i + R_{perm} + R_S) \cdot I + U_{br} + 2 \cdot \pi \cdot c_2 \cdot B \cdot l_{Fe} \cdot d \cdot n,$$

from which it follows that the speed can be represented by

$$n = \frac{U_L - U_{br} - (R_i + R_{perm} + R_S) \cdot I}{2 \cdot \pi \cdot c_2 \cdot B \cdot l_{Fe} \cdot d} \quad \text{Eq. 2}$$

The product of torque and speed is the power output, P_i, which is given by

$$P_i = 2 \cdot \pi \cdot M \cdot n$$

$$= \frac{c_1}{c_2} \left[(U_L - U_{br}) \cdot I - (R_i + R_{perm} + R_S) \cdot I^2 \right]$$
Eq. 3

Thus it is equal to the battery power output, $U_L \cdot I$, minus the losses at the brushes, $U_{br} \cdot I$ and the combined impedance losses

$$(R_i + R_{perm} + R_S) \cdot I^2.$$

At the specified nominal voltage, the only technical parameters that can be varied in order to determine the power output are thus the impedance of the battery, the power-supply cable, and the starter motor itself. The actual power output, P_m, delivered at the starter motor pinion is reduced even further by the mechanical friction, V_R, and magnetization-reversal losses, V_{Fe}, in the iron core so that

$$P_m = P_i - V_R - V_{Fe}$$

Permanent-magnet motor
Electric motors with permanent-magnet stators (Figure 3) are at present the most commonly used design for car starter motors worldwide. New types of magnetic material such as strontium ferrite are able to achieve an air-gap induction of B ≈ 400 mT. As it is imposed by the permanent magnet

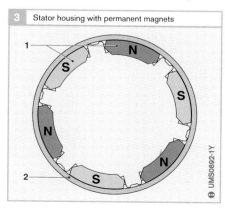

Fig. 3
1 Permanent magnets
2 Stator housing

(1), it is independent of the armature current. Such motors are also said to have a shunt characteristic. The effective magnetic flux is reduced at high armature currents by the "armature crossfield"[1], however.

As is evident from Equation 1, the torque increases in linear relationship with the current, whereas the motor speed decreases in linear fashion (Equation 2). As a result of the two linear relationships, the power output characteristic is represented by a convex parabolic curve.

Real starter-motor characteristics are marginally different from the theoretical graphs. That is due in particular to the retroactive effect of the armature magnetic field at high currents, but also because of nonlinear material and friction-related effects. The characteristic graphs are shown in Figure 4, with losses taken into account.

---
[1] "Elektrische Maschinen", Rolf Fischer, Hanser Verlag 2001, ISBN 3-446-21810-6.

Series-wound motor

In a series-wound motor, the stator magnetic field is electromagnetically created by excitation windings (Figure 5, Item 2). They are connected in series with the armature winding so that the same current is flowing through them. Consequently, the induction, B, is not constant and is dependent instead on the starter-motor current. An initial approximation using the machine constant c_3 can be represented by

$$B = c_3 \cdot I$$

so that the torque is given by Equation 1 thus:

$$M = c_1 \cdot c_3 \cdot l_{Fe} \cdot d \cdot I^2$$

It increases quadratically relative to the current. The motor speed drops hyperbolically according to Equation 2 thus:

$$n = \frac{1}{2 \cdot \pi \cdot c_2 \cdot c_3 \cdot l_{Fe} \cdot d} \left[\frac{U_L - U_{br}}{I} - (R_i + R_{perm} + R_S) \right]$$

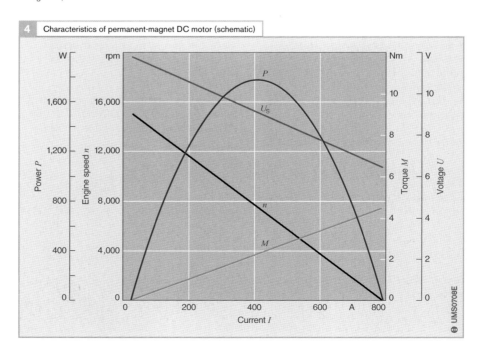

4 Characteristics of permanent-magnet DC motor (schematic)

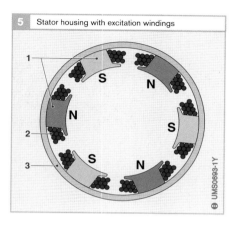

5 Stator housing with excitation windings

Similarly, in the case of the torque, the theoretical quadratic characteristic is evident only at low currents. Due to magnetic saturation of the iron, the magnetic flux ceases to increase proportionally at high currents, and instead, virtually proportional torque increase results.

Figure 6 shows the characteristic graphs.

Due to their high peak transient torque combined with adequate speed at low loads, series-wound motors are used mainly as starter motors for commercial vehicles.

Fig. 5
1 Pole shoes
2 Excitation winding
3 Stator housing

This means that the series-wound motor characteristically has a high peak transient torque and a high no-load speed. Theoretically, the speed would increase to infinity in response to decreasing current. In practice, however, the ever-present friction losses constitute the limiting factor.

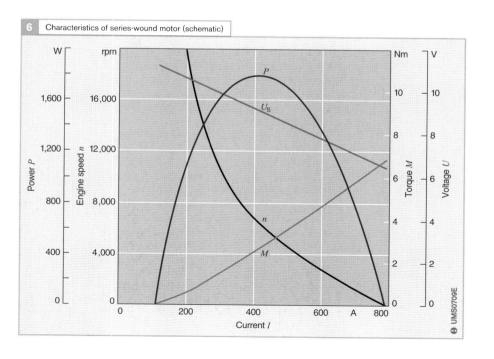

6 Characteristics of series-wound motor (schematic)

Permanent-magnet motor with flux concentrators

The torque and speed characteristics of permanent-magnet motors can be manipulated with the aid of flux concentrators (Figure 8, Item 2). If part of the leading edge of the magnet (1) is replaced by a piece of soft iron, the shape of the shunt characteristic curve can be changed to a certain degree to more closely resemble the series-wound characteristic (Figure 7).

As the size of the magnet is smaller and the soft-iron piece short-circuits part of the magnetic flux, the primary flux is reduced under no-load conditions. In accordance with the equation for the induced voltage, the noload speed is then increased. In the vicinity of the short-circuit, the soft-iron pole edge offers the flux resulting from the stator flux and the armature flux a path of high magnetic conductivity, as a result of which the overall flux is greater than for a motor without flux concentrators and produces higher torque in accordance with the torque equation. The flux concentrators only marginally increase the maximum power output. Flux concentrators are primarily a common feature of medium-power starter motors for car diesel engines.

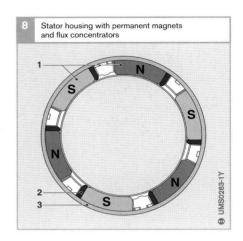

Fig. 8
1 Permanent magnets
2 Flux concentrators
3 Stator housing

8 Stator housing with permanent magnets and flux concentrators

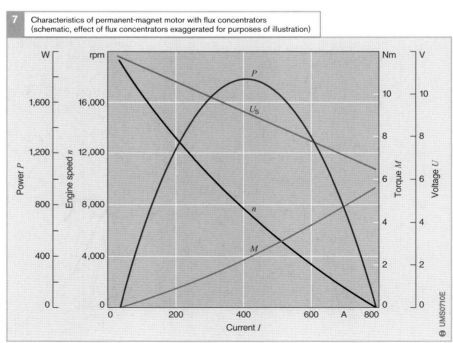

7 Characteristics of permanent-magnet motor with flux concentrators (schematic, effect of flux concentrators exaggerated for purposes of illustration)

Compound motor

The compound motor has two separate excitation windings: the series winding and the shunt winding. This purpose of design is to obtain a defined no-load speed while retaining the advantages of the series-wound motor. The shunt winding is connected in parallel with the armature winding. Compound motors are used as heavy-duty starter motors for commercial vehicles. Figure 9 shows the typical characteristics.

Direct-drive and reduction-gear starter motors

In a conventional direct-drive starter motor, the pinion rotates at the same speed as the armature. The one-way clutch and the pinion are mounted directly on the motor's armature shaft. In order to be able to deliver the high torque required for cold starting, the motor has to be relatively large – and therefore heavy – with this type of design. Consequently, it is now only commonly used in cars for power ratings < 1 kW.

If planetary gearing is used (Figure 10), the same torque can be obtained from a smaller and faster-running electric motor with the result that a weight saving of 30...40 % can be achieved, depending on model. Furthermore, less weight ultimately means lower fuel consumption by the vehicle concerned.

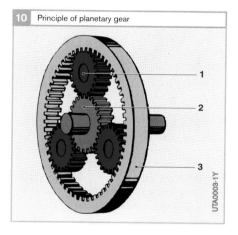

Fig. 10
1. Planet gear
2. Sun gear
3. Internal gear

Figure 10: Principle of planetary gear

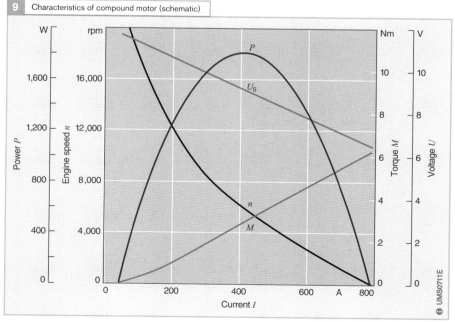

Figure 9: Characteristics of compound motor (schematic)

11 Starter-motor reduction gear

$$i = 1 + \frac{\text{Number of teeth on internal gear}}{\text{Number of teeth on sun gear}}$$

On standard-design starter motors for cars, the planet gears are made of sintered steel while the internal gear is constructed in glass-fiber reinforced polyamide. Where there are particularly demanding requirements regarding smoothness, temperature-resistance and durability, the internal gear may also be made of sintered steel. Such designs are then use three elastic rubber supports in the intermediate bearing.

The transmission ratio between the armature and the pinion is variable across a broad range from roughly 3.3 : 1 to 6 : 1 depending on the design of the planetary gear. This allows optimum matching of the starter motor to the characteristics of the engine on which it is fitted and to the vehicle's electric "Dimensioning starting systems".

Higher transmission ratios enable higher engine speeds for warm starting, while lower transmission ratios allow the starter motor to cope with extreme cold-starting conditions and result in lower power consumption by the starter motor when turning the engine over.

The characteristics of a reduction-gear starter motor also offer other advantages.

The high armature speed obtained as a result of the low gearing ratio creates a "flywheel" effect which helps to turn the engine through the cylinder top-dead-center positions and smooth out engine-speed fluctuations. As a result, the engine has a higher momentary speed precisely at the point when fuel is injected, and this has a positive effect on the fuel-injection pattern and therefore the ease with which the engine starts and the exhaust emission levels.

On engines with smaller numbers of cylinders, this "flywheel effect" also has the advantage that the high peak-torque levels of the individual cylinders can be reliably overcome with a relatively low amount of starting power.

Smaller starter motors allow the vehicle manufacturer more scope in the design of the engine compartment and/or the placement of other equipment within the vehicle.

Planetary gearing systems generally offer the advantage of combining compact overall dimensions with low gearing ratios. The gearing geometry allows the achievement of high torque output without generating high levels of noise. Furthermore, they do not generate any external transverse forces so that the armature-shaft and pinion-shaft bearings are not subjected to high stresses even at high power outputs.

The planetary gearing system used in starter motors has a fixed internal gear, that is the outer, internally toothed ring gear. The drive input is carried by the sun gear, which is attached to the armature shaft of the electric motor. The planet gears (of which there are normally three) are thus in engagement with both the sun gear and the internal gear. As the planet gears "orbit" around the sun gear, their bearing-shaft journals drive the output shaft which carries the familiar helix and one-way clutch (Figure 11).

The transmission ratio, i, for this type of planetary gearing system is given by the equation:

Solenoid switch

The purpose of the solenoid switch is to be able to switch a high current by means of a relatively low control current. The starter-motor current can be as high as 1,500 A on cars and as much as 2,500 A on commercial vehicles. Due to the diminishing effect of the supply-cable impedance on the starter-motor power output, the length of the power cable from the battery must be kept as short as possible and the switch impedances as low as possible. In addition, the switch contacts for such currents are subjected to high loads. The use of a power relay in the form of the solenoid switch is therefore absolutely essential. A simple mechanical switch (ignition/starter switch, starter button) or a mini-relay operated by the engine control unit can then be used to switch the relatively low control current.

The solenoid switch built into the starter motor is the combination of a solenoid and a relay switch. As previously explained in the section "Operating sequence of the starter motor", it performs the following two functions:
- It moves the drive pinion outwards so that engages in the engine's ring gear and
- It closes the switch which completes the starter motor's primary electric circuit

The design of a solenoid switch is shown in Figure 12.

The solenoid core (4) protrudes into the solenoid coil from one side, while the movable coil (1) protrudes from the other side. The distance between the core and the armature when at rest represents the total travel of the armature. The solenoid housing, core and armature together form the magnetic circuit.

The magnetic field created when the solenoid is energized draws the armature into the coil. That armature movement is utilized firstly to move the pinion along its longitudinal axis, and secondly to close the contacts (6, 8) of the relay switch.

The solenoid coil in most solenoid switches consists of a pull-in winding and a hold-in winding (Figure 12, Items 2 and 3, and Figure 13, Items 4a and 4b). This arrangement is particularly effective with regard to thermal stress capacity and the achievable magnetic forces. When the coil first starts to draw in the armature, the air gap between armature and the core is relatively large. Only a high magnetomotive force can overcome meshing resistance.

As the air gap closes up as the armature retracts, the magnetic force increases significantly. When the armature is fully retracted, i.e. when there is only a minimal residual air gap, the force of the hold-in winding on its own is sufficient to hold the armature in position until the starting sequence is completed. The pull-in winding is thus short-circuited when the main-circuit switch and the ignition

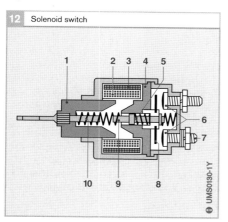

Fig. 12 Solenoid switch

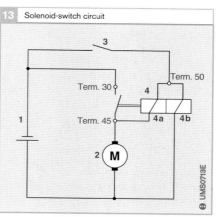

Fig. 13 Solenoid-switch circuit

Fig. 12
1 Solenoid armature
2 Pull-in winding
3 Hold-in winding
4 Solenoid core
5 Contact spring
6 Switch contacts
7 Electrical connection
8 Switch contact
9 Armature shaft (split)
10 Return spring

Fig. 13
1 Battery
2 Starter motor
3 Ignition/starter switch
4 Solenoid switch
4a Pull-in winding
4b Hold-in winding

switch are closed. It is important that the two coils have the same number of turns. Otherwise, the solenoid switch might adopt a self-locking characteristic due to the supply of the coils, which would then be connected in series, in reverse through Terminal 45. Having the same number of turns in each coil ensures that that the magnetic fields of the two coils, through which the current then flows in opposite directions, cancel each other out and the solenoid switches off reliably.

Since the starter motor causes a substantial voltage drop in the vehicle's electrical system when its main circuit is closed, the hold-in winding must be dimensioned so as to be capable of reliably holding the armature in place even at supply voltages significantly below half the nominal battery voltage. Otherwise, the solenoid switch might under certain circumstances repeatedly switch on and off in rapid succession, resulting in damage to the switch contacts.

Return springs between the individual components ensure that the switch opens again when the solenoid is switched off and the armature returns to its resting position.

All starter-motor terminals (Terminals 50, 30 and 45) are usefully incorporated within a single terminal block on the cap of the solenoid switch. The solenoid-switch cap is made of a high-quality duromer material which provides the required resistance to mechanical stress even at high temperatures (up to 180 °C for short periods) and thus ensures that the solenoid switch and all its components are held together reliably.

In order to ensure reliable retraction and switching of the solenoid even at high temperatures, there are special coil designs that can also withstand temperatures up to 180 °C.

The primary-circuit switch contacts on car starter motors generally consist of steel bolts with riveted copper contacts that provide reliable switching characteristics and minimal switch impedance. The relay switches for commercial-vehicle starter motors are manufactured as one-piece components from a high-strength copper alloy in order to provide high current capacity.

Depending on the type of wiring harness used, the control-circuit terminal (Terminal 50) may take the form of a screw terminal or some type of push-fit connector which may or may not be insulated.

Modern relay switches for cars have a round plug connector to connect the cap to the solenoid coil. This allows a closed-surface cap design without riveted or soldered connections or the holes they would require. In addition, the standard arrangement provides for a seal between the cap and the solenoid housing.

For situations requiring special sealing arrangements, the solenoid armature can be protected by a flexible rubber boot (Figure 14, Item 2) in order to prevent the ingress of humidity from the starter motor's pinion-engaging mechanism. This measure is primarily adopted for applications in which it is not possible to entirely prevent the entry of water into the drive-end shield. Another situation where the armature boot may be used is one where the fitted position of the motor is

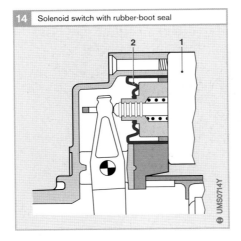

Fig. 14
1 Solenoid switch
2 Rubber boot

such that the solenoid switch is underneath, with the result that moisture can collect in the drive-end shield cowl.

Some large commercial-vehicle starter motors do not have an integral solenoid switch; instead the solenoid for engaging the pinion and the relay switch for controlling the electric circuits are separate from one another (refer to the section "Starter motors for commercial vehicles").

Overrunning clutches

On all types of starter motor, the drive is transmitted via an overrunning or one-way clutch. That one-way clutch is positioned between the motor and the pinion. Its job is to transmit the drive to the pinion when the starter motor is driving the ring gear, but to disengage the pinion from the pinion drive shaft as soon as the ring gear is moving faster than the pinion. The one-way clutch therefore prevents the armature of the starter motor from being accelerated to excessive speeds once the engine has started.

Overrunning clutches for starter motors can be either frictional (roller and multiplate types) or positively interlocking (spur-gear type) in design.

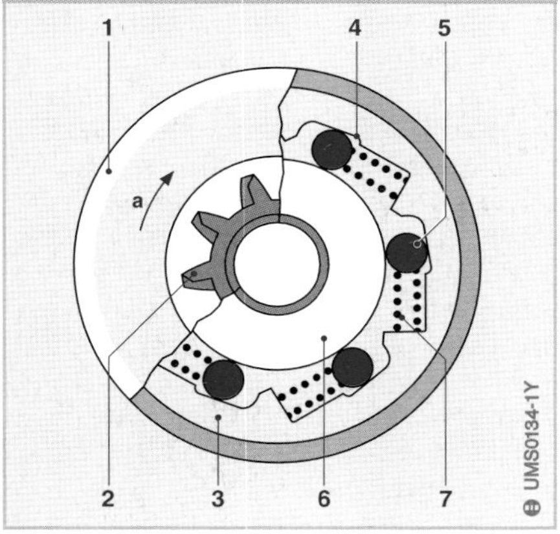

Roller-type overrunning clutch
Pre-engaged starter motors normally have roller-type overrunning clutches (Figure 15). The central component of this type is a clutch shell (3) with a roller race (4). The clutch shell forms part of the driver (3) and is thus connected to the pinion drive shaft by a helix. The frictional link between the central cylindrical shaft of the pinion and the clutch shell surrounding it is formed by cylindrical rollers (5) that are able to move within the roller race.

When the mechanism is at rest, the springs (7) force the rollers into the constricted space between the roller race in the clutch shell and the pinion shaft (6). Because of the constriction, the rollers jam between the surfaces of the roller race and the pinion shaft, so that when the motor turns the clutch driver, the pinion shaft is forced to turn as well.

As soon as the pinion starts to overrun, the friction between the pinion shaft and the rollers pushes the rollers into the wider part of the roller race; the force of the springs ensures that the rollers are held in contact with the pinion shaft and roller race at all times. The resulting overrunning torque is relatively small and has very little effect as far as the no-load speed of the starter-motor armature is concerned.

Multiplate overrunning clutch
The multiplate overrunning clutch is used on larger pre-engaged starter motors. Its job is:
- To establish the frictional link between starter-motor pinion and armature
- To prevent overload by limiting the torque transmitted to the pinion by the armature shaft and
- To disengage the pinion from the armature shaft as soon as it is overrunning

15 Roller-type overrunning clutch

Fig. 15
1 Cap
2 Pinion
3 Driver and clutch shell
4 Roller race
5 Cylindrical roller
6 Pinion shaft
7 Springs

a Direction of rotation

Starter motors — Starter-motor design

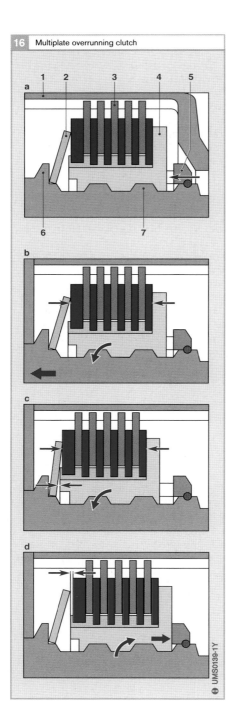

Fig. 16
1. Driver flange
2. Disc spring
3. Laminated core
4. Clutch race
5. Stop ring
6. Stop collar on output shaft
7. Helical spline on output shaft

Explanation of Figure 16:
a Resting position
Disc spring compresses laminated core with initial force.
Transmission of drive to clutch race ensured by frictional effect.
b Transmission of drive
Pinion engaged.
Clutch race presses against disc spring.
Compression increases.
Laminated core fully locked.
c Torque limitation
Clutch race presses disc spring against collar.
Forces in equilibrium at design rating.
Plates slip if design rating exceeded.
d Overrunning
Direction of force transmission reversed.
Clutch race runs up against stop ring, thus relieving plate pressure/disengaging drive.

Transmission of drive

So that the multiplate overrunning clutch can transmit the drive from the motor to the pinion, there has to be a certain amount of pressure holding the plates against each another (Figure 16a). When the clutch is in the resting position, the laminated core (3) is compressed by the force of a disc spring (2) to such as extent that it is able to drive the clutch race (4).

Once the pinion has reached the limit of its outward travel, the clutch must be capable of transmitting the full power of the motor. As the pinion is held stationary while the armature shaft is rotating, the clutch race slides outwards on the helical spline (7) of the output shaft and presses against the disc spring. This further compresses the laminated core and increases the pressure between the individual plates. Compression continues to increase until the friction between the plates is sufficient to transmit the torque required to start the engine. The force transmission path as follows:
Armature shaft – driver flange – outer plates – inner plates – clutch race – output shaft – pinion (Figure 16b).

Torque limitation

The increasing plate pressure due to the helix effect of the clutch race, and the resulting torque transmission, are limited by the fact that, on reaching the maximum load, the end face of the clutch race comes into contact with the disc spring. In doing so, it presses the disc spring against the stop collar (6) of the output shaft (Figure 16c).

This prevents the laminated core from being compressed any further. The multiplate overrunning clutch then functions as an overload clutch since the plates will slip if the maximum power transmission rating, and therefore the maximum torque, is exceeded.

Disengagement of drive (overrunning)

When the engine flywheel accelerates due to ignition pulses or the engine starting, the pinion starts to turn faster than the starter motor. As a result, the direction of force transmission is reversed and the clutch race is thrown inwards along the helical spline until it meets the stop ring (5, Figure 16d). The disc spring is released and no longer presses against the laminated core. As a result, the plates separate from one another and the drive is no longer transmitted between them.

Radial-tooth overrunning clutch

A radial-tooth overrunning clutch (Figure 17) is fitted on Type JE and KE pre-engaged starter motors (described in more detail in the section "Starter motors for commercial vehicles") in combination with a mechanical two-stage pinion-engaging mechanism. This type of overrunning clutch transmits the torque by a positively interlocking mechanism in the form of a system of helical teeth.

During the overrunning phase, the ring gear on the engine's flywheel is driving the pinion (1) which is connected to the clutch race (4) by helical teeth. Due to the special shape of the helical teeth, the clutch race is forced inwards along the helical spline towards the motor when the pinion is overrunning. Separation of the meshing components (pinion and clutch race) of the radial-tooth overrunning clutch is assisted by three flyweights (2) as the latter generate a longitudinal force via an internally tapered thrust ring (3).

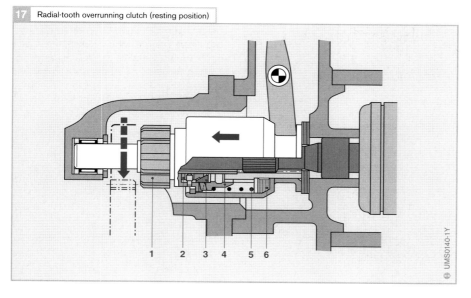

17 Radial-tooth overrunning clutch (resting position)

Fig. 17
1 Pinion
2 Flyweights
3 Internally tapered thrust ring
4 Clutch race with spur gearing
5 Spring
6 Rubber bumper

Pinion-engaging mechanism

The pinion-engaging mechanism is the system of components that engages the pinion with the ring gear. The lever-and-helix system (Fig. 18) has established itself as the global standard for such mechanisms. For this reason, only this system is described in this section. Other types of mechanism are detailed in the section "Starter-motor design variations" in connection with the relevant starter motor types.

On starter motors with a lever-and-helix engaging mechanism, the engaging sequence is made up of a lever-travel phase and a helical-travel phase.

Lever-travel phase

When the ignition/starter switch is operated, the solenoid armature is drawn inwards against the force of the return spring. It is connected to the pinion-engaging lever in a manner that allows for a certain amount of play. The pinion-engaging lever pushes the overrunning clutch against the ring gear on the engine flywheel with the aid of the guide collar and the meshing spring. The starter-motor armature is not rotating at this stage as the primary electric circuit has not been closed. If the pinion is in such a position that its teeth are aligned with the gaps in the ring gear, it will mesh immediately and the lever will move it outwards as far as the solenoid-armature travel allows. This marks the end of the lever-travel phase.

Helical-travel phase

The helical travel corresponds to the distance between the end of the lever-travel phase and the limit of the continued axial travel of the overrunning clutch (stop ring) due to the effect of the helical spline. This mechanism ensures that the maximum possible overlap between the teeth of the pinion and the ring gear is utilized. The helix prevents transmission of torque to engine until the pinion is fully engaged in the ring gear.

The solenoid armature is connected to the pinion-engaging lever in such a way that there is a certain amount of free travel or "play". If the engine on which the starter motor is mounted fails to start (e.g. due to lack of fuel), the starting sequence has to be aborted. At this point, the pinion is fully engaged, and pinion and ring gear are under maximum load.

When the power supply to the solenoid is switched off, there must be sufficient armature free travel to allow it to open the primary-circuit switch. If this were not the case, the pinion-engaging lever would hold the armature in place. The primary-circuit switch would remain closed and it would not be possible to stop the starter motor.

On commercial-vehicle starter motors, the geometrical constraints demand that rather that a "cutout spring" (Figure 19) is used instead of the "free-travel" arrangement explained above. The force of that spring when the pinion is in the resting position overcomes the force of the solenoid-armature return spring and so pushes the armature back until it is resting against the lever. When the solenoid armature is in the retracted position, on the other hand, the force of the armature return spring must be sufficient to compress

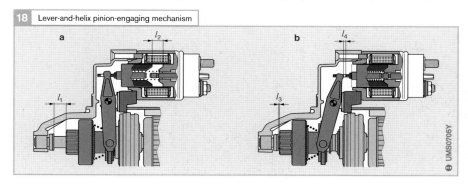

18 Lever-and-helix pinion-engaging mechanism

Fig. 18
a Resting position
b Extended position, shown at end of lever travel

l_1 Overall pinion travel
l_2 Solenoid-armature travel
l_3 Helical travel
l_4 Free travel

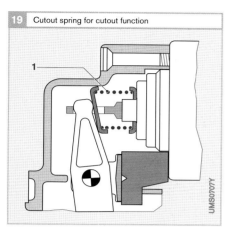

Fig. 19
1 Solenoid with cutout spring

the cutout spring enough for the primary-circuit switch contacts to be separated.

There are two possible arrangements for the pinion end bearing as illustrated in Figure 20.
- Most starter motors have a drive-end shield which has a "mouth" where it fits over the flywheel. The drive-end shield holds the pinion end bearing and has an opening facing the flywheel. The bearings used can be plain or rolling bearings (in some cases sealed). The pinion-engaging mechanism is exposed to the ring gear.
- For applications where the pinion-engaging mechanism would be subjected to high levels of dirt, moisture or clutch-plate dust, the closed-shield design is preferable. The pinion bearings are completely enclosed within the drive-end shield. Special sealing rings provide such effective protection that these starter motors are suitable for even the most unfavorable conditions. In order to be able to cope with the higher bearing loads and last the full service life of the motor, the pinion bearings are suitable reinforced.

Closed-shield starter motors also have advantages in terms of fitting because they do not require a mating flange that is shaped to match the "mouth" of the end shield. Consequently, they allow more scope for choice of fitted position and the space requirements in the bell housing are smaller.

Nevertheless, the following features prevent the general use of this design:
- The bearing forces in the drive-end shield and the effective radius of friction are about twice as great as for an open-shield starter motor, with the result that friction-related power losses are greater (the use of rolling bearings can help to alleviate this problem and also improves wear and noise characteristics).
- Bearings, seals and elongated pinion design result in higher costs.
- The overall length of the starter motor is greater.
- The gearing arrangements, and therefore the noise characteristics, are less favorable.

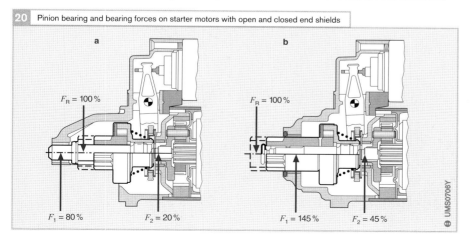

Fig. 20
a Open-shield starter motor
b Closed-shield starter motor

Starter-motor design variations

Starter-motor type designations

Bosch starter motors are classified according to the type designation key shown in Figure 1.
- The first letter indicates whether the starter motor is a direct-drive model, a reduction-gear model or a starter motor for commercial vehicles.
- The second letter indicates the method of excitation; the third letter shows whether the motor has a closed end shield.
- The figure that comes next indicates the stator-housing diameter in millimeters. This identifies the starter-motor family or type range.
- A class code letter follows to indicate the size of the armature core which can vary within the individual motor series. "S" is for short, "M" for medium, "L" for long, and "E" for extended.
- An additional numerical suffix in the type designation indicates other features which affect the motor characteristics.
- Details of the voltage rating and the direction of rotation complete the designation code where necessary.

Starter-motor types and applications

Bosch manufactures starter motors for a wide range of internal-combustion engines The suitability of the various starter-motor types for the different types of engine is shown in Figures 2 (cars), 3 and 4 (commercial vehicles). The starter motors specified in Figure 4 have been in production for many years and therefore have a different type-designation format from newer models.

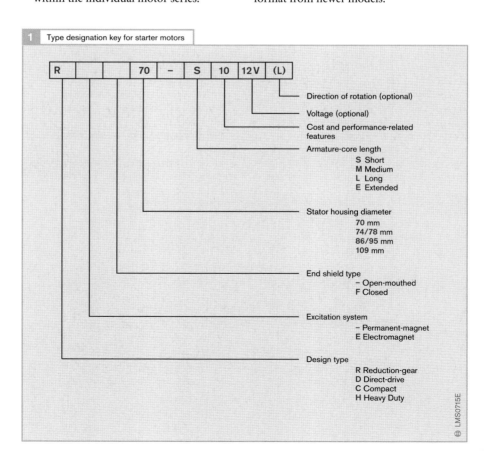

1 Type designation key for starter motors

R   70 - S 10 12V (L)

- Direction of rotation (optional)
- Voltage (optional)
- Cost and performance-related features
- Armature-core length
  S Short
  M Medium
  L Long
  E Extended
- Stator housing diameter
  70 mm
  74/78 mm
  86/95 mm
  109 mm
- End shield type
  – Open-mouthed
  F Closed
- Excitation system
  – Permanent-magnet
  E Electromagnet
- Design type
  R Reduction-gear
  D Direct-drive
  C Compact
  H Heavy Duty

Starter motors Starter-motor design variations

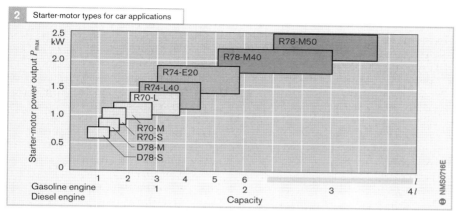

2 Starter-motor types for car applications

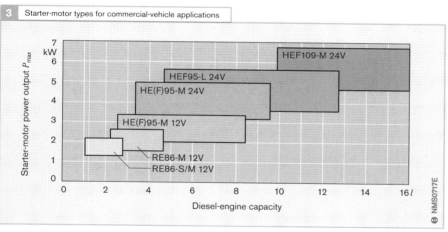

3 Starter-motor types for commercial-vehicle applications

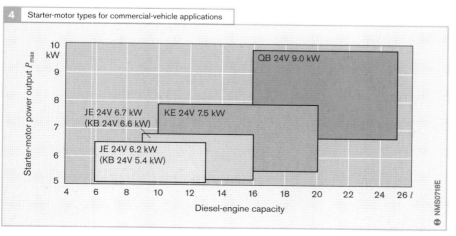

4 Starter-motor types for commercial-vehicle applications

Starter motors — Starter-motor design variations

Direct-drive starter motors for cars

Application
The Series D78 pre-engaged starter motors have a permanent-magnet excitation field. They are suitable for use in cars with small gasoline engines up to approx. 1.6 l capacity.

Design
The design and internal configuration of the Series D78 can be seen in Figure 5.

This model has a DC motor with six-pole permanent-magnet excitation and four carbon brushes. Power transmission is by way of a roller-type overrunning clutch which is driven directly by the armature shaft.

The two basic models available (D78-S and D78-M) differ primarily by virtue of the length of the electric motor. In addition, the D78-M has flux concentrators which increase the short-circuit torque.

Reduction-gear starter motors for cars

Application
For gasoline engines with capacities above 1.4 l and for car diesel engines, it is advisable due to considerations of weight and it is preferable to fit the R70, R74 and R78 series of reduction-gear start motors for reasons of weight and space restrictions. They are available with open or closed end shields.

Design
Design and internal configuration can be seen in Figure 6. These starter motors have a tough planetary reduction gear of the type previously described with three planet gears. They provide the high starting torque required by gasoline engines up to a capacity of approx. 6 l and for diesel engines up to a capacity of approx. 3 l.

In order to cover the range of power outputs required by modern car engines, there

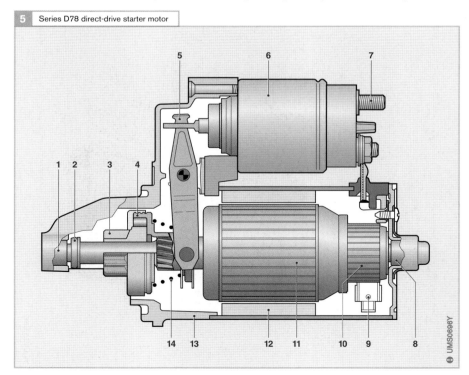

5 Series D78 direct-drive starter motor

Fig. 5
1 Drive shaft
2 Stop ring
3 Pinion
4 Roller-type overrunning clutch
5 Pinion-engaging lever
6 Solenoid switch
7 Electrical connection
8 Commutator bearing
9 Brush holder
10 Commutator
11 Armature
12 Magnet
13 Stator housing
14 Meshing spring

are three type ranges with varying stator housing diameters available. Common to all models are the integral solenoid switch, six-pole permanent-magnet excitation field and four carbon brushes. The permanent magnets used ensure that demagnetization does not occur even under unfavorable conditions. The more powerful types also have flux concentrators that increase the short-circuit torque and no-load speed.

The use of planetary reduction gears with transmission ratios between 3.38 : 1 and 5.67 : 1 enables optimum starting-system compatibility with the requirements of the engine and vehicle electrical system in question. The standard internal gear is made of glass-fiber reinforced polyamide plastic, while an alternative planetary gear with a sound-insulated internal gear made of sintered steel is also available.

The bearings in the planetary gear and the drive end shield are fitted with sintered bushes on the smaller models (up to R70-M, R74-M) but are also available with needle-rolling bearings as an option for higher performance.

Starter motors for commercial vehicles

Starter motors for commercial vehicles are nowadays generally fitted with a reduction gear in order to utilize the weight and size advantages compared with direct-drive starter motors.

The use of lighter materials and optimized manufacturing methods has made it possible to reduce the overall weight by as much as 40% compared with conventional direct-drive starter motors.

The area of application for reduction-gear starter motors with electromagnetic excitation is primarily diesel engines with capacities between 2.5 and approx. 16 l.

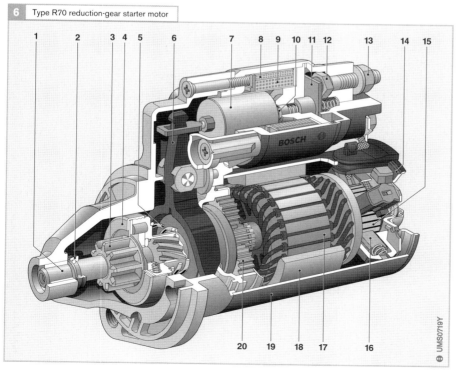

6 Type R70 reduction-gear starter motor

Fig. 6
1 Drive shaft
2 Stop ring
3 Pinion
4 Roller-type overrunning clutch
5 Meshing spring
6 Pinion-engaging lever
7 Solenoid switch
8 Hold-in winding
9 Pull-in winding
10 Return spring
11 Switch contact
12 Switch contact
13 Electrical connection
14 Commutator end shield
15 Commutator
16 Brush holder
17 Armature
18 Magnet
19 Stator housing
20 Planetary gear

Type RE86 and HE(F)95 starter motors for commercial vehicles

Application

These models are suitable for diesel engines with a capacity of between 2.5 and 9.5 *l*.

Design

The basic design is the same as the starter motors for cars described above. They are available with the conventional open-mouthed end shield (Figure 7a) or with the closed-type end shield (Figure 7b).

In contrast with the permanent-magnet car starter motors described previously, these models use electromagnets. Accordingly the motor is of the DC series-wound variety.

Special features

Due to the high torque loads, the intermediate bearing, which supports the drive shaft and planetary-gear carrier as well as the rubber-insulated internal gear, is made of aluminum. The internal and planet gears are constructed from high-strength sintered metal. The overrunning clutch can be modified both in terms of size and choice of materials to suit specific requirements.

Type HEF95-L starter motor for commercial vehicles

Application

The HEF95-L (Figure 8) is the next larger model in the present range of reduction-gear starter motors.

The area of application for this model covers diesel engines with a capacity of between 9 and approx. 12 *l*.

7 Type RE86/HE(F)95 starter motors for commercial vehicles

Fig. 7
a Conventional design with open end shield (Type RE86/HE95)
b Closed-shield design (Type HEF95)

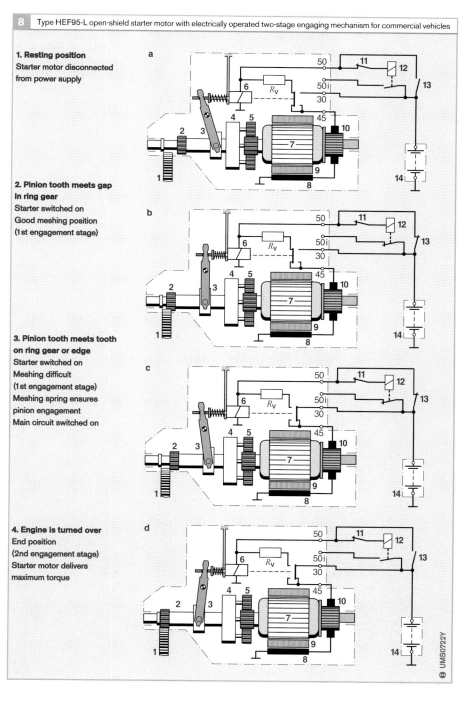

Fig. 8
1 Ring gear
2 Pinion
3 Pinion-engaging lever
4 Roller-type overrunning clutch
5 Planetary gear
6 Solenoid switch
7 Armature
8 Excitation winding
9 Pole shoe
10 Commutator
11 Thermostatic switch
12 Type IMR pilot solenoid
13 Ignition/starter switch
14 Battery

R_V Series resistor

Design and method of operation
The HEF95-L starter motor is generally a closed-shield model in order to utilize the benefits of the design (e.g. greater resistance to environmental effects and greater adaptability with regard to fitting). In contrast with the less powerful Series HE95 starter motors, this model has an electrically operated, two-stage pinion-engaging system (Figure 8) in order to enable reliable pinion engagement and minimize ring-gear wear over the full service life of the motor despite the high dynamic forces at play.

The two-stage pinion-engagement sequence is controlled jointly by the Type IMR pilot solenoid (Figure 9) and the solenoid switch. The meshing spring is also integrated in the solenoid switch (Figure 10). A cutout spring ensures that the starter motor switches off if the starting sequence is aborted.

1st stage (preliminary stage)
A voltage signal (e.g. from the ignition switch or a control unit) activates the starter motor. It initially switches the IMR pre-control relay which is capable of switching the high current of approx. 150 A required for a two-stage electrical system.

That current then flows through a series resistor and a combined retracting and hold-in winding. This firstly moves the over-running clutch and pinion shaft outwards by operating the pinion-engaging lever. Secondly, a sufficiently high current can flow via the series resistor through the electric motor to start it turning before the main circuit is closed. As a result, the pinion can normally be fully engaged in the ring gear before the maximum current is applied to the motor by the closure of the primary-circuit switch.

In exceptional cases, the pinion may not engage with the ring gear if the two gears are not aligned. In such cases, the meshing spring on the solenoid armature makes sure that the main circuit is completed before the pinion is fully engaged with the ring gear. The pinion-engaging sequence is then completed within a single stage.

2nd stage (main stage)
When the pinion has successfully engaged with the ring gear, the solenoid armature disconnects the circuit through the series resistor by means of a break switch shortly before it reaches its fully retracted position. Immediately after that, the main circuit is completed and the starter motor begins to generate its full torque output.

9 Type IMR pilot solenoid (external and cutaway views)

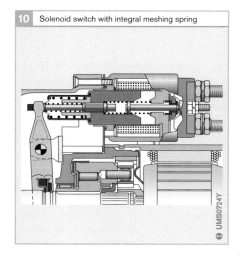

10 Solenoid switch with integral meshing spring

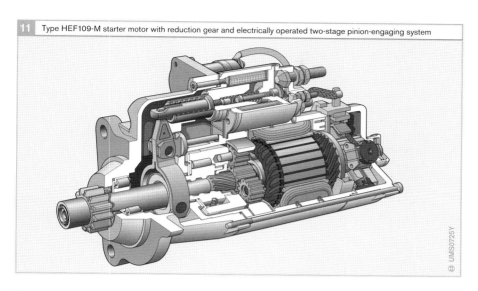

11 Type HEF109-M starter motor with reduction gear and electrically operated two-stage pinion-engaging system

Type HEF109-M starter motor for commercial vehicles

Application

The HEF109-M starter motor is the most powerful model in the current range for commercial vehicles. It is designed for use on diesel engines with a capacity of between 9 and approx. 16 l.

Like many of the other models, it utilizes the advantages of the reduction-gear concept to reduce weight and size, improve short-circuit torque and minimize speed fluctuations when turning the engine over (Figure 11).

Design

The HEF109-M is a closed-shield model in order to utilize the benefits of the design (such as greater resistance to environmental effects and greater adaptability with regard to fitting). In addition, it has an electrically operated two-stage pinion-engaging system which ensures reliable pinion engagement and significantly longer life both on the part of the starter motor and the ring gear.

The electrical functions, in particular those of the electric motor and the two-stage engagement system, are the same as described previously for the HEF95-L starter motor. In order to further significantly reduce the dynamic forces that occur during the pinion-engaging sequence, the masses moved along the motor axis are kept as small as possible. The roller-type overrunning clutch is combined with the planetary reduction gear to form a single unit which does not move in an axial direction. The pinion-engaging lever moves the pinion shaft via a thrust plate; the overrunning-clutch driver does not move in an axial direction (Figure 12). The pinion itself is mounted on the pinion shaft by means of straight-tooth gearing incorporating a degree of spring-cushioned axial travel which further contributes to load reduction during the pinion-engaging sequence.

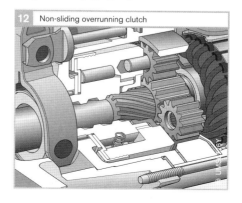

12 Non-sliding overrunning clutch

Starter motors with pre-engaged starter pinion engagement mechanism incorporating mechanical pinion rotation

Application

Starter motors with pre-engaged starter pinion engagement mechanism incorporating mechanical pinion rotation (Types JE and KE) are used to start large-scale internal-combustion engines. They have a two-stage mechanical pinion-engagement sequence in order to reduce the loads on the pinion and the ring gear. The most important functional component of the mechanism is the radial-tooth overrunning clutch described in the section "Overrunning clutches".

Design and method of operation

The Type KE has a series-wound motor, while the Type JE has a compound motor in order to limit speed.

1st pinion-engagement stage
When the ignition/starter switch is operated, the solenoid switch initially moves the pinion- engaging lever against the action of the return spring. The pinion-engaging lever slides the radial-tooth overrunning clutch assembly outwards along the straight-tooth gearing until it meets the ring gear. If the pinion happens to be in a position where it can mesh with the ring gear, it will continue to move outwards to the full extent of its linear travel.

2nd pinion-engagement stage
If the pinion is not in a position where it can mesh with the ring gear, the remaining components of the radial-tooth overrunning clutch continue to move outwards on a linear path towards the ring gear. The action of the clutch-race helix forces the pinion to rotate in its working direction while simultaneously compressing the meshing spring of the radial-tooth overrunning clutch.

The pinion teeth are brought into alignment with the ring gear so that the pinion slides into engagement with the ring gear under the force of the compressed meshing spring.

The pinion and ring-gear teeth can sometimes meet in such a way (edge-to-edge) that the pinion can neither rotate nor slide outwards into engagement with the ring gear. In such cases, the starting sequence must be aborted and repeated. When this happens, the helix in the radial-tooth overrunning clutch forces the starter-motor armature to rotate in the opposite direction to its normal direction of rotation ("armature reversal") when the overrunning clutch is being moved by the pinion- engaging lever. The next time the starting sequence is performed, the starter-motor armature/pinion is then in a better position relative to the ring gear and is able to mesh.

The overrunning-clutch travel and the solenoid- armature travel are precisely matched to each another so that the switch contacts do not close until the pinion is fully engaged in the ring gear. In other words, the motor main circuit is not completed until the pinion is fully engaged in the ring gear. Consequently, the starter motor cannot develop its full torque until that point.

Overrunning and pinion retraction
As soon as the engine is running faster than it is driven by the starter motor, the ring gear starts to drive the pinion. Due to the special design of the radial-tooth overrunning clutch, the pinion pushes the clutch race inwards on the helix and simultaneously compresses the spring in the overrunning clutch. The helical teeth disengage. Three flyweights that produce an axial force by way of an internally tapered thrust ring effect separation of the gears.

As the speed drops, the flyweights lose their effect and the helical teeth engage again.

Resting position
When the ignition/starter switch is switched off, the radial-tooth overrunning clutch is moved back into its resting position with the aid of the return spring and pinion-engaging lever.

Starter motors with pre-engaged starter pinion engagement mechanism incorporating motor-assisted pinion rotation

Application
Starter motors with pre-engaged starter pinion engagement mechanism incorporating motor-assisted pinion rotation (Types KB, QB, TB and TF) are used to start large-scale internal-combustion engines.

Design and method of operation
This type of starter motor has a motor-assisted two-stage operating sequence in order to reduce the loads on the pinion and ring gear.

The first stage of the sequence only involves engagement of the pinion; at this stage the pinion does not start to turn the engine over. Only in the second stage of the sequence, just before the pinion reaches the limit of its outward travel, is the full excitation and armature current connected. The characteristic feature of the KB/QB and TB/TF starter-motor designs is that the solenoid is positioned co-axially with the other components of the motor. Figure 13 shows the motor design, Figure 14 the method of operation.

1st stage (preliminary stage)
When the ignition/starter switch is operated, current flows through the control-relay winding and the hold-in winding of the pinion-engaging solenoid. As a consequence, the control relay also immediately completes the engaging-solenoid pull-in-winding circuit.

The engaging-solenoid armature then pushes the pinion against the engine ring gear by means of the pinion-engaging rod and the pinion drive shaft. At the same time, the shunt winding (which initially is connected in series with the armature winding) is energized. It acts in conjunction with the engaging-solenoid pull-in winding as a series resistor for the motor-armature winding. This circuit limits

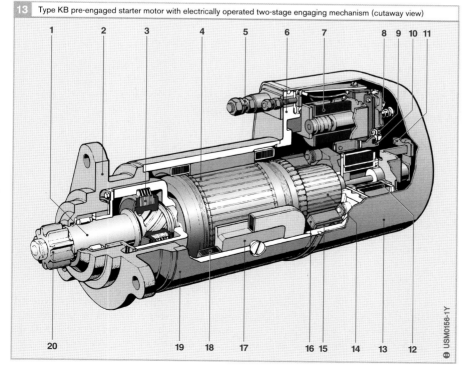

13 Type KB pre-engaged starter motor with electrically operated two-stage engaging mechanism (cutaway view)

Fig. 13
1 Output shaft
2 Drive end shield
3 Multiplate overrunning clutch
4 Armature
5 Electrical connection
6 Switch contact
7 Control relay
8 Switch contact
9 Stop
10 Tripping lever
11 Release lever
12 Pinion-engaging solenoid
13 End cap
14 Commutator
15 Carbon brush
16 Brush holder
17 Pole shoe
18 Excitation winding
19 Stator housing
20 Pinion

Starter motors Starter-motor design variations

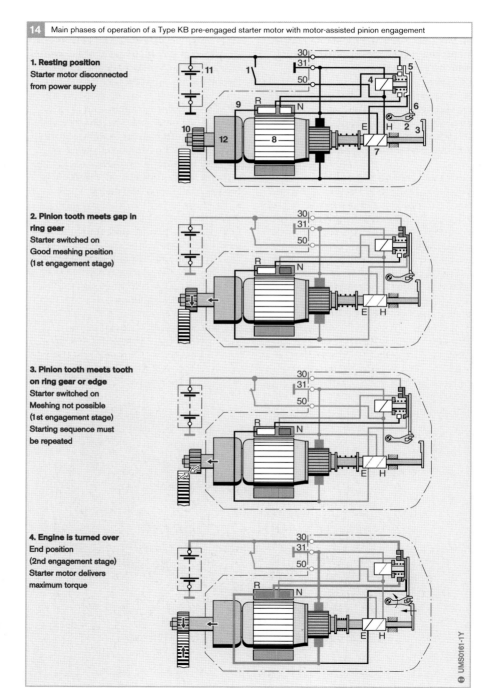

14 Main phases of operation of a Type KB pre-engaged starter motor with motor-assisted pinion engagement

1. Resting position
Starter motor disconnected from power supply

2. Pinion tooth meets gap in ring gear
Starter switched on
Good meshing position
(1st engagement stage)

3. Pinion tooth meets tooth on ring gear or edge
Starter switched on
Meshing not possible
(1st engagement stage)
Starting sequence must be repeated

4. Engine is turned over
End position
(2nd engagement stage)
Starter motor delivers maximum torque

Fig. 14
1 Ignition/driving switch
2 Tripping lever
3 Release lever
4 Control relay
5 Switch contact
6 Stop
7 Pinion-engaging solenoid
8 Armature
9 Excitation winding
10 Pinion
11 Battery
12 Multiplate overrunning clutch

E Pull-in winding
H Hold-in winding
N Shunt winding
R Series winding

the armature current to such a degree that the motor can only produce a small amount of torque and, therefore, only turns very slowly.

In the first stage, therefore, the starter-motor pinion is moved outwards along the motor axis and simultaneously slowly rotated in order to allow smooth engagement with the ring gear. If the pinion teeth are not in a position where they can mesh with the ring gear when the two gears meet, the pinion continues to rotate until it is able to mesh with the ring gear. The pinion and ring-gear teeth can sometimes meet in such a way (edge-to-edge) that the pinion can neither rotate nor slide outwards into engagement with the ring gear; in such cases, the starting sequence must be aborted and repeated.

2nd stage (main stage)
Immediately before the pinion reaches the end of its outward travel, a release lever lifts up a tripping lever and releases the control-relay switch contacts. A pre-loaded spring can then abruptly close the switch contacts. The full current is then applied to the starter motor so that it can generate its maximum torque output in order to turn over the engine by means of the multiplate overrunning clutch and the pinion.

Overrunning and pinion retraction
Once the engine has started and the speed of the starter-motor pinion exceeds the no-load speed of the motor armature, the direction of force transmission is reversed. With the aid of the helix in the multiplate overrunning clutch, the drive between the pinion and the motor armature is disengaged. The pinion remains engaged in the ring gear as long as the ignition/starter switch remains on. Only when the ignition/starter switch is released, and the starter motor therefore switched off, can the pinion retract from the ring gear and return to its resting position.

Type TB/TF pre-engaged starter motor
Application
This type of starter motor is used on very large diesel engines such as those used in railway locomotives, ships, specialized vehicles and fixed installations.

Design and method of operation
The basic design and method of operation are to a large degree the same as for the Type KB/QB models. Figure 15 shows the design of a Type TB starter motor. The Type TF differs from the Type TB essentially only by virtue of a reduction gear, i.e. its pinion is offset from

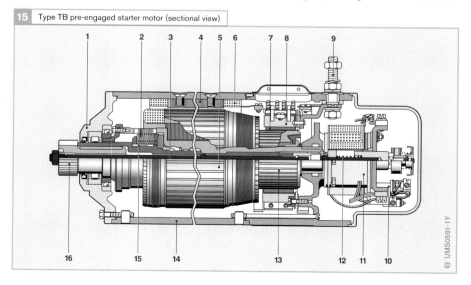

15 Type TB pre-engaged starter motor (sectional view)

Fig. 15
1 Mating collar for engine flange
2 Multiplate overrunning clutch
3 Pinion-engaging rod
4 Pole shoe
5 Armature
6 Excitation winding
7 Carbon brush
8 Brush holder
9 Electrical connection
10 Control relay
11 Pinion-engaging solenoid
12 Cutout spring
13 Commutator
14 Stator housing
15 Helical spline
16 Pinion

the motor-armature axis. This offset configuration often facilitates fitting of the starter motor to the engine. In addition, the reduction gear is significant in terms of the dimensioning of the starting system (speed/torque).

Technology of electrical starting systems

Starter-motor control

Conventional method of control
In a conventional starting system, the driver connects the battery voltage to the starter-motor solenoid switch (by turning the ignition key to the starting position). The current flowing through the solenoid switch (approx. 30 A in cars, up to approx. 70 A in commercial vehicles) produces the electromagnetic force which firstly moves the pinion towards the engine ring gear and secondly switches on the starter-motor main circuit (200...1,000 A on cars, approx. 2,000 A on commercial vehicles).

The starter motor is switched off when the ignition key is released and moves back from the starting position to the "drive" position, i.e. the starting circuit is broken and the power supply to the solenoid switch is disconnected.

In order to ensure that the solenoid switch functions reliably, the effective impedance of the supply cable between the positive terminal of the battery and Terminal 50 on the starter motor must be within a specific, predefined range (approx. 50 mΩ for 12V systems and approx. 100 mΩ for 24-V systems).

In addition to the impedance of the cables, the contact impedance across terminals, connectors, switches and resistors must also be taken into account by fuses. Furthermore, the power consumption of electrical devices connected in parallel with the solenoid which use sections of the control lead must also be considered.

If the permissible limits are exceeded, the reliability of the starting system may be impaired. The control line to the solenoid switch on the starter motor must be kept as short as possible. Where this is not possible (e.g. articulated buses, construction machinery, etc.) an intermediate precontrol relay is often used.

Automatic starting systems
The demanding requirements placed on modern vehicles in terms of convenience, safety, quality and low noise output have resulted in the increasing popularity of automatic starting systems.

An automatic starting system differs from a conventional one by virtue of two additional components:
- A pilot control solenoid (Figure 1, Item 2) and
- A control unit in the vehicle (e.g. an engine control unit, Item 3) which controls the starting sequence

The driver then no longer directly controls the starter-motor solenoid-switch circuit; instead the ignition key is used to send a request signal to the control unit, which then performs a series of checks before initiating the starting sequence. A wide range of such checks is possible and is aimed at establishing such things as:
- Is the driver authorized to start the vehicle (theft-deterrence feature)?
- Is the engine already running (in quiet vehicles or even in buses or coaches, the driver can hardly hear the engine inside the vehicle)?
- Is the battery sufficiently well charged for starting taking account of the engine temperature?

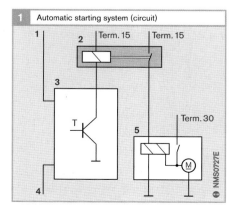

Fig. 1
1 Start signal from driver
2 Pilot control solenoid
3 ECU
4 Selector-lever/clutch-position signal
5 Starter motor

- Is the selector lever in the neutral position (automatic transmission) or the clutch disengaged (manual transmission)?

On successful completion of the checks, the control unit initiates the starting sequence by switching on the transistor, T. This connects the coil of the precontrol relay to equipment ground so that a current flows from Terminal 15 (connected to battery power supply when ignition is on). The pilot control solenoid closes the circuit between the battery (Terminal 30) and the starter-motor relay. The starting sequence then proceeds as previously described.

During the starting sequence, the engine speed (which is in any case detected by the control unit) is compared with a reference figure for the speed at which the engine is running under its own power (the reference speed may also be variable according to engine temperature). Once the engine reaches the reference speed, the control unit switches off the starter motor by switching off the transistor, T. As a result, the starting time is always as short as possible, noise output is reduced and starter-motor wear is minimized.

This system can also be modified to perform a stop-start function. This involves automatically switching the engine off when the vehicle is stationary and restarting it when required. This can achieve substantial reductions in fuel consumption, particularly in urban traffic. At present, reduction-gear starter motors used in this type of system in conjunction with a high-quality ring gear can perform over 200,000 starts.

Due to its inherent characteristics, when it is switched on the starter motor causes a significant, transient voltage drop in the vehicle's electrical system. For this reason, the control units and starting system must be mutually compatible. The voltage drop must be limited in terms of size and duration, and the control units must be capable of performing their functions even at a substantially reduced power-supply voltage.

Power circuit
Starter-motor power cables
As explained in the section "DC motors", the power-supply cable to the starter motor has a fundamental effect on the power of the starting system. Under normal conditions, it has to carry the starter-motor short-circuit current for a short period (<50 ms) and the current required to turn the engine over for the duration of the starting sequence. The voltage drop over the length of the cable and the permissible temperature rise must also be taken into consideration.

In exceptional circumstances, the starter-motor power lead must also be capable of carrying currents close to the short-circuit current for relatively long periods without suffering damage. This would be the case if the vehicle had to be driven by the starter motor in an emergency, for example.

The starting point for determining the permissible temperature rise is a permissible transient current density of $J = 30$ A/mm². Together with the short-circuit current (at +20 °C) for the starter motor in question, this can be used to calculate the cross-sectional area, q_w, of the power cable thus:

$q_w = I_k/J$
where I_k is the short-circuit current of the starter motor at +20 °C.

The power-cable impedance including the internal impedance of the battery must not fall below or exceed the permissible level.

If it *exceeds* the permissible level, the power of the starting system will be reduced, leading to problems starting the engine at low temperatures and malfunctions in the vehicle's electrical system.

If it *falls below* the permissible level, thermal overload, demagnetization of the permanent magnets, gear damage or pinion engagement problems may occur.

Starter-motor batteries
Starter-motor batteries are nowadays almost exclusively lead-acid accumulator batteries. The nominal voltage per cell for such batteries is

2 V, so that a conventional 12-V battery has six cells connected in series. Accordingly, a 24-V commercial-vehicle battery has 12 cells.

The capacity of the battery, i.e. the amount of current that can be drawn from the battery over a specific period expressed in ampere-hours [Ah], or the size of the accumulator plates essentially determines the size of the battery.

The most important battery characteristic as far as starting the engine is concerned is its power, that is the product of the current supplied and the voltage between the terminals, U_K. The battery-terminal voltage decreases as the load current increases. It is useful to define an internal battery impedance, R_i, which reflects that phenomenon.

The internal impedance, R_i, of the starter battery has a diminishing effect on the starter-motor power output in addition to the impedance of the power cable, switches and contacts. However, the internal impedance of the battery is not a fixed quantity but a variable which is dependent not only on the battery design but also the temperature, the battery charge level, battery age, and usage history. Figure 2 shows an ideal battery characteristic graph.

When the battery is fully charged, the voltage drops steeply at low levels of current draw due to capitative discharge. At higher levels of current draw, the gradient of the graph is shallower. This is the range in which the battery is operating during the starting sequence. It makes sense, therefore, to define the internal impedance of the battery for starting within this range. The basis for that determination is the low- temperature test current, I_{CC}, as defined by EN 60 095. According to that definition, the battery-terminal voltage, U_K, when discharging at I_{CC} and −18 °C measured 10 s after commencement of discharge, must be at least 7.5 V (1.25 V per cell). The internal impedance of the battery is defined as the gradient of the shallower section of the graph. It is determined by extrapolation of the shallower section of the graph to the Y-axis. That point of intersection represents the steady-state voltage, U_0, of the battery, which is not the same as the no-load voltage, U_L (the latter is the voltage of the battery when no load is applied). Accordingly, the internal impedance, R_i, of the battery is given by
$R_i = (U_0 - U_f)/I_{CC}$

The ideal battery characteristic is given by
$U_K = U_0 - R_i \cdot I$

The maximum power that a battery can deliver is produced at the current at which the voltage between the terminals has dropped to half the steady-state voltage:
$P_{Kmax} = U_0^2/(4 \cdot R_i)$

The internal impedance of the battery decreases as the battery capacity increases so that maximum power increases along with greater capacity. Starter motors are designed for a maximum battery size.

If powered by a *smaller battery*, the actual power of the starting system will be less than the rated power. As long as the cold-starting requirements are satisfied, this is technically allowable.

If the starter motor is powered by a *larger battery*, the power will be the corresponding amount above the rated power. This can result in overloading of the mechanical components, increased wear and thermal overload. In the case of permanent-magnet starter motors, it can cause partial demagnetization of the magnets and the associated irreversible loss of torque output. Consequently, the specified battery size should not be exceeded.

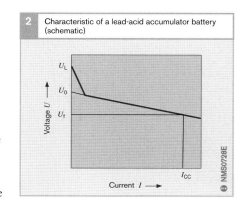

2 Characteristic of a lead-acid accumulator battery (schematic)

Fig. 2
I_{CC} Low-temperature test current
U_f Voltage at low-temperature test current
U_0 Steady-state voltage
U_L No-load voltage

Development and production of alternators and starter motors

Quality management

Alternators and starter motors are high-quality, highly developed products. The failure rates encountered nowadays are of the order of a few per million. This means that of a million products manufactured, only a few fail to achieve the intended service life. This is the result of a systematic quality management process that follows the product from the initial phases of development right through to the end of its useful life.

The starting point is a set of specifications agreed with the customer which define the requirements for the alternator/starter motor as precisely as possible. In the course of product development, a series of regular design reviews are undertaken in order to compare the design results with those specifications. This method of working ensures that, when the product goes into volume production, all external requirements as well as all internal requirements are met.

Beyond the testing of the specifications defined at the start, there are other procedures adopted in the course of the development process. Thus a system known as Failure Modes and Effects Analysis (FMEA) is used to determine the possible failure risk of the product as a whole and of each individual component and to eliminate that risk by design modifications or testing procedures during production (Figure 1).

A similar procedure is adopted when planning and designing the production processes. The production processes must also demonstrate their suitability for volume production in short-term and long-term viability studies. This applies not only to internally produced parts but also to pasts procured from external suppliers.

At the end of the development process (i.e. before the start of volume production), the product has to be approved by the vehicle manufacturer and by Bosch.

Alternators and starter motors are generally developed for a specific application, i.e. for use on a specific engine. The start of volume production is therefore coordinated and synchronized with the start of engine/vehicle production by the vehicle manufacturer.

1 Testing area for endurance testing of starter motors

Even after the start of full production, the product must continually demonstrate its quality, i.e. its compliance with specified requirements, in a series of "requalification tests". For starter motors, those include tests relating to:
- Torque/speed characteristics,
- Service life incorporating specific testing of
  – number of switching operations
  – resistance to corrosion
  – resistance to vibration/shock
- Dimensions
- Weight and
- Electrical data

Quality management therefore also involves the analysis of returned products from the field. The knowledge gained from such analysis is used to improve the design of the product and/or the production process. Quality documentation and "design rule-books" are used to record the knowledge gained and apply it to new developments. In this way, a process of continual improvement is maintained.

The entire system of maintaining and improving quality is also constructed in collaboration with European and American automobile producers and regularly checked for correct application and effectiveness by means of internal and external quality audits. Certificates are issued to confirm that the required standards are being maintained.

The "Alternator" and "Starter Motor" divisions of Robert Bosch GmbH are certified in accordance with all applicable motor-industry requirements catalogs (VDA Vol. 6.1; QS9000).
 In 2003, certification according to the international environmental management standard, ISO 14001, is due to follow.

Development

Computer simulation identifies the optimum design

In order to improve electrical systems, Bosch uses computer programs that calculate such things as the charge-balance equation. This involves calculating a system's battery charge level after a simulated 14-day winter urban driving cycle.

The following data is incorporated in the charge-balance equation:
- Road and traffic conditions
- Vehicle type
- Driving style
- Temperatures
- Daytime and nighttime journeys
- Characteristics of alternator, battery and starter motor, and
- Power consumption of electrical consumer units

If the battery has an adequate charge level (usually above 50%) at the end of the testing cycle, the charge balance is acceptable. If the result of the charge-balance equation is negative or if the system is overdimensioned, the optimum combination of alternator, starter motor and battery is recalculated.
 Such calculations help to determine whether the system can be improved with existing products or whether the development of a new custom-designed product would provide a better solution.

Using CAD systems to reduce development time

Bosch engineers use computers to test out all possible solutions right from the start of the development process. The Bosch Technical Center for Automotive Equipment at Schwieberdingen near Stuttgart in Germany has CAD systems for this purpose. They allow designs and technical or scientific calculations such as magnetic-field and temperature distribution to be completed much more quickly.

Finite-element calculations for materials testing

The power, weight, dimensions and strength of a product such as an alternator is defined even before the first specimen is produced. An important role in that process is played by finite-element calculations on a computer. They simulate the behavior of components and materials under a wide range of conditions (Figure 2). If unacceptable deficiencies exist, the design engineers can identify them immediately on the computer, determine the causes and develop suitable remedies. This method of working saves having to perform costly tests.

From endurance test to volume production

For testing fitted position, function and endurance (Figures 3 and 4) the use of test specimens is unavoidable. They are produced by staff who have gained qualifications in a wide range of specializations at Bosch factories. Their experience is brought to bear on the process of specimen production and is also incorporated in subsequent series production.

Requirements of the fitted location

The location in which the product is fitted has a decisive influence on its design. As part of a "simultaneous engineering" process, Bosch consults with the vehicle manufacturer on questions of available space for fitting, ambient conditions of the location and possibilities for reducing space requirements. In the early stages of a development project, CAD models of the product are used, whereas in the later stages testing specimens are required.

There is hardly any other location in which electrical, electronic and mechanical components are subjected to such high stresses as in the engine compartment of a motor vehicle. Temperatures can rapidly change from extreme cold to searing heat; shocks and vibration demand enormous strength and resistance; components are under attack from saltwater and dust. The consequences of exposure to such stresses are investigated in climatic and endurance-testing facilities according to specified testing schedules.

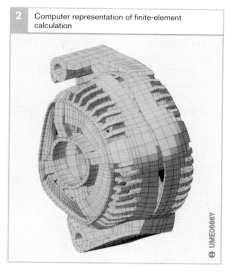

Fig. 2
Finite-element calculations enable simulation of material response to various types of stress at an early stage of the development process (example: simulation of natural oscillation of an alternator showing areas of high amplitude)

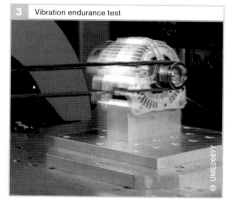

Fig. 3
Simulation of the resonant frequencies encountered on the vehicle on vibration tables and spatial-vibration test benches (example: alternator testing specimen)

Fig. 4
Product trials also involve testing the product's resistance to splash-water (example: alternator testing specimen)

Speed capacity and vibration-resistance

Comprehensive laboratory and practical tests provide data on vibration and speed-related stresses on components in the vehicle. From that information, the testing conditions for the test specimens can be defined.

Electrodynamic vibration tables and spatial-vibration test benches are used to simulate the frequencies that occur on the vehicle. This demonstrates whether the material characteristics meet the high quality standards demanded. Rotating components must prove their strength in overspeed and fluctuating-speed tests.

Cold-starting tests in the cold room

For detailed and reproducible cold-starting tests at temperatures down to −35 °C, Bosch has three cold cells for complete vehicles. When joined together, they can even accommodate commercial vehicles up to 18 meters long.

One of the cold cells is equipped with a chassis dynamometer and another has an engine test bench.

The first practical trials (conducted in collaboration with the vehicle manufacturer) take place concurrently with the cold-room tests. The results enable precise adaptation of products to the engine at an early stage.

Global quality standards

All Bosch production facilities for alternators and starter motors – whether in Germany, the UK or Spain – work to the same exacting and internationally applicable Bosch standards – in other words, the manufacturing and testing methods are universally standardized.

Spare parts too are produced by the same methods as the original components. A worldwide dealer network ensures that those parts are universally available.

Production (starter motors)

The manufacture of a starter motor under volume-production conditions is to a large extent a fully automated process (Figure 5). The large number of custom designs places particular demands on the design of the production facilities.

The drive-end shield (Figure 5a) and reduction-gear and overrunning-clutch assemblies (Figure 5b) are manufactured in a preproduction phase on extensively automated production lines.

The solenoid switch is assembled from the individual components "casing", "coil" and "cap" (Figure 5c).

The armature is the central component of the motor and therefore the essential determinant of power output as well as motor noise. The commutator is produced by a complex press-forming process. The armature shaft, laminated core and armature conductor are combined with the commutator on the armature assembly line (Figure 5d). Once assembled, the armature is balanced (Figure 5e) in order to ensure that the starter motor runs as smoothly as possible.

The precision with which the components are manufactured is the essential prerequisite for reliable operation over the full service life of the starter motor.

Final assembly completes the involved process of producing a starter motor. In this phase, all subassemblies are combined to produce the finished product (Figure 5f), which then undergoes final inspection.

5 Highly automated starter-motor production

Fig. 5
a Drive-end-shield production line
b Planetary-gear assembly line
c Solenoid-coil production
d Armature assembly
e Armature balancing
f Final assembly

Service technology

Important!
This chapter provides general descriptions of service technology, and is *not* intended to replace repair and instruction manuals! Repairs should always be performed by qualified professional technicians!

Over 10,000 Bosch service agents in 132 countries are standing by to provide car owners with assistance. As Bosch service agents do not represent the interests of any one vehicle manufacturer, this help is neutral and unbiased. Fast assistance is always available, even in sparsely populated countries in South America and Africa. The same quality standards apply everywhere. It is no wonder that the Bosch service warranty is valid throughout the world.

Overview

The specifications and performance data of Bosch components and systems are precisely matched to the requirements of each individual vehicle. Bosch also develops and designs the test equipment, special tools and diagnosis technology needed for tests and inspections.

General-application test equipment from Bosch – extending from basic battery testers to comprehensive vehicle inspection lines – is used by automobile repair workshops and official inspection agencies throughout the world.

Service personnel receive training in the efficient use of this test technology as well as information focusing on a range of automotive systems. Meanwhile, feedback from our customers flows into the development of new products.

The service AWN
Test technology
It is still possible to test mechanical systems in motor vehicles using relatively basic equipment, but mastering the increasingly

---

[1]) Bosch service technology stems from development activities carried out by the Bosch AWN service network. The "asanetwork GmbH" is responsible for advanced development and marketing under the "AWN" name.

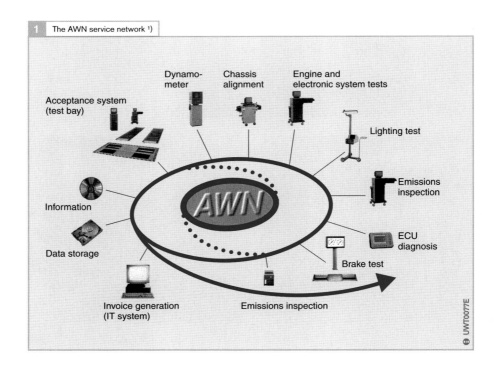

1 The AWN service network [1])

a battery is then "sulfated". Sulfation is a result of careless maintenance. It causes an increase in the battery's internal resistance which impedes the chemical processes and makes charging more difficult.

A sulfated battery gets very hot when charged with a "W" charging characteristic. The charge voltage rises sharply as soon as charging starts. If sulfation is only slight, the lead sulfate is broken down slowly and the charge voltage drops steadily. As soon as the lead sulfate has been broken down completely (regenerated), the voltage climbs again the same as it does when charging a serviceable, non-sulfated, battery (Fig. 2).

Regeneration of lead batteries
If the sulfation is not severe, the battery can be regenerated by applying a minimal charge current of approx. 25 mA per Ah battery nominal capacity ($\approx 0.5\ldots2$ A) for about 50 hours.

On the other hand, if the acid has converted completely it will be impossible to regenerate the battery. It is unusable.

Troubleshooting
If the battery fails to start the engine, this can be due to insufficient charge or to a battery defect. The state of charge and starting power can be measured using the Bosch battery tester which only needs a few moments to provide clear information on the battery's condition. Such details are needed for defining the steps to be taken for remedying the problems.

The battery's exact state of charge can be determined by measuring the electrolyte specific density and also by precise measurement of the open-circuit voltage. Excessive water consumption, and/or charging immediately before the test, will falsify the results in the direction of a more favorable verdict. The charge on the plate surfaces which falsifies the results can be compensated for by pausing briefly in the test sequence so that the plate charge can reduce.

Assessment of starting capability is only possible by loading the battery with a standard current for a stipulated time, e.g. 30 seconds, and measuring the voltage drop. Taking into account the battery's initial state of charge, the measured value is compared with a standard value stored in the tester. The results of this comparison are displayed as the percentage starting capability of the battery compared to that of a fully functional battery.

The complex test procedure means that its sequence must follow a program in the tester. Test results are only possible on batteries which have a residual charge. Due to the generation of a mean value for a number of cells, and external influences, this test only provides reliable results in approx. 90 % of the cases. In borderline cases, in order to stabilise the chemical reactions an intermediate charge is required or the test must be repeated.

Charging with the battery charger
Safety requirements
In order to avoid the risk of accidents, the battery charger's exposed terminal clips must be safely electrically isolated from the 230-volt mains power supply and there must be a facility for switching off the charging current before disconnecting the terminal clips. An additional safety circuit prevents sparking if the terminal clips are connected the wrong way round.

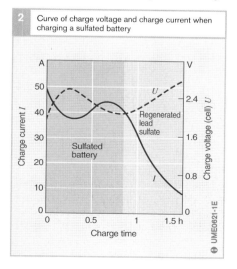

2 Curve of charge voltage and charge current when charging a sulfated battery

Battery chargers
If it is impossible for the alternator to charge the battery adequately, a battery charger (Fig. 3) must be used. This applies when the battery has been out of use for a long time, or directly before it is removed from the vehicle and put into storage.

Battery chargers with voltage limitation must be used on maintenance-free batteries. Otherwise overpressure is generated during charging and the battery dries out. A fully charged modern battery can be stored for 6 months without difficulty. A battery can only be charged completely with a relatively low current (max. 1 A).

Electronic battery charger
The LW- and BML-electronic battery chargers allow the battery to be charged in the vehicle without disconnecting it from the electrical system. The charge voltage is free from voltage peaks, and controlled electronically so that the battery cannot be overcharged. During charging, the electronic components, such as engine ECUs and airbag triggering units, etc., are protected against damage. These battery chargers are designed for trickle charging and floating operation, particularly with maintenance-free batteries.

Rapid-start battery charger
The rapid-start charger with start-assist stages provides starting help for trucks and passenger cars. Charging times are very short even for large batteries. These chargers are provided with a charge-monitoring facility which ensures that charging is efficient and does not overload the battery.

Home battery chargers
So-called home battery chargers are particularly suitable for charging the batteries in small-power drives. They are also useful for home and hobby requirements.

Charging methods
Normal charging
Generally speaking, normal charging takes place using the I_{10} charge current. This corresponds to 10% of the battery's nominal Ah capacity:
$I_{10} = 0.1 \cdot K_{20} \cdot$ A/Ah.

Depending upon the process used, the charging time can be up to 14 hours.

Boost charging
Boost charging can bring an empty battery back up to about 80% of its rated capacity so that it can withstand the loading which is typical for automotive applications. Boost charging has no negative effects. Below gassing voltage, high charge currents present no problems. Such currents can be in the vicinity of the number for the nominal Ah capacity (relative charge current $I_{100} = K_{20} \cdot$ A/Ah). Once the gassing voltage is reached, the charging current must either be switched off or a change made to normal charging.

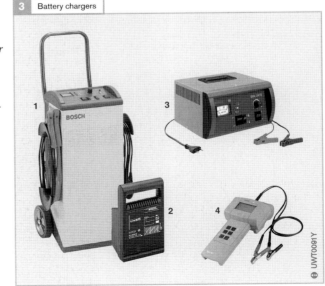

3　Battery chargers

Fig. 3
1　Rapid-start battery charger
2　Home battery charger
3　Electronic battery charger
4　Battery tester

Gassing voltage:
The gassing voltage at ≤ 15 °C is approx. 14.4 V (2.4 V/cell) and at > 15 °C approx. 13.8 V (2.3 V/cell). If it is exceeded when charging, the battery will start to produce noticeable amounts of gas. This will result in water loss from the battery and the risk of gas detonation.

Trickle-charging
To compensate for the self-discharge losses in stored batteries (for instance when caravan or mobile-home batteries are stored during the winter), the battery is left connected to a battery charger for an extended period of time, whereby charging current is limited to 1 mA/Ah.

Floating operation
With this type of operation, battery charger and load are permanently connected to the battery. Depending upon the load, a changeover continuously takes place between charging and discharging, whereby the battery charger's electronic circuitry prevents battery overcharge.

Charging curves
There are various methods for charging the battery, each of which is characterized by its own charging curve:

W	Constant resistance (charge current drops when the voltage increases)
U	Constant charge voltage
I	Constant charge current
a	Automatic switch-off
e	Battery charger switches on again automatically
o	Automatic switch-over to other charging curves

Whereby, it is possible to combine the various charging curves. For instance:

WU	As W curve but charge voltage remains constant above a given value (e.g. gassing voltage)
IU	Constant charge current up to a value above which voltage remains constant and the charge current falls
WoW	Switch-over from one charging curve to the other. For instance from a high to a low charge current

The commonly used W charging curve (on home and workshop battery chargers for instance) is defined by the charger's internal resistance. These chargers respond to increasing battery voltage by steadily lowering the charge current. Using the normal charge method as per Fig. 5, with a relatively low initial 20 A charge current, charging takes about 12 hours. Since reduced current continues to flow above the gassing voltage, the battery charger must be switched off once full charge has been reached.

Since the battery-charging process must be monitored, this technique is only applicable to a limited extent on maintenance-free batteries.

On IU-curve battery chargers, charge current and charge voltage are automatically controlled to a constant level. This means that the current remains constant, irrespective of mains-voltage fluctuations, until the gassing voltage is reached, after which it drops

4 Rapid-start charger in use in the workshop

sharply due to constant voltage limiting (Fig. 6). This means that the charge current can be considerably higher (in the example given: 3 h charging time with 50 A initial charge current). This method permits a high level of charge in a short time whereby overcharge is avoided. Similar results are obtained from battery chargers whose charge voltage is limited (WU curve), or which automatically switch to a lower current upon reaching a given limit voltage (WoW curve), or which terminate the charging process completely upon reaching this limit voltage (Wa curve).

Charge-voltage setting

The nominal voltages of the battery and the charger must be identical. Since the charge current results solely from the difference in voltage between the battery and battery charger, it is highly affected by the magnitude of the voltage differential. Undervoltage can lead to inadequate charging, and overvoltage on battery chargers without charge-voltage limitation can lead to over-charging. Maintenance-free batteries therefore should only be charged using voltage limiting, and if they are to be charged over a longer period, charging is to take place with a reduced voltage limit (e.g. 2.3 V/cell instead of 2.4 V/cell). On the other hand, if voltage limiting is applied, very cold batteries cannot be charged. Here, a higher charge voltage is necessary. This usually presents no problems, and can be implemented for instance by a Summer/Winter changeover facility.

Direct charging in the vehicle

In modern vehicles, more and more electrical and electronic "aids" contribute to a high level of safety and comfort. Prime examples are the airbag, car telephone, car radio, and ECU, to name but a few. These highly sensitive devices though must be protected against voltage peaks when charging the battery. Previously, for charging, the battery had to be disconnected from the vehicle electrical system. Electronic battery chargers make this step a thing of the past. Since charging is faster and there is no danger to the vehicle's electronic equipment, this is of course a considerable help in the vehicle repair shop because:

- The time-consuming disconnection/reconnection and removal/installation of the battery is no longer necessary.
- Data memories in the car radio, ECUs, car telephone, trip computer etc. are no longer erased due to battery disconnection.
- Electrical consumers (airbag, ECUs etc.) are protected.
- No danger of damage due to incorrect handling.

Fig. 5
Battery charger without closed-loop control. Charge-current switch-off is therefore necessary on reaching full charge.

Fig. 6
Battery charger with closed-loop control, automatic reduction of charge current upon reaching full charge.

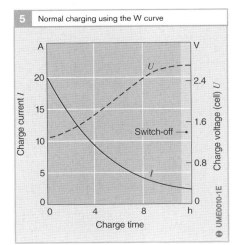

5 Normal charging using the W curve

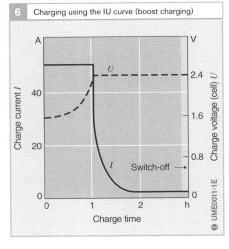

6 Charging using the IU curve (boost charging)

- No dangerous battery gassing during trickle charging.
- Batteries can be recharged without disconnecting the current consumers/loads (floating operation).
- Brief charge times due to the high power reserves inherent in the IU/IWU charging curves.

Backup mode

When replacing the battery, the LW 30/50 E battery charger is operated in backup mode to retain the data in the data storages of such equipment as car radio, car telephone, etc. In this mode, current output is limited to approx. 2 A.

Incorrect-polarity protection

In case of falsely connected battery terminals, this protection facility prevents battery short-circuit and destruction of the battery charger.

Start-assist using the battery charger

The battery charger's start-assist function supports the battery when starting the engine. Its increased short-time power output makes it possible to supply the required high current.

Caution! Start-assist is only permitted on vehicles for which the manufacturer has not imposed limitations or bans on start assist in the operating instructions.

Start-assist with jumper cables

The battery from another vehicle can be used to provide start assistance. In such cases, the battery of the vehicle needing assistance must remain installed and connected, and manufacturer instructions must be complied with. For efficient starting assistance, only standardized jumper cables are to be used (DIN 72 553) with a conductor cross-section of at least 16 mm² for SI engines and 25 mm² for diesel engines. Both of the batteries (or battery chargers) which are to be connected must have the same nominal voltage.
Working steps:
- Determine the cause of battery dis-charge. If the fault is in the electrical system, do not use start assistance since this can damage the battery charger, or the battery and alternator of the vehicle providing the assistance.
- Connect the positive terminal of the empty battery to the positive terminal of the outside source of power.
- Connect the negative terminal of the outside source of power to a bare metallic surface (remote from the battery) on the vehicle receiving assistance, for instance to the engine ground braid.
- Check that the battery jumper cables are firmly attached (good electrical contact).
- Start the vehicle which is providing the assistance. After a brief pause also start the disabled vehicle.
- Once starting has been successful, disconnect the cables in the opposite order to that given above.

Safety instructions

Before starting work on the vehicle's electrical system, or near the battery, the ground cable is to be disconnected provided all consumers/loads have been switched off. This is necessary in order to prevent short circuits (for instance with tools) which can cause sparks and burns.

For the same reasons, particular care must be taken when connecting/disconnecting battery-charger cables or battery jumper cables. The following basic rules concerning safety must be followed when working with batteries:
- When handling sulfuric acid or when topping-up with distilled water, always wear protective goggles and rubber gloves.
- Do not fill up with electrolyte to above the MAX mark.
- Do not tip batteries to extreme angles from the vertical and do not keep them tilted for long periods.
- Due to the danger of electrolytic-gas detonation, smoking and naked flames are forbidden and sparks must be avoided when batteries are being charged (connect and disconnect cables in the prescribed order with the battery charger switched off).
- Battery-charging rooms are to be well ventilated.

Testing technology for alternators

Average driving cycles and operating times can be determined for all types of vehicles (passenger cars, trucks, buses, construction machinery, etc.) and their typical operating conditions. There are therefore a wide variety of alternator designs available which comply with these different requirements.

In case of malfunction in the vehicle's electrical power-generating system, a check should first of all be carried out directly in the vehicle. If, in the process, an alternator defect is located, the alternator is either replaced or repaired using the service-information and service-instructions documentation. Before being installed in the vehicle again, the alternator must be tested on the combination test bench.

Testing directly in the vehicle

A visual inspection is first of all carried out to check the V-belt, the wiring, and the charge-indicator lamp. Basically, an engine analyzer and a volt-ammeter are required for the electrical tests.

The following tests can be performed using these two testers:
- Oscilloscope display of the DC voltage with low harmonic ripple (between B+ and B−)
- Voltage (between B+ and B−), with and without load
- Charging current under load
- Current without load
- Quiescent current
- Battery voltage
- Short-circuit of lines to ground or plus (+)
- Line open-circuit and
- Contact resistance of the lines

Alternator repair

For the various alternator types, there are specific service instructions available which describe the alternator repairs.

These instructions also contain the relevant test and adjustment values.

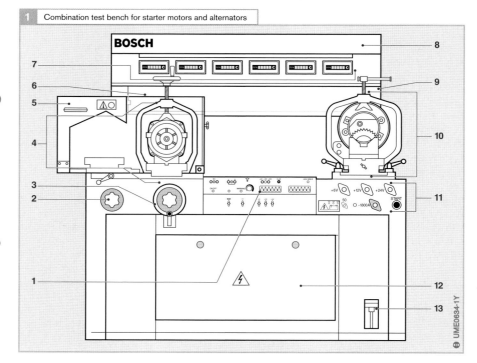

Fig. 1
1 Operator panel for alternator and starter-motor tests
2 Adjustable loading resistor (alternator test)
3 Handwheel for adjusting clamping-table height (alternator test)
4 Alternator test setup
5 Protective hood
6 Tool tray
7 Display unit
8 Lighting unit
9 Socket connection for rotational-speed sensor (alternator test)
10 Starter-motor test setup
11 Connection terminal for starter motor
12 Battery compartment with cover
13 Pedal for starter-motor loading (drum brake)

1 Combination test bench for starter motors and alternators

A number of different testers (e.g. alternator tester and interturn-short-circuit tester) are used for alternator repairs. In addition, in order to localize defects inside the alternator and repair them efficiently, special tools are required for each alternator type.

Checking the alternator on the combination test bench

Once the alternator has been repaired, it is clamped in the relevant test setup on the combination test bench (Fig.1).

Depending upon the version concerned, the alternator can be driven directly up to speeds of approx. 6,000 rpm). At higher speeds, the alternator is driven through a V-belt (Fig. 2).

The alternator is bolted to the clamping device using a swivel arm. The rotational-speed sensor is calibrated after aligning and tensioning the V-belt. The alternator is then connected electrically.

Two points on the power curve are run up to when testing an alternator:
Using an adjustable loading resistor, the alternator is loaded with the maximum attainable current at two different test speeds (e.g. 1,500 or 1,800 rpm and 6,000 rpm). The alternator voltage must remain above the stipulated limit value (e.g. 13.5 V).

If these desired values are reached, the alternator can be installed in the vehicle.

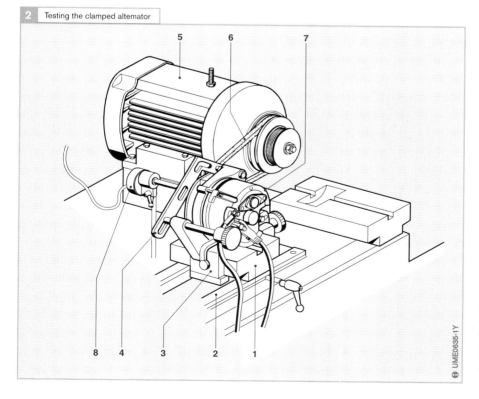

2 Testing the clamped alternator

Fig. 2
1 Clamping table
2 Guide
3 Clamping device
4 Swivel arm
5 Drive
6 V-belt
7 Alternator
8 Rotational-speed sensor

Testing systems for starter motors

Starter motors for cars are designed to last for the average service life of the engine and therefore do not require any special maintenance.

In the case of commercial vehicles with duty profiles that involve frequent short journeys or vehicles that are subjected to other exceptional stresses (particularly taxis, parcel delivery vans, etc.), regular inspection of the starting system is advisable.

This also applies to commercial vehicles that are designed for much greater mileages than the average car.

If there is a fault in the starting system, the equipment is usually first *tested in the vehicle*. If the fault can be traced to the starter motor, it is either replaced or repaired with the help of service information bulletins and servicing instructions.

Before it is refitted on the vehicle, the starter motor should then be tested on the combination test bench (Figure 1).

Testing in the vehicle

Before the starter motor can be tested, the function of the battery (voltage under load, electrolyte level and electrolyte specific gravity) must be checked.

The following problems can be identified by listening to the starter motor:
- Unusual noises during the starting sequence
- Starter motor engages but turns the engine very slowly or not at all
- No sound of pinion engagement
- Starter motor disengages too slowly or fails to disengage

In the case of unusual noises during the starting sequence, the fault can be traced to the starter motor, the way it has been fitted or the ring gear. Other problems require selective electrical testing of the starting system (e.g. using an engine analyzer). The following tests are carried out with the starter motor at rest:
- Line short to equipment ground or positive
- Voltage at Terminal 30
- Line continuity, and
- Line contact impedance

1 Combination test bench for starter motors and alternators

The following are tested during the starting sequence:
- Voltage at Terminal 50
- Voltage at solenoid-switch output, and
- Starter-motor current (up to 1,000 A)

Repairing the starter motor

First of all, the starter-motor pinion is checked for damage (broken teeth, excessive wear, etc.) and replaced if necessary. Then various instruments (e.g. alternator testers and coil-winding short-circuit detectors) are used in conjunction with the relevant servicing instructions. In addition, each individual type of starter motor requires special tools for carrying out the repairs; only in this way can faults be isolated within the starter motor and properly rectified.

Testing the starter motor on the combination test bench

Once the starter motor has been repaired, it is fixed on the starter-motor testing table of the combination test bench (Figure 1) – either by means of a flange (Figure 2) or wedges and a hold-down clamp, depending on design.

The handwheel and the fixing table are used to adjust the play between the teeth of the pinion and ring gear and the distance of the pinion from the ring gear (if the backlash is not correct, excessive tooth wear will result). The speed sensor is then adjusted and the electrical connections to the starter motor completed.

The starter-motor testing procedure essentially consists of two parts:
- Testing the starter motor under no-load conditions. The criteria for this part of the test are that the starter-motor current remains below a specified threshold and the motor speed reaches a minimum level when not under load.
- The short-circuit test involves braking the starter motor to a standstill using the drum brake built into the test bench. During this procedure, the starter motor may only be held stationary for a short period (no more than 2 s). The starter-motor current and voltage are measured under those conditions. The test results must be within the specified limits.

2 Starter motor fixed to testing table

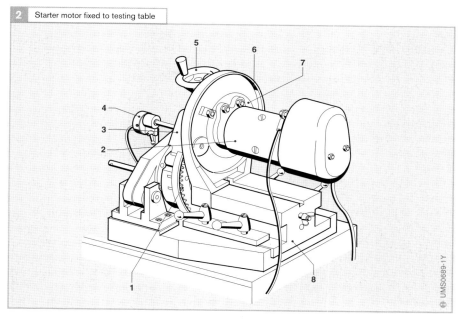

Fig. 2
1 Ring gear
2 Starter motor
3 Safety cover
4 Speed sensor
5 Handwheel
6 Clamping bracket
7 Mounting flange
8 Fixing table

Automotive lighting technology

The vehicle's lighting equipment makes a vital contribution to safety on the roads. Increasing traffic density combines with high vehicle speeds and a wide spectrum of driving conditions to confront headlamp and lighting-system designers with an increasingly exacting array of challenges which are defined by legal regulations.

Technical demands

The challenge of mastering the extensive range of potential driving situations is the decisive factor behind the engineering and layout of automotive lighting equipment. Effective illumination of the road surface under a variety of conditions not only serves as a vital reinforcement for human vision; in some cases sight would be completely impossible without the help of the vehicle's lighting system. Priorities include brightness, color and spatial perception, while also embracing shape and motion along with recognition of luminance and contrasting hues. This is why automotive headlamps are subject to such stringent technical demands.

High-performance headlamps and other lamps at the front and rear of the vehicle form the foundation of an effective "see and be seen" policy.

Requirements in the realm of illumination technology are now being joined by the proliferating stylistic demands that headlamps and vehicle lights are expected to satisfy. One essential condition for meeting these demands is partial or total deletion of optical profile elements from the external lens to produce a "glass-smooth" finish. The inside of the headlamp or lamp unit can then be designed to reflect stylistic requirements. The ultimate result is powerful and attractively styled headlamps and lights.

Front-end lighting

The primary function of the front headlamps (Figure 1) is to illuminate the road, making it possible for the driver to monitor road conditions and recognize obstructions in time to react. The image resulting from switched-on headlamps also serves as a vehicle-recognition signal for oncoming traffic. The turn signals alert other drivers to changes of direction and, when they flash together, to potentially critical situations.

The headlamps and other lamps installed on the front end of the vehicle include:
- High-beam/low-beam headlamps
- Fog lamps
- Auxiliary driving lamps
- Turn-signal lamps
- Parking lamps
- Side-marker and clearance lamps (wide vehicles) and
- Daytime running lamps (as prescribed in some countries)

1 In combination: Low beam, high beam, fog lamp and turn signal form a single unit for harmonious integration in the vehicle's front end (example)

Rear-end lighting

Rear-mounted lamps (Figure 2) are switched on in response to inclement weather and darkness to indicate the vehicle's position, and also to signal the vehicle's current and intended direction. The stop lamps show whether a vehicle's brakes are being applied or not. The turn-signal lamps indicate an intended change of direction, and when both flash at once they warn of a hazardous situation. The backup lamps provide illumination when the vehicle is reversing.

The lamps and lights on the vehicle's rear end include:
- Stop (brake) lamps
- Tail lamps
- Fog warning lamps
- Turn-signal lamps
- Parking lamp

2 Rear lamps and backup lamp integrated to form a compact and efficient assembly at the rear of the vehicle (example)

- Clearance lamps (wide vehicles)
- Backup lamps, and
- License-plate lamp

Vehicle interior

Regarding the inside of the vehicle, the vital factors include safe operation of controls and switchgear along with an adequate flow of information on operating conditions (all with a minimum of driver distraction). These priorities dictate effectively illuminated instrument panels and discrete lighting for various control clusters (such as the sound and navigation systems), where they satisfy a prime requirement for relaxed and safe vehicle operation. Visual and acoustic signals should be transmitted to the driver in accordance with their priority.

Legal framework

Overview

Numerous regulations govern highway traffic both within Germany and in the EU as a whole.

These legal mandates regulate such areas as traffic safety, driver conduct and homologation approval for vehicles intended for operation on the public high-ways.

Automotive lighting equipment is defined in the Design and Operation Regulations, while official certification is subject to yet other stipulations.

Approval codes and symbols

Lighting equipment on motor vehicles is subject to national and international design and operation regulations that govern both manufacturing processes and test procedures.

A special approval code is specified for each kind of lighting device. This code or the corresponding symbol must be clearly visible on the lens of the headlamp or other lighting device (Figure 3, next page). This provision also applies to type-tested replacement headlamps and lights.

The presence of a code or symbol attests that the light unit has absolved testing with an official technical appraisal institute to

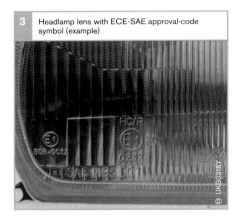

3 Headlamp lens with ECE-SAE approval-code symbol (example)

earn official government type approval (i.e., from the Kraftfahrzeug-Bundesamt in Germany). The specifications of all units from series production that display a particular code or symbol must correspond to those of the type-test specimens. Examples of certifications symbols are:

- ⏧K   German national test symbol,
- Ⓔ1   ECE test symbol,
- ⌐e1⌐ EU test symbol.

The number 1 following each letter indicates that type approval has been granted in Germany (Figure 3).

Moving beyond the individual national ordinances, installation of all automotive lighting and visual signalling equipment within Europe as a whole is governed by European regulations (ECE: All of Europe, Japan, EU). These pan-European ordinances assume legal priority over national laws. Continuing steps toward European unity have been accompanied by harmonization of regulations and ordinances. This, in turn, has allowed progressive relaxation of export restrictions. Because headlamps from Bosch comply with the applicable ECE and EU regulations, they can be used in all ECE and EU nations, regardless of the country of acquisition.

A completely different regulatory framework applies in the USA. The self-certification principle compels each manufacturer, in its role as importer of technical lighting equipment, to verify (and prove when required) 100 % compliance with the specifications defined in FMVSS 108 as entered in the Federal Register. As this implies, there is no specific individual homologation certification in the European sense. The FMVSS 108 specifications are partly based on SAE industrial standards.

This means that vehicles being reimported to Europe must be modified to comply with the European standards.

Development of lighting technology

Introduction of lighting equipment

When the first motor vehicles appeared at the end of the 19th century lighting played only a minor role; because at that time the complete lack of any vehicle safety at all meant that night driving was too dangerous in any case to become common. The first lamps were candle-powered hurricane lanterns, followed by petroleum lanterns and, finally, acetylene lamps.

In 1908, the only electric lamps on the vehicle were auxiliary devices incorporated in the rear and side lights, as no means was available for recharging the batteries while the vehicle was actually being driven. The introduction of the dynamo, or "Lichtmaschine" (the German term) fostered the gradual rise of electric headlamps. The year 1913 also witnessed the first deliveries of Bosch lighting equipment, consisting of the following components (Fig. 4):

- Headlamps and side lanterns
- Number-plate and splash-wall lanterns
- Generator
- Battery and
- Switch unit

The march toward mass motorization, higher traffic density and stylistic demands have all combined to propel lighting technology to its current level (Figure 5).

Physical principles

Visual perception

The human eye
The eye relies on various mechanisms to adapt to different lighting conditions. One of these is an adjustment in the sensitivity of the retina, while another is the adaptive response of the iris. The iris, which encircles the lens like a shutter, reacts to changes in light intensity by expanding and contracting to enlarge or reduce the pupil's effective aperture.

The eye reacts to sudden glare with contraction, marked by muscular tension and lid closure. This leads to a temporary but substantial loss of vision.

This is why the avoidance of glare is one of the essential priorities of headlamp technology, the aim being to ensure consistent visual performance.

Evaluating visual acuity
Sensitivity
The human visual apparatus assumes a vital role as a source of sensory information, of which it supplies roughly 90% of the total. The primary perceptual classifications for all visual information are brightness, color and shape, while sensitivity to brightness, in particular, is also marked by pronounced spectral sensitivity.

In 1924, efforts to quantify "perceived" light based on technically reliable data led to the definition of a spectral luminous efficiency function $V(\lambda)$ for a standard observer. The corresponding spectral sensitivity peaks at a wavelength of $\lambda = 555$ nm (in the yellow-green range).

The boundary separating UV radiation and visible light (violet) lies at the short-wave end of the spectrum, at 380 nm. In the long-wave range, the red sector extends to 780 nm, which marks the start of the "invisible" infrared range.

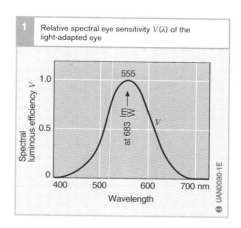

1 Relative spectral eye sensitivity $V(\lambda)$ of the light-adapted eye

Visual range and signal identification range
The visual range is the maximum distance at which objects (such as vehicles) and persons on the road or on the road shoulder remain visible.

Among the factors which influence visual range are the shape, size and reflectivity of the person or object, the road-surface material, the technical characteristics and cleanliness of windshields and headlamps, and the visual acuity of the driver.

This extensive array of influencing factors renders it impossible to define visual range in purely numerical terms. Under extremely favorable conditions it can extend forward beyond 100 m (along the verge of the road), while negative factors (the other side of the road being wet) can reduce it to below 20 m.

The signal identification range is the distance at which a light signal (such as that generated by the brake lights on another vehicle) can still be perceived in fog, mist or other inclement weather conditions.

Safety
Light sensitivity is particularly significant for "perceptibility efficiency", a factor of vital importance in road safety.

Satisfying the demands of perceptibility efficiency entails total exploitation of all available means for ensuring that the driver retains the unobstructed ability of perceiving and recognizing all visual and acoustic impressions and signals generated or transmitted by road traffic.

Yet another significant factor is "control efficiency", which focuses on fostering road safety by minimizing certain demands on the driver. Here the object is to position all instruments and control elements in highly visible locations within convenient reach of the driver, thereby facilitating their operation and use. Lighting for instruments and displays joins dashboard control configuration and general interior illumination as an essential contributor to achieving this end.

Impairment of vision

The lighting system can support the driver's perceptiblity and control efficiency while driving under restricted visibility (dawn, dusk, darkness, driving in a tunnel or on roads surrounded by dense forest). Even the surface of a dry, brightly illuminated road absorbs 70 % of the incident light, leaving only 30 % for visible roadway illumination.

Inclement weather and the accompanying effects (fog, rain, snow, etc.) also impinge upon the driver's clear field of sight. When wet, a dark road surface absorbs 85 % of the light directed toward it. Under these conditions, perceptibility efficiency can be substantially enhanced by lighting equipment such as headlamps on low or high beam as well as by front fog lamps.

Among other factors with the potential to directly limit visual perception are condensation, contamination and damage (scratches and fissures) on the wind-shield. Yet another danger source is sudden glare (from approaching traffic at night, the setting sun, etc.).

Direct impairment of the driver's vision may arise from dirty headlamps, on which the dirt prevents more than a fraction of the available light from reaching the road surface.

Various countermeasures are available for responding to these negative influences:
- Wipe/wash systems for windshields and headlamps
- Sun visors
- Variable light distribution and
- Optimal road illumination (Figure 2, good low-level illumination, extended side illumination, concentration on specific road sectors)

> **Luminous intensity**
>
> The "brightness" of light sources can vary. Luminous intensity serves as an index for comparing them. It is the visible light radiation that a light source projects in a specific direction. The unit for defining levels of luminous intensity is the candela (cd), roughly equivalent to the illumination emitted by one candle. The "brightness" of an illuminated surface varies according to its reflective properties, the luminous intensity and the distance separating it from the light source.
> Samples of approved intensities (in cd):
> Stop (brake) lamp   (single)   60...185
> Tail lamp   (single)   4...12
> Fog warning lamp   (single)   150...300
> High beam   (max. total)   225,000

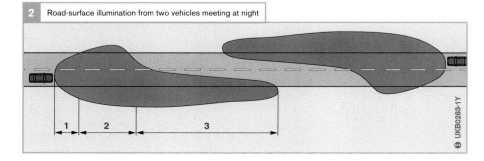

Fig. 2
1 Wide-ranging side illumination
2 Targeted road-surface illumination
3 Effective low-level illumination

Road-surface illumination from two vehicles meeting at night

Light sources

Thermal radiators
Thermal radiators generate light from heat energy. This means that luminous-intensity levels in these devices are proportional to the heat that can be generated at the source.

The major liability of the thermal radiator is its low working efficiency (below 10%) which, relative to the gaseous-discharge lamp, leads to very low potential for luminous efficiency.

Incandescent (vacuum) bulb
The bulb is a thermal-radiator device featuring a tungsten filament that glows when electrical energy flows through it. The luminous power of a standard bulb is modest, and service life is limited by vaporized tungsten particles from the filament, which deposit on the inner surface of the bulb. These considerations led to the virtual extinction of the incandescent bulb in headlamps, where it has been replaced by the halogen bulb. Other lamps, including backup lamps, are still equipped with incandescent bulbs (Figure 8).

Halogen bulbs
H1, H3, H7, and HB4 halogen bulbs have only *one* filament. These and similar bulbs are used in low-beam headlamps and fog lamps. The H4 halogen bulb (Figs. 3 and 5), is a dual-filament unit capable of alternating between low and high beam. A cap covers that portion of the low beam with the highest glare potential and forms a light-dark

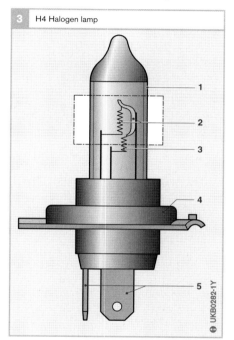

H4 Halogen lamp

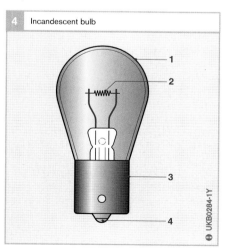

Incandescent bulb

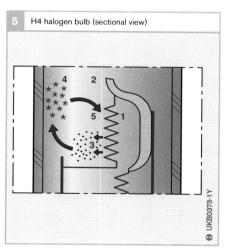

H4 halogen bulb (sectional view)

Fig. 3
1 Glass bulb
2 Low-beam filament with cap
3 High-beam filament
4 Lamp base
5 Electric connection

Fig. 4
1 Glass bulb
2 Filament
3 Lamp socket base
4 Electric connection

Fig. 5
1 Tungsten filament
2 Halogen charge (iodine or bromine)
3 Evaporated tungsten
4 Halogenated tungsten
5 Tungsten deposits

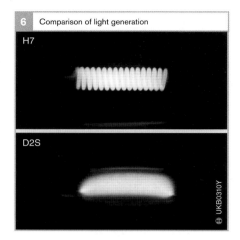

Fig. 6
H7 Halogen bulb (filament)
D2S Gaseous-discharge lamp (high-intensity discharge arc)

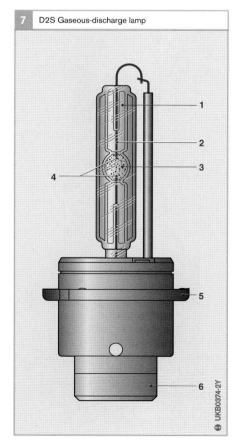

Fig. 7
1 Glass capsule with UV shield
2 Electrical lead
3 Discharge chamber
4 Electrodes
5 Lamp base
6 Electric connection

cutoff line within the light beam. A halogen unit rated at 60/55 W radiates approximately twice as much light as a comparable 45/40 W Bilux unit, and the inner surface of the bulb resists clouding, remaining clear throughout the H4's service life. The halogen-gas charge (iodine or bromine) makes it possible to use filament temperatures approaching tungsten's melting point (around 3,400 °C) for commensurately high levels of luminous power. The tungsten vapor adjacent to the hot bulb walls combines with the atmospheric halogen to form a translucent gas (halogenated tungsten) that remains stable through a temperature range of roughly 200 ... 1,400 °C. Tungsten particles approaching the filament respond to the high temperatures at the filament by dispersing to form a consistent tungsten layer (Fig. 10). As external bulb temperatures of approximately 300 °C are needed to maintain this cycle, clearances between quartz bulb and filament must be minimal. A further advantage of this layout is the fact that higher fill pressures can be used which contribute to inhibiting tungsten's inherent tendency to evaporate.

Even minute traces of oils and grease on the bulb's surface, of the kind caused by handling with bare fingers, lead to harmful desposits that can attack and destroy the glass during high-temperature operation.

Gaseous-discharge lamps

Gaseous discharge describes the electrical discharge that occurs when an electrical current flows through a gas and causes it to emit radiation (Examples: sodium-vapor street lamps and fluorescent lamps for interior lighting).

Light sources relying on the gaseous-discharge concept acquired new significance for automotive applications with the advent of the "Litronic" electronic lighting system. This kind of system is better adapted to human vision, as well as providing for longer ranges and brighter and more consistently distributed illumination of the road surface.

Durability is such that the units usually last for the entire life of the vehicle, with no need for replacement. The concept also makes it possible to design more compact headlamps for smooth front-end styling. The gaseous-discharge lamp is filled with the noble gas xenon and a mixture of metal halides, and an electronic ballast unit is required for ignition and operation. Application of an ignition voltage in the 10...20 kV range ionizes the gas between the electrodes, producing an electrically conductive path in the form of a luminous arc (Fig. 6, D2S). A regulated supply of alternating current (400 Hz) heats the lamp body so that the metallic charge is vaporized and light is radiated.

Under normal circumstances the lamp requires several seconds to ionize all of the particles and generate full illumination. This process is accelerated by feeding a high "start-up current" into the unit until it reaches maximum luminous power, at which point limitation of lamp current commences. A sustained operating voltage of only 85 V is then sufficient to maintain the arc.

This concept features several decisive assets compared with conventional bulbs:
- Extended service life. Because no solid metal evaporates, the lamp is not subject to mechanical wear
- High levels of luminous efficiency owing to the extreme temperature of the gaseous mixture (hotter than 4,000 K)
- Higher overall efficiency levels, due to high luminous efficiency combining with the low power consumption resulting from the lower operating temperatures

Gaseous-discharge bulbs for automotive applications are designed around a socket and a glass element that serves as a shield against ultraviolet radiation:
- For PES headlamps, the D2S bulb (Fig. 7)
- For reflection headlamps, the D2R bulb with an integral light shield designed to generate a pronounced light-dark cutoff (comparable to the low-beam bulb shield found in H4 units (Fig. 8)

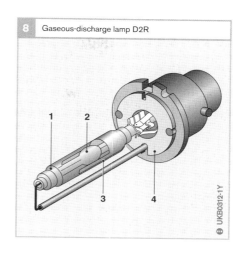

Fig. 8
1 UV inert-gas bulb
2 Discharge chamber
3 Shield
4 Lamp holder

Automotive lamps
ECE-R37 defines lamps for vehicle lighting systems in 6 V, 12 V and 24 V versions (Table 1 on next page). Various socket configurations are employed to ensure that the correct bulb types are installed at individual locations, while the operating voltages are also given on the bulb units to further distinguish bulbs of different voltage ratings that share a single socket design. The correct bulb type must be indicated on the assembly.

The luminous efficiency (in lumens per watt) indicates photometric efficiency as a function of the electrical power supplied to the unit. Thus the luminous efficiency of incandescent vacuum bulbs is 10 ... 18 lm/W, while halogen bulbs operate at higher levels in the 22 ... 26 lm/W range, primarily due to higher filament temperatures. At 85 lm/W, the D2S and D2R gaseous-discharge lamps (Litronic) make substantial contributions to further improving low-beam illumination.

▶ Luminous flux

Luminous flux is that light emitted by a light source that falls within the visible wavelength range. It is quantified in lumens (lm). For comparitive data refer to Table 1: Automotive lamps and bulbs.

Table 1. Specifications for motor-vehicle bulbs (2-wheeled vehicles not included)

Application	Category	Voltage rating V	Power rating W	Luminous flux Lumen	IEC Base type
High beam/ Low beam	R2	6 12 24	45/40[1] 45/40 55/50	600 min/ 400-550[1]	P 45 t-41
Fog lamp, High, low beam in 4-HL systems	H1	6 12 24	55 55 70	1,350[2] 1,550 1,900	P14.5 e
Fog lamp, High beam	H3	6 12 24	55 55 70	1,050[2] 1,450 1,750	PK 22s
High beam/ Low beam	H4	12 24	60/55 75/70	1,650/ 1,000[1], [2] 1,900/1,200	P 43 t – 38
High beam, Low beam in 4-HL systems, Fog lamp	H7	12	55	1,500[2]	PX 26 d
Fog lamp	H10	12	55	1,250[2]	PY 20 d
Low beam in 4-HL systems	HB4	12	55	1,100	P 22 d
High beam/ Low beam	HIR2	12	55	1,875	PX 22 d
High beam in 4-HL systems	HB3	12	60	1,900	P 20 d
High beam/ Low beam	HIR1	12	65	2,500	PX 20 d
Stop, flasher, rear fog, reversing lamp	P 21 W PY 21 W[6]	6 12 24	21	460[3]	BA 15 s
Stop lamp	P 21/5 W PY 21 W[7]	6 12 24	21/5[4] 21/5 21/5	440/35[3], [4] 440/35 440/40[3]	BAY 15 d
Side-marker lamp, tail lamp	R 5 W	6 12 24	5	50[3]	BA 15 s

Table 1: Specifications for motor-vehicle bulbs (continued)

Application	Category	Voltage rating V	Power rating W	Luminous flux Lumen	IEC Base type
Tail lamp	R 10 W	6, 12, 24	10	125[3]	BA 15 s
Stop, flasher, rear fog, reversing lamp	P 27 W	12	27	475[3]	W 2.5 x 16 d
Stop lamp/ tail lamp	P 27/7 W	12	27/7	475/36[3]	W 2.5 x 16 q
License-plate lamp, tail lamp	C 5 W	6, 12, 24	5	45[3]	SV 8.5
Reversing lamp	C 21 W	12	21	460[3]	SV 8.5
Side-marker lamp	T 4 W	6, 12, 24	4	35[3]	BA 9 s
Side-marker lamp, License-plate lamp	W 5 W	6, 12, 24	5	50[3]	W 2.1 x 9.5 d
Side-marker lamp, License-plate lamp	W 3 W	6, 12, 24	3	22[3]	W 2.1 x 9.5 d
Low beam in 4-HL systems	D1S	85, 12[5]	35, ca. 40[5]	3,200	PK 32 d-2
Low beam in 4-HL systems	D2S	85, 12[5]	35, ca. 40[5]	3,200	P 32 d-2
Low beam in 4-SW systems	D2R	85, 12[5]	35, ca. 40[5]	2,800	P 32 d-3

[1]) High/low beam. [2]) Specifications at test voltage of 6.3; 13.2 or 28.0 V.
[3]) Specifications at test voltage of 6.75; 13.5 or 28.0 V
[4]) Main/secondary filament. [5]) With ballast unit. [6]) Yellow-light version.

Headlamp elements

The headlamps installed in most vehicles still consist of the bulb, reflector and lens as their primary elements.

The bulb serves as the light source by emitting a concentric beam (actual geometry may vary according to configuration).

That portion of the light beam which is not emitted directly along the target path impacts against the reflector, which concentrates it to form a roughly parallel projection pattern (Figure 9).

A refraction lens then redirects the light waves toward the desired target region on the road surface (Figure 10).

In a number of newer headlamp designs it is the reflector that redirects the light toward the target area, thus assuming the function earlier performed by the lens. In these units the lens merely seals and protects the interior of the headlamp against external influences and contamination.

Reflectors
Purpose

The purpose of the reflectors installed in automotive headlamps is to collect as much of the bulb's light as possible in order to maximize beam range. As a basic rule, a headlamp's potential illumination range is proportional to its lens aperture, while luminous efficiency grows as the included angle embraced by the reflector increases (reflector depth).

Further demands on headlamp design arise from such requirements as dictated by vehicular styling considerations (for installation in flat-surfaced front ends or for general adaptation to body shape, etc.).

Earlier reflectors were almost always parabolic. Current designs respond to various imperatives by using a variety of configurations, as found in stepped (graduated) reflectors, transitionless free-form reflectors, and headlamp designs based on optical imaging technology (PES, or Poly-Ellipsoid System).

Reflection

High levels of reflection are encountered whenever a beam of light impacts against a mirrored surface. The light's angle of incidence is identical to its angle of reflection. As this also applies to irregular mirror surfaces, parabolic headlamp reflectors are designed so that the light beams leaving the focal point are emitted parallel to the reflector axis.

The intensity of the emerging light relative to the incident illumination impacting against the reflector (reflection factor) is determined by the angle of incidence and

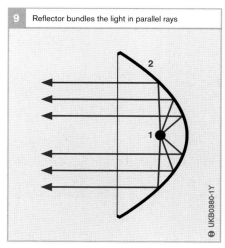

9 Reflector bundles the light in parallel rays

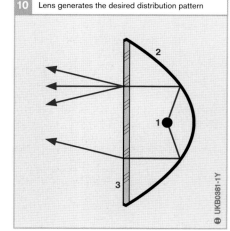

10 Lens generates the desired distribution pattern

Fig. 9
1 Lamp in focal point f
2 Reflector

Fig. 10
1 Lamp in focal point f
2 Reflector
3 Lens

the refractive indices of the materials in contact with each other. Reflectors with aluminum applied in an evaporative coating process have reflection factors in the 90 % range. This figure can drop to below 50 % for a reflector surface that has been damaged by corrosion, which is why rust protection is a major consideration with headlamp reflectors. Reflector-surface quality and corrosion protection are prime determinants of headlamp quality.

Reflector focal length
Reflectors with shorter mean focal lengths f or short mean focal length $\bar{f}$ promote effective exploitation of the bulb and achieve high efficiency levels. On these units the reflector encloses the lamp bulb to a large extent, making it possible to redirect a high proportion of the incident light into the ultimate beam pattern.

The focal length (also known as the focal depth) is the distance between the vertex of the parabola and the focal point. It is between 15 and 40 mm on conventional headlamp reflectors. A lens redistributes the light to provide the desired pattern on the road surface (Figure 10).

The geometries of modern reflectors sometimes display quite substantial departures from the purely parabolic form. Special mathematical (HNS, or Homogenous Numerically Calculated Surface) processes are employed to calculate optimal configurations. The mean focal length $\bar{f}$, as defined relative to the distance between the reflector's vertex and the center of the filament, is a given figure. The actual value is between 15 and 25 mm. Reduced focal lengths make it possible to install three separate reflectors (Fig. 11) – for low-beam, high-beam and fog lamp (H1 and H2 bulbs) – within the same space needed by a conventional H4 parabolic reflector, all with a simultaneous increase in luminous efficiency.

In the case of the graduated (stepped) or complex-surface (segmented) reflectors PD2 (**Partition Design 2**), each partition can be produced with its own specific focal length f.

Reflector materials
Bosch reflectors are precision-manufactured in sheet-metal or plastic.

The first step in the production process for sheet-metal reflectors is deep-draw molding to produce a paraboloid or one of the more complex geometrical configurations described above. This is followed by galvanization or application of a powder coating to protect against corrosion. The reflector is then painted to produce a smooth and consistent surface, after which evapora-

11 HNS reflectors (Example): Using numerically-calculated reflector surfaces to modify light distribution

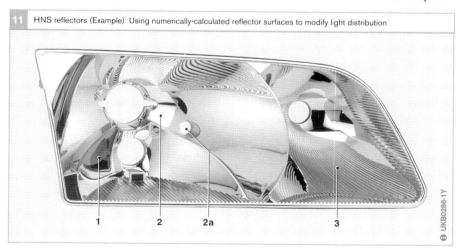

Fig. 11
1   Fog lamp
2   Low beam
    (2a opening for the side-marker lamp)
3   High beam

tive or sputter-spray techniques are employed to apply the aluminum reflective layer and the special protective coating.

This process hermetically seals the sheet-metal while providing an extremely smooth and consistent surface featuring residual unevenness of no more than 1/10,000 mm.

Plastic reflectors are also components of extremely high quality, and high-precision spray (Thermoplast) or compression molding (Duroplast) techniques are employed to endow them with their ultimate geometry. This manufacturing strategy facilitates production of reflectors with special graduated patterns as well as multichamber units. Although no special anti-corrosion treatment is needed to protect the base material, the reflective layer is guarded by a special coating.

Lenses
(outer lens with light-dispersion optics)
Purpose
The purpose of the lens is to produce the desired illumination pattern on the road surface by refracting, dispersing and focussing the light emitted by the reflector. During the lens molding process, high priority is assigned to surface quality. Flaws must be avoided, as a lens that emits stray light beams could pose a glare hazard to oncoming traffic.

Refraction and reflection
Headlamps and lamps in general rely on transparent substances and mediums such as air, glass, plastic, etc., to project light. Light rays entering the transition zone between the air and the transparent lens material splits into refracted and (relatively minute) reflected elements. Refraction results from variations in light-wave propagation velocities in different mediums such as air, glass and plastic. Both the transition from air (refractive index n of roughly 1) to glass (refractive index n of roughly 1.52) and the passage from glass to air produce reflection rates of around 4.3% at a 90% angle of incidence. The orientation of the reflected component is defined by the fact that the angle of incidence is equal to the projection (reflection) angle.

"Total" reflection, in which all light rays are reflected with no loss, occurs once the angle of incidence in a high-density substance exceeds a specified level. This effect is exploited in devices such as reflectors, road-side-post light reflectors and optical-fiber cables (fiber-glass or plastic) as a means of relaying light.

On the inside of the lens there is a precisely defined arrangement of focal and prismatic elements, which combine with planar sectors

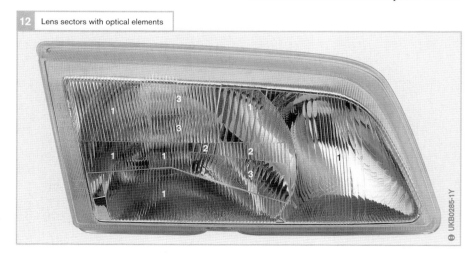

12 Lens sectors with optical elements

Fig. 12
1 Lens elements
2 Prismatic lens sector
3 Combined lens sector

to provide an extended high-beam range as well as good luminous distribution when the headlamps are dimmed. The specific type and configuration of these optical elements on the projection surface of the lens varies according to the reflector's focal length and the desired light pattern (Figs. 12 and 13).

The outside of the lens is always smooth to prevent the accumulation of dirt.

Optical element (Fig. 13)
Light refraction disperses parallel rays impacting against a *lens element* at right angles to the axis of the cylindrical lens. The scatter effect is inversely proportional to the radius of the lens.

Light rays encountering a *prismatic lens sector* at a given angle are redirected at an angle defined by the geometry of the prismatic element. Parallel incident rays remain parallel as they emerge from the prism.

Combined lens sectors unite focal and prismatic elements in a single field.

"Clear" outer lenses
Modern free-form reflectors are developed specifically for use in combination with "clear" outer lenses, in which no optical elements of any kind are required.

Lens materials
Conventional lenses are manufactured from high-purity glass that must be absolutely free of bubbles or streaks. Today, plastic is often specified as the lens' material in order to reduce weight. Within Europe the legal framework allowing official certification of plastic lenses is in place, and the first plastic lens entered series production in 1992. A varnish coat is applied to the outside surface of the lens to protect it against aging and scratches. While plastic lenses can be specified as a weight-reduction measure, they also open new vistas in headlamp design by placing practically no limits on geometrical configuration. Yet another significant factor in automotive-engineering applications is the fact that plastic units are less expensive.

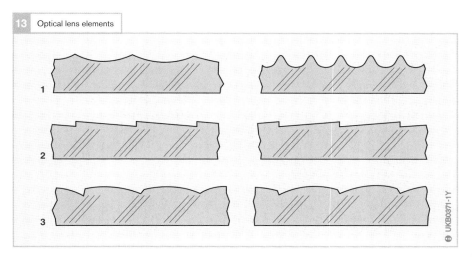

Fig. 13
1   Lens elements
2   Prismatic lens sector
3   Combined lens sector

Lights and lamps

Purpose
Lights and lamps are intended to enable recognition of the vehicle's presence and its current or intended state of motion.

Regulations
Minimum and maximum luminous intensity levels are prescribed for the light emerging along the central horizontal axes of all lamps. The intent of the regulations is to ensure highly-visible and conspicuous signal images while simultaneously preventing dazzle for other road users. Luminous intensity may be lower in light emerging from above, below and to the side of the horizontal axis.

Versions
All vehicle-mounted lamps can be designed based on either one of two basic concepts, with the final choice being dictated by such factors as available space.

The design regulations are defined in ECE R6, R7, and R38, and the installation/mounting in 76/756/EEC, ECE-R48, and in StVZO §49A (CUR/FMVSS)

Lamps with reflector optics
A reflector in any of a wide variety of different shapes (often parabolic) redirects the light emitted by the bulb to produce a roughly axial projection angle. A lens featuring optical dispersion elements then defines the ultimate projection pattern as required (Figure 14).

Lamps with fresnel optics
The light from the bulb is projected directly against the lens, with no refraction by the reflector. A lens based on fresnel refraction technology then redirects the beam in the desired direction (Figure 15).

Fresnel optics devices are usually less efficient than reflector-optics lamps.

Lamps combining fresnel and reflector optics
Combination lamps incorporating elements from both concepts are being successfully applied. As an example, the GP (**G**edrehte **P**arabel, or rotated parabola) reflector (Fig. 16) can be employed to derive undiminished luminous flux from volumetrically compact units with both shallower reflectors and smaller lenses. This design employs a specially designed GP reflector to capture the light beam emanating from the bulb at the largest-possible peripheral angle. The fresnel optics then homogenize the beam for projection along the specified path.

Free-form lamps with fresnel cap (Figure 17) combine good phototechnical efficiency with substantial scope for stylistic variation. The reflector refracts and redirects the rays from the bulb while at the same

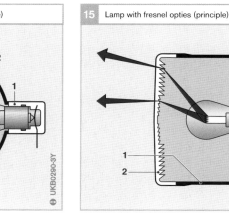

14 Lamp with reflector optics (principle)

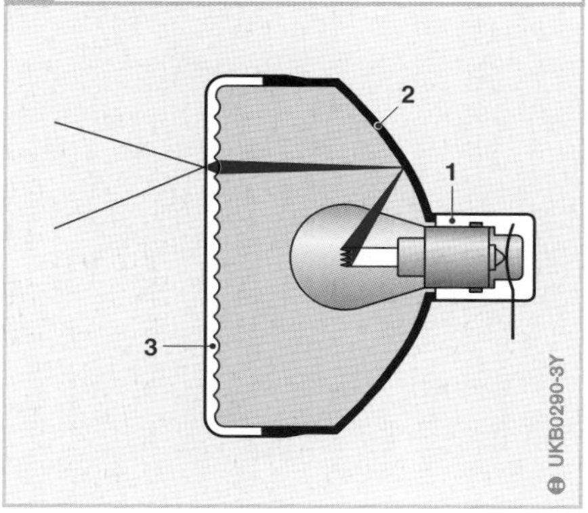

15 Lamp with fresnel opties (principle)

Fig. 14
1 Housing
2 Reflector
3 Lens with cylindrical fluting

Fig. 15
1 Housing
2 Lens with fresnel optics

time contributing all or part of the required diffusion to the final distribution pattern. As a result, the cover lens can be clear and unfluted, or it can incorporate cylindrical dispersion elements to generate vertical or horizontal scatter. The fresnel lens provides greater photometric efficiency by capturing light that would otherwise be wasted, and reprojecting it in the desired direction.

The primary application for both designs is in front turn signals. Here and in other applications, the ultimate considerations are vehicle shape (which defines space availability), stylistic requirements and the required level of light intensity.

Color filters
Specific red and yellow-based colors have been assigned to automotive lamps according to their intended function (brakes, turn signals, fog warning lamps, etc.). These colors are defined on the basis of a standard color spectrum (color location). Because white light is actually composed of numerous colors, it is possible to filter luminous radiation to partially or completely remove undesired spectral elements from an emerging beam. The color filter can be in the form of a dyed lens or a colored layer on the bulb's glass surface (e.g., amber bulbs in turn signals with neutral lens tones).

This filter technology can also be employed to produce color-coordinated lamp lenses. These adapt to the vehicle's finish when not in operation, but still comply with the certification regulations when switched on.

The color locations specified within the jurisdiction of the ECE include "amber" with a wavelength of 592 nm for turn signals, and "red" at 625 nm for stop (brake) lamps and tail lamps.

Designs
The term *"grouped design configuration"* refers to components that share a single housing but still have individual lenses and bulbs (example: multichamber rear lamp assembly comprising individual light units).

"Combined design configuration" indicates a component assembly in which a single housing and bulb are employed together with more than one lens (example: combined tail lamp and license-plate lamp in commercial-vehicle applications).

The term *"nested design configuration"* is applied to component assemblies that share a common housing and lens, but have individual bulbs (example: nested design of headlamp unit with built in side-marker lamp). A frequent form of nested design features a bulb in which several functions are combined (example: tail and stop (brake) lamps both relying on a single dual-filament bulb).

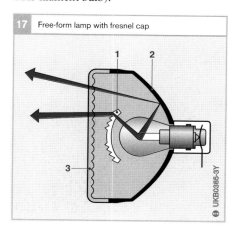

Fig. 16 Lamp with GP (rotated parabola) reflector

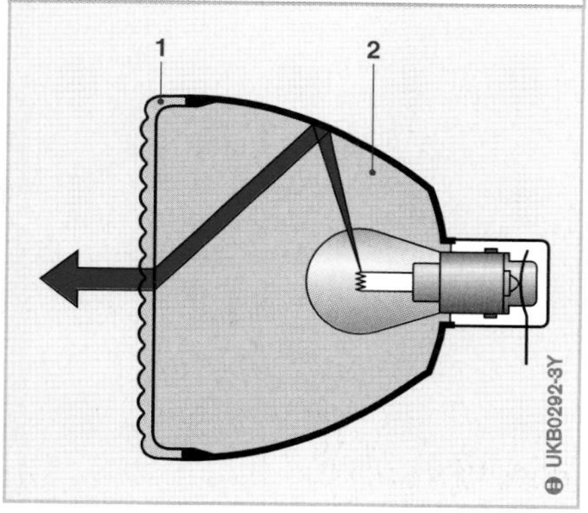

Fig. 17 Free-form lamp with fresnel cap

Fig. 16
1 Fresnel lens
2 GP rotated parabola reflector

Fig. 17
1 Fresnel cap
2 Reflector
3 Lens with cylindrical fluting

Front lighting system: Components

Main headlamps (Europe)

Purpose

The vehicle's main headlamps must satisfy several – sometimes mutually antagonistic – imperatives. On the one hand, they must provide maximum visual range while at the same time ensuring that light distribution immediately in front of the vehicle remains in line with the requirements of safe operation. Although blinding glare for approaching road users must be minimal, it is vital to provide the lateral illumination needed to safely negotiate curves, i.e., the light must extend outward to embrace the verge of the road. Although it is impossible to achieve absolutely consistent luminance across the entire road surface, it is possible to avoid sharp contrasts in light density.

High beam

The high beam is usually generated by a light source located at the reflector's focal point, causing the light to be reflected outward along a plane extending along the reflector's axis (Fig. 1). The maximum luminous intensity which is available during high-beam operation is largely a function of the reflector's mirrored surface area.

In four and six-headlamp systems, in particular, purely parabolic high-beam reflectors can be replaced by units with complex geometrical configurations for simultaneous use of high and low beams.

In these systems the high-beam component is designed to join with the low-beam's light (simultaneous operation) to produce a harmonious overall high-beam distribution pattern. This strategy abolishes the annoying overlapping sector that would otherwise be present at the front of the distribution pattern.

Low beam

Because the high traffic density encountered on modern roads severely restricts the use of high beams, the low beams serve as the primary source of light under normal conditions. Basic design modifications implemented within recent years are behind the substantial improvements in low-beam performance. Developments have included:

- Introduction of the asymmetrical low-beam pattern, affording extended visual range along the road shoulder.
- Official certification for various types of halogen bulbs which increase the luminous intensity on the road surface by 50 ... 80 %.
- Introduction of innovative new headlamp concepts with complex geometries (PES, HNS) with efficiency levels increased by up to 50 %.

▶ Luminous intensity and range

The luminous internsity is the luminous flux arriving at a given surface. It increases proportionally along with the light intensity, and decreases with the square of the distance. It is measured in Lux (lx):
1 lx = 1 lm/m²

The range is defined as the distance at which the luminous intensity in the light beam still has a given value (e.g. 1 lx). The geometric range is the distance at which the horizontal part of the light-dark cutoff is shown on the road surface with the headlamps on low beam.

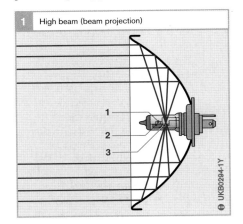

Fig. 1
1 Low-beam filament
2 Shield
3 High-beam filament at local point

High beam (beam projection)

Fig. 3 Testing low-beam illuminance and light-dark cutoff in the light tunnel

- Headlight leveling control (also known as vertical aim control) devices adapt the attitude of the headlamps to avoid blinding oncoming traffic when the rear of the vehicle is heavily laden. Headlamp washing systems are mandatory equipment on motor vehicles fitted with headlight leveling control.
- "Litronic" gaseous-discharge headlamps supply more than twice as much light as conventional halogen units.

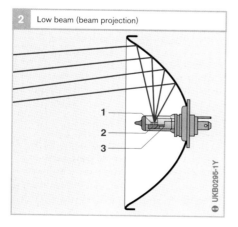

Fig. 2 Low beam (beam projection)

Method of operation
Virtually all of the headlamp designs employed up to 1988 (with H4 bulbs) used a low-beam light source mounted forward of the parabolic reflector's focal point which, following reflection, endowed the beam with a post-reflective inclination toward the reflector's axis (Fig. 2).

Low-beam headlamps need a light-dark cutoff in the light pattern. In the case of H4 halogen headlamps, or Litronic headlamps with D2R bulbs, this is achieved by the image from the shield (H4) or the shutter (D2R). On headlamps for all-round use (H1-, H7-, HB4 bulbs), the light-dark cutoff is achieved by the special imaging of the filament.

The resulting "dark above/bright below" distribution pattern provides acceptable visual ranges under all driving conditions. This layout keeps the glare projected toward approaching road users within reasonable limits while at the same time supplying relatively high levels of luminous intensity in the sector below the light-dark cutoff line (Fig. 3).

Automotive headlamp performance is subject to technical assessment to verify compliance with the regulations governing

Fig. 2
1 Low-beam filament
2 Shield
3 High-beam filament

official certification. Precise test points are specified for measuring high and low-beam light. Among the requirements to be satisfied at these locations are maximum luminous-intensity levels – to ensure adequate road-surface visibility – and maximum intensity levels, to prevent glare.

Regulations
The following ordinances apply for main headlamps, their installation and use.
76/761/EEC and ECE-RI and R2:
High and low-beam headlamps and their bulbs.
ECE-R8: Headlamps with HI, H2, H3, H7, H8, H9, H11, HIR1, HIR2, HB3 and HS4 bulbs.
ECP-R20: Headlamps with H4 bulbs.
StVZO § 50 (FMVSS/CUR): High and low-beam headlamps.
76/756/EEC and ECE-R48: Mounting and application.
ECE-R 98/99: Headlamps with gaseous-discharge bulbs.

High beams: Installation
A minimum of two and a maximum of four headlights are prescribed for the high-beam mode.

Prescribed color for the high-beam indicator lamp in instrument cluster: Blue or yellow.

High beams: Illumination technology
High-beam light distribution is defined in the regulations and guidelines together with stipulations governing low beams.

The most important specifications are: Symmetrical distribution relative to the central vertical plane and maximum illumination along the headlamp's central axis.

The maximum approved light intensity, a composite of the intensity ratings for all high-beam headlamps installed on a vehicle, is 225,000 cd. The corresponding rating is indicated by a reference code located adjacent to the official certification code on each headlamp. The code for 225,000 cd is 75. The light intensity of the high beam is indicated by a figure, such as 20 (as an example), stamped next to the round ECE test symbol.

If these are the only headlamps on the vehicle (no auxiliary driving lamps), then this sample system's composite luminous intensity would be in the range of 40/75 of 225,000, viz., 120,000 cd.

Low beams: Installation
Regulations prescribe 2 white-light low-beam headlamps for all multiple-track vehicles.

Low beams: Illumination technology
The relevant ordinances governing symmetrical low-beam headlamps within Germany are contained in the Technical Specifications (TA) of the German StVZO (FMVSS/CUR). The only applicable regulations for asymmetrical low beams are the international regulations and guidelines; these contain precise definitions of the photometric test procedures prescribed for use on various low-beam units (with incandescent, halogen or gaseous-discharge bulbs).

Certification testing is carried out using test lamps manufactured to more precise tolerances than units stemming from standard series production.

Headlamp glare is assessed based on StVZO § 50 (6). Glare is considered to be eliminated when the illuminance at a height equal to that of the headlamp's center does not exceed 1 lx at a distance of 25 m. This test is conducted with the engine running at moderate rpm.

Low/high beam: Switching
Selection of the low-beam mode must simultaneously extinguish all high-beam illumination. Gradual dimming is permissible within a maximum period of 5 seconds. A 2-second response delay is required to prevent the dimming mode switching on when the high-beam flashers are used. The low beams may continue in operation when the high-beam mode is selected (simultaneous operation). H4 bulbs are generally designed for limited use with both filaments in operation.

Design configurations
Lens (movable with respect to the bodywork)
This design concept, now superseded, featured a lens and reflector that combined to form a single headlamp module, with beam adjustment being performed by pivoting the complete assembly. Under unfavorable circumstances this type of layout can result in the lens being slightly tilted relative to the vehicle's body. The headlamp assembly was usually equipped with seals in the area immediately surrounding the bulb as well as a special ventilation system.

Lens (fixed with respect to the bodywork)
There is no direct connection between the lens and the reflector, which is mounted in a housing that moves relative to the lens during adjustment (housing concept). Because the lens is not attached to the housing it remains stationary in one invariable position relative to the body, an asset allowing total integration within the front end. The complete headlamp is sealed or provided with ventilation elements.

Headlamp systems
Dual-headlamp systems rely on a single shared reflector for low and high-beam operation, e.g., in combination with a dual-filament H4 bulb (Fig. 4a).

In *quad headlamp* systems one pair of headlamps may be switched on in both modes or during low-beam operation only, while the other pair is operated exclusively for high-beam use (Fig. 4b).

Six-headlamp systems differ from the quad configuration by incorporating a supplementary fog lamp within the main headlamp assembly (Fig. 4c).

Main headlamps (North America)

High beam
As in the European system, the light source is usually located at the focal point of a parabolic reflector (Fig. 5, Pos. 3, next page).

Low beam
Headlamps with a light-dark cutoff that rely on visual/optical adjustment procedures have been approved in the USA since 1 May, 1997. These units reflect the ECE guidelines in effect for Europe, thus making it possible to equip vehicles for Europe and the USA with the same headlamps.
 The "classical" sealed-beam design in use prior to this date featured a low-beam light source located just above (Fig. 5, Pos. 1) and somewhat to the left of the reflector's focal point (as viewed looking in the direction of vehicle travel). This arrangement directs virtually the entire luminous flux downward onto the road surface. This is accompanied by a rightward offset producing an asymmetrical beam. Because there is no shield beneath the light source, the entire reflector can be used. However, the absence of a clearly defined light-dark cutoff limits visual ranges on the driver's side of

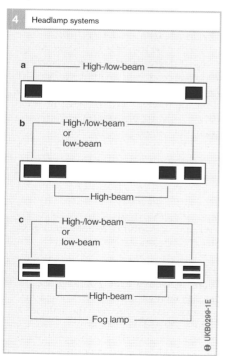

Fig. 4
a Dual-headlamp system
b Quad headlamp system
c Six-headlamp system

the lane, and also produces higher glare for oncoming traffic than the European system.

Legal regulations
Federal Motor Vehicle Safety Standard (FMVSS) No. 108 and the SAE Ground Vehicle Lighting Standards Manual (Standards and Recommended Practices).

The regulatory framework governing installation and switching arrangements for main headlamps is comparable to that in Europe.

As mentioned above, headlamps with a light-dark cutoff as defined in ECE Directives have been approved in the USA since 1 May, 1997.

Prior to this date, there were major differences in the respective headlamp systems. Until 1983 only sealed-beam units in the following dimensions were approved in the USA:
Dual-headlamp systems:
- 178 mm diameter (round)
- 200 x 142 mm (rectangular)
Quad headlamp systems:
- 146 mm diameter (round)
- 165 x 100 mm (rectangular)

An amendment to FMVSS 108 that entered effect in 1983 made it possible to start using headlamp modules of various shapes and sizes with replaceable bulbs. These were known as the RBH, or Replaceable Bulb Headlamps.

Headlamps which are in line with ECE Regulations or EU Directives can be used in the USA on dual-track vehicles provided they comply with the US stipulations on lighting technology and adjustment (visual). There are no restrictions regarding motorcycles.

Design configurations
Sealed beam
This design concept has disappeared from common use. Because the light source is exposed, the glass reflector with its vapor-coated aluminum reflective layer must be hermetically sealed with the lens. After sealing, the entire unit is filled with an inert gas. A ruptured filament necessitates replacement of the entire unit. Units relying on halogen light bulbs are also available.

The limited range of sealed-beam headlamps on the market severely restricted the designers' latitude in front-end styling.

Replaceable Bulb Headlamp (RBH)
Starting in 1983, European advances in the realm of replaceable bulb technology began making inroads into American system technology. The RBH headlamp's geometry and dimensions can be adapted for improved design (styling); these units usually feature plastic lenses and housings.

Vehicle Headlamp Aiming Device (VHAD)
This is a mechanical device used with RBH headlamps. The integral mechanism relies on a concept using spirit levels for vertical adjustment, while a system using a pointer and scale is used for horizontal adjustment in an "on-board aiming" process.

Headlamp systems
North America mirrors European practice in employing dual, quad and six-head-lamp systems.

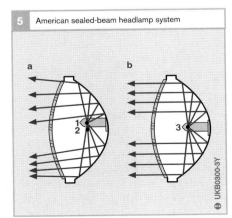

5 American sealed-beam headlamp system

Fig. 5
a Low beam
b High beam

1 Filament for low beam
2 Focal point
3 High-beam filament (at focal point)

Headlamp versions

Conventional headlamps

In conventional headlamp systems, the quality of the low-beam light improves as the reflectors get larger, while geometrical range is proportional to installation height. A contrary consideration is that vehicle front ends should remain low in response to the dictates of aerodynamics.

Taking these considerations into account, when the reflector is increased in size this leads to wide headlamps. If the reflector size remains unchanged, the focal length becomes a dependent factor, whereby short focal lengths generate a wider light beam with better side and close-range illumination. This is of particular advantage in curves.

Headlamps with stepped reflectors
Stepped, or graduated, reflectors are segmented reflectors consisting of paraboloid and/or parelliptical (combined parabola and ellipse) sectors designed to provide various focal lengths. These units retain the advantages of a deep reflector in a more compact unit suitable for installation in shallower apertures (Fig. 6).

Homofocal reflector
The homofocal reflector is composed of base and supplementary reflector units (Fig. 6, Pos. 1a and 1b). Because the homofocal reflector's supplemental sectors share a single focal point to achieve a shorter focal length than can be derived from a base reflector alone, they make a substantial contribution to effective luminous flux. While the light from the supplementary reflectors improves side and close-range illumination, there is no increase in range.

Multifocal reflector
The basic concept behind the multifocal reflector is similar to that employed in its homofocal counterpart. Based on mathematical definitions, parelliptic reflector sections designed for horizontal light diffusion provide a large number of focal points.

Headlamps with stepless reflectors
Specially-developed computer programs (CAL, **Computer Aided Lighting**) assist in designing stepless VFR (**Variable-Focus Reflectors**) with non-parabolic sectors. The focal points of the various reflector zones can change their position with respect to the light source. By applying this principle it is possible to make full use of the whole of the reflector surface.

Headlamps without focal lenses
(clear outer lens)
Today's expanded HNS (**Homogeneous Numerically Calculated Surface**) technology makes it possible to achieve headlamp effi-

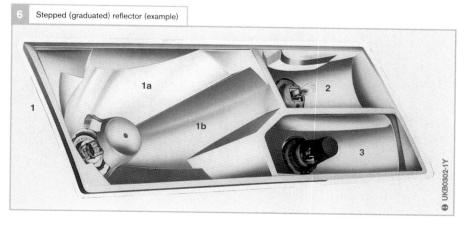

6 Stepped (graduated) reflector (example)

Fig. 6
1 Homofocal reflector
1a Base reflector
1b Supplementary reflector surfaces
2 High-beam reflector
3 Fog-lamp reflector

ciency levels of up to 50%. In addition, the entire pattern of light distribution can be defined by the reflector, meaning that fluted lens patterns have become redundant. Headlamps with clear, unfluted lens covers are also expanding the horizons of headlamp styling (Fig. 6, BMW 3 Series).

Headlamps with complex-surface reflectors
Such reflectors are partitioned into a number of segments. The PD2 program applies the CAL and HNS modules for the individual optimization of each segment. The salient factor with the surfaces developed using the PD2 program is that graduations and inconsistencies can be accepted at the boundary areas. This results in freely designed reflector surfaces which high levels of homogeneity and sideways beam spread.

PES headlamps
The headlamp designated PES (Poly-Ellipsoid System) as shown in Figure 4 (VW New Beetle) is distinguished by reliance on imaging optics to improve on the technical performance provided by conventional headlamps. This system celebrated its world premiere with the start of series production at Bosch in 1985.

While the conventional headlamp relies on a diffusion lens for light distribution, the PES light-distribution pattern is defined by the reflector and then projected through the lens and onto the road surface. The underlying concept is related to that used for overhead slide projectors, with optical reproduction of an image being the basic objective in both cases. While this image (or object) is the slide itself in an overhead projector, in the headlamp it is the light-distribution pattern generated by the reflector and modified by the screen that produces the light-dark cutoff needed for low-beam operation (Fig. 7).

Thanks to the eliptical (CAL designed) reflector and to optical projection technology, the PES needs a light-emission surface of no more than 28 cm² to project the light-distribution patterns formerly available only from large-surface conventional lamps. An imaging screen projects precisely defined light-dark cutoffs. This concept offers tremendous latitude for defining these transition areas, with either high or low contrast, and there is also virtually unlimited range for accomodating special geometries (Fig. 8a).

PES headlamps can be combined with conventional upper beam, side-marker lamps, and PES fog lamps to form a lighting-strip unit with a height of only 80 mm.
The PES PLUS concept relies on light projection directed toward a section of the reflector beneath the imaging screen to enhance close-range illumination (Fig. 8b). The signal image is also enlarged to reduce psychological glare. This effect is enhanced by using a supplementary ring-shaped reflector element (Fig. 8c) to provide special benefits in case of oncoming traffic.

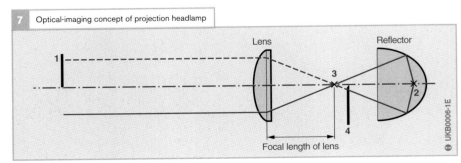

Fig. 7
1 Screen image
2 Rear reflector focal point
3 Front reflector and lens focal point
4 Screen

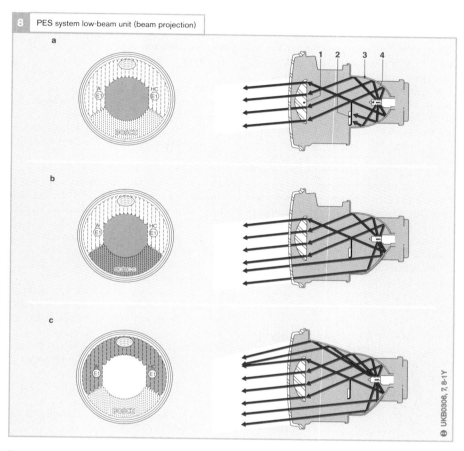

Fig. 8
a PES
b PES-PLUS
c PES-PLUS with ring-shaped reflector

1 Lens
2 Shield
3 Reflector
4 Bulb

"Litronic"

Assignments and requirements

The Litronic (**Light Elec*tronic*s**) headlamp system from Bosch features a xenon gaseous-discharge bulb to satisfy the most exacting performance demands in the field of lighting technology. The catalog of new requirements concerns the type of light and its intensity while also embracing a call for compact designs.

A service life in excess of 1,500 hours means the unit can be be expected to last for as long as the car itself. The illumination furnished by Litronic headlamps also represents a substantial improvement over that provided by conventional halogen units (Fig. 9, next page).

Design

The components of the Litronic headlamp system are:
- Optical unit with gaseous-discharge lamp
- Electronic ballast unit with ignition unit and ECU

Principle of operation

Compared to halogen units, Litronic headlamps produce a higher luminous flux with specifically adapted light distribution. This results in well-illuminated road shoulders, while significant improvements in both visibility and general orientation are valuable in potentially hazardous situations and in bad weather. Litronic headlamps comply with ECE-R48 by always incorporating automatic headlight leveling and a headlamp washer. These features combine to ensure consistently optimal exploitation of the unit's very long range and unobscured light projection.

An electronic ballast unit – comprising the ignition unit and the ECU – is responsible for igniting, operating and monitoring the gaseous-discharge lamp. The ignition unit generates the high voltage required to initiate arcing in the gaseous-discharge lamp. The ECU regulates the current supply in the warm-up phase before reverting to maintenance of a consistent 35 W for static operation. The current flowing to the lamp is raised in the initial seconds after ignition to accelerate the lamp's progress to 100 % illumination.

The system also compensates for fluctuations in the vehicle's battery voltage to maintain highly consistent levels of luminous flux.

Should the lamp go out owing to a momentary lapse or collapse in the vehicle's on-board voltage supply, reignition is automatic. The electronic ballast unit responds to defects (such as a damaged lamp) by interrupting the power supply to prevent personal injury in the event of inadvertent contact.

Versions
Headlamps with gaseous-discharge lamps are installed in quad systems in combination with conventional high-beam lamps (Fig. 10).
Two different optical systems are available:

PES projection headlamps
The world's first Litronic headlamp debuted in 1991 in a unit combining PES projection headlamps with DI gaseous-discharge units. Current new vehicles equipped with PES projection headlamps are supplied exclusively with DS2 units (Figs. 7 and 11).

Reflection headlamps
Provided that larger areas are available for the light-emission surface, Litronic can also be in the form of a reflection headlamp. Either a lens with integral optical focal elements or a clear lens cover may be found within this substantially larger projection surface.

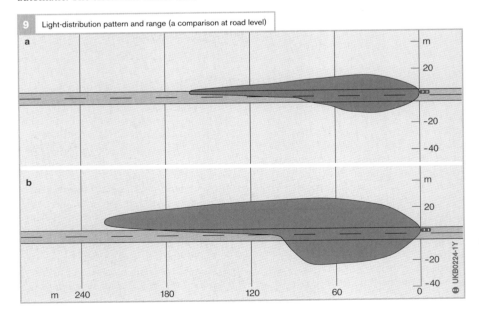

Fig. 9
a PES HI lamp
b Litronic PES D2S lamp

The aiming device is aligned with the vehicle axis using a sighting device such as a mirror with an orientation line. The device is rotated to bring the orientation line into uniform contact with two external vehicle-reference marks. The imaging chamber can be adjusted vertically to the level of the vehicle's headlamps prior to being locked in position.

Headlamp testing
Headlamp aim can be assessed once the equipment's optical system has been correctly positioned in front of the lens. An image of the light distribution pattern emitted by the headlamp appears on the collector screen. Some test devices are also equipped with photodiodes and a display to facilitate measurement of luminous intensity.

On headlamps with asymmetrical low-beam patterns, the light-dark cutoff must contact the horizontal delineation line with the horizontal and vertical components intersecting on a vertical line extending through the center mark (Fig. 18).

Following the stipulated adjustment of the low-beam light-dark cutoff, the center of the high beam (when both settings are being adjusted simultaneously) should be within the boundary marks that encompass the center mark (Fig. 19).

Headlight levelling control
Purpose
The function of the headlight levelling control (also known as vertical aim control) is to compensate for changes in vehicle load, and maintain consistently satisfactory visual ranges while at the same time preventing oncoming traffic being blinded by excessive glare. The headlight levelling control executes this function by adjusting the inclination of the low-beam headlamp. Without headlight levelling control, visual range is subject to continuing variations due to changes in vehicle load (Fig. 20).

1	Geometrical range for horizontal component of low-beam light-dark cutoff (Headlamp installed at height of 65 cm)					
Inclination of light-dark cutoff (1% = 10 cm/10 m)		1%	1.5%	2%	2.5%	3%
Aiming dimension e		10 cm	15 cm	20 cm	25 cm	30 cm
Geometrical range for horizontal component of low-beam light-dark cutoff		65 m	43.3 m	32.5 m	26 m	21.7 m

Table 1

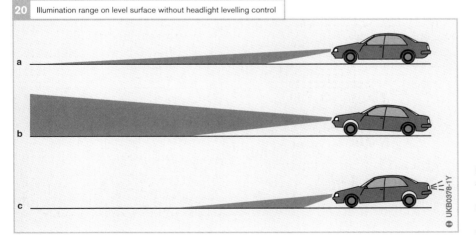

Fig. 20
a Unladen steady-state operation
b Under acceleration or with rear load
c During braking

Regulations

Table 1 gives the geometrical ranges for various angles of inclination for headlamps installed at a height of 65 cm. The inspection tolerance extends to include inclinations of up to 2.5% (1.5% below standard setting). The EU specifications governing the basic setting for headlight levelling controls stipulate:

The basic setting relative to dimension e is 10..15 cm at a distance of 10 meters with one person in the driver's seat. The specifications for this setting are provided by the vehicle manufacturer.

Inside Europe, since 1 Jan. 1998, an automatic or manually-operated headlight levelling control device is mandatory on all new vehicles as a condition for initial registration. The sole exception to this requirement is when the vehicle is equipped with some other device capable of ensuring that the light beam remains within specified vertical tolerances (such as automatic self-levelling suspension). Although this equipment is not mandatory in other countries, its use is permitted.

Designs

All headlight levelling controls feature actuators that move the headlamp reflector (housing-type design) or headlamp unit up and down. Automatic systems rely on sensors that monitor suspension travel as the basis for generating proportional signals for transmission to the aiming actuators. Manual layouts employ a driver-operated switch to control height adjustment.

Automatic headlight-levelling control

Automatic headlight-levelling control systems fall into two categories: static and dynamic. While static systems compensate for load variations in the luggage and passenger compartments, dynamic systems also correct headlamp aim during acceleration – both from standing starts and when underway – and when braking.

The components of a typical headlight-levelling control system include (Fig. 21):
- Sensors on the vehicle axles to precisely measure the vehicle's inclination or tilt
- An ECU that uses the sensor signals as the basis for calculating the vehicle's pitch angle. The ECU compares these data with the specified values and responds to deviations by transmitting appropriate control signals to the headlamps' servomotors
- Servomotors to adjust the headlamps to the correct angle

Static system

In addition to the signals from the suspension sensors, the static system's con-troller also receives a speed signal from the electronic speedometer. The controller relies on this signal to decide whether the vehicle is stationary, undergoing a dynamic change in speed, or proceeding at a constant speed. Automatic systems based on the static concept always feature substantial response inertia, so the system corrects only those vehicle inclinations that are consistently registered over relatively long periods.

21 Schematic diagram of an automatic headlight-levelling control system (dynamic system)

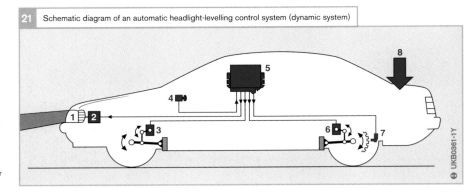

Fig. 21
1. Headlamp
2. Actuator
3. Front-suspension travel sensor
4. Light switch (on/off)
5. ECU
6. Rear-suspension travel sensor
7. Wheel-speed sensor
8. Load

Each time the vehicle moves off, the system corrects the headlamp angle to compensate for load variations, with a second correction cycle being initiated once the vehicle goes into steady-state operation. The static system employs the same servomotors found in manual systems to compensate for the deviation between the specified and monitored vertical angles of the headlamps.

Dynamic system
The dynamic automatic system relies on two distinct operating modes to ensure optimal headlamp orientation under all driving conditions. Supplementary capabilities in speed-signal analysis endow the system with the ability to differentiate between acceleration and braking; this represents a distinctive advance beyond the static aim-control concept.

During steady-state operation, the dynamic system operates in the same manner as its static counterpart with large rates of response inertia, but as soon as the controller registers acceleration or braking, the system immediately switches to its dynamic mode. This mode's distinctive feature is its faster signal processing and the higher servomotor adjustment speed, thus allowing beam range to be readjusted within fractions of a second. The ultimate result is that the driver always enjoys the visual range needed for safe and efficient monitoring of traffic conditions. Following acceleration or braking, the system automatically reverts to operation in its delayed-response mode.

Manual headlight-levelling control
This type of control is activated by the driver. A detent is prescribed for the standard setting, which also serves as the reference point for setting the beam to its basic position. Regardless of whether they feature infinitely-variable or graduated control, all manual units must incorporate hand switches in the close vicinity of which there are visible markings corresponding to the various vehicle-load conditions for which different correction settings are prescribed.

Adjustment mechanisms
Hydromechanical units use a fluid medium to initiate motion in the headlight levelling control actuators, and *vacuum systems* exploit the negative pressure in the intake manifold. Both methods rely on an interior-mounted switch for control.

Electric stepper motors are used as the actuators in *electrical systems*. These can be activated either by hand or by signals from the suspension-travel sensors.

Front fog lamps

Purpose
Fog lamps are designed to improve road-surface illumination under conditions marked by limited visibility (fog, snow, heavy rain and dust).

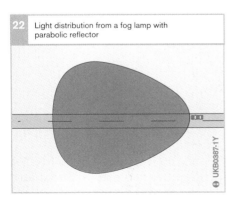
22 Light distribution from a fog lamp with parabolic reflector

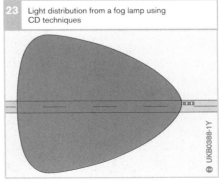
23 Light distribution from a fog lamp using CD techniques

Optical concept

Paraboloid

A parabolic reflector with a light source located at the focal point reflects light outward along a plane parallel to its axis (as with a high beam). The lens extends the beam to form a horizontal band (Fig. 22), while a screen prevents the emerging light rays from being projected upward.

Free-form technology (CD techniques)

With the help of CAL (Computer Aided Lighting) calculation methods, reflectors can be so designed that they scatter the light directly. That is, without optical profiling (fluting) of the lens. At the same time, they produce a sharp light-dark cutoff (delineation line between rays of light projected in different directions) without the necessity for separate screening. This multiple orientation of light rays forms the basis of CD (Converging/Diverging) technology.

The fact that the lamp bulb is extensively enveloped leads to an extremely high volume of light as well as maximum scatter width (Fig. 23, previous page).

24 "Compact 100" upright-mounted fog lamp

PES fog lamp

This concept minimizes backglare for drivers in fog. The lens projects the imageof a screen onto the road to furnish maximum contrast along the light-dark cutoff line.

Designs

External-fitted fog lamps incorporating their own optical elements can be mounted above the bumper (Fig. 24) or suspended below it (Fig. 25). Stylistic and aerodynamic considerations are leading to increased use of built-in fog lamps adapted to the shape of the bodywork, or fog lamps integrated within a combined lighting unit (with adjustable reflectors when the fog lamps are combined with the main headlamps).

Most fog lamps project a white beam; there is no substantive evidence that yellow lamps provide any special physiological benefits. A fog lamp's effectiveness depends upon the size of the illuminated area and the focal length of the reflector. Assuming identical illuminated areas and focal points, from the lighting-engineering viewpoint any differences between round and rectangular fog lamps is insignificant.

Regulations

Design is governed by ECE-R19, while mounting/installation specifications are set

25 "Pilot" pendant-mounted fog lamp

forth in 76/756/EEC, ECE-R48, StVZO (FMVSS/CUR) § 52.

Two fog lamps projecting white or yellow light are permitted.

The fog-lamp circuit must be separate frorn the circuits for low and high beams, i.e., it must be possible to switch the fog lamps on and off independently. The (German) StVZO allows fog-lamp installation in positions more than 400 mm from the widest point on the vehicle's periphery provided that the switching circuit ensures that operation is possible only in conjunction with the low beams. The adjustment procedures for front fog lamps mirror those used for the main headlamps. The setting dimensions e are contained in the specifications.

Auxiliary driving lamps

Purpose
Auxiliary driving lamps enhance long-range visibility with dual, quad and six-headlamp assemblies by generating a concentrated – and thus far-reaching – beam of light.

Optical concept
The optical principle is similar to that used for fog lamps, although from the lighting-engineering viewpoint, the lens is aligned to the high-beam requirements. Often, the size and shape of auxiliary driving lamps and fog lamps are identical.

Mounting and installation, regulations
Installation, lighting engineering, and adjustment procedures all correspond to those for for high beams. Auxiliary driving lamps are also subject to the same regulations defining maximum levels of luminous intensity, according to which the sum of all reference numbers for all the vehicle's driving lamps is not to exceed 75. The number 10 is used for general assessment purposes on older lamps for which no approval number is available.

Turn signals and flashers

Purpose
These lamps generate signal images to alert other road users to changes in direction (turn signals) and potential danger (hazard warning flashers).

Regulations
The StVZO (FMVSS/CUR) and EU Directive 76/756/EWG stipulate that the basic illumination equipment will be supplemented by optical signaling devices to indicate changes in direction and to serve as hazard flashers on all vehicles with a maximum speed in excess of 25 km/h (Fig. 26).

Flasher signals
The flasher signals are generated at a frequency of 60...120 pulses per minute and a relative illumination period of 30...80%. Light must be emitted within 1.5 sec of initial switch on. Should one bulb fail the remaining lamps must continue to produce visible light.

Turn signals
Turn signaling is in the form of synchronized signals which are generated by all

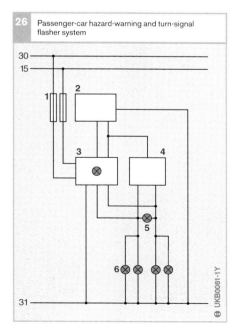

26　Passenger-car hazard-warning and turn-signal flasher system

Fig. 26
1　Fuses
2　Hazard-warning and turn-signal flasher
3　Hazard-warning switch with indicator
4　Turn-signal switch
5　Indicator lamp
6　Turn-signal lamps

flashers on one side of the vehicle. Lamp performance is monitored by an electronic circuit. Malfunctions are indicated by a warning lamp or a substantial change in the signal's flash rate.

Hazard-warning flashers
This mode is in fact synchronized flashing of all the vehicle's turn-signal lamps, and must remain available when the engine is switched off. An operation indicator is mandatory.

Turn-signal and flasher systems for vehicles without trailers
The electronic hazard-warning and turn-signal flasher includes a pulse generator designed to switch on the lamps via relay, and a current-controlled monitoring circuit which modifies the flash frequency in response to bulb failure. The turn-signal stalk controls the turn signals, while the hazard flashers are switched on with a separate switch.

Turn-signal and flasher systems for vehicles with/without trailers
This type of hazard-warning and turn-signal flasher differs from that employed on vehicles without trailers in the way that turn-signal operation is monitored.

Single-circuit monitoring
Tractor/towing vehicle and trailer share a common monitoring circuit that triggers two indicator lamps designed to flash at system operating frequency. If the first turn signal on the tractor or trailer fails, the first indicator lamp remains off. If the second turn signal also fails, the first and the second indicator lamp remain off. This configuration does not indicate the location (tractor/towing vehicle or trailer) of defective lamps. The flashing frequency remains unaltered.

Dual-circuit monitoring
Tractor/towing vehicle and trailer are equipped with separate monitoring circuits. Faulty flasher lamps can be located depending upon which indicator lamp remains off. The flashing frequency remains unchanged.

Fig. 27 Front turn-signal lamp integrated in a front headlamp assembly

Fig. 27
1 Turn-signal lamp

Front and side turn-signal lamps

Purpose

Turn-signal lamps indicate intended changes in direction (turn-signal function) and potentially dangerous situations (hazard-warning function). Design and location must be selected to ensure the lamps can always be clearly seen by other road users regardless of lighting and operating conditions (Fig. 27).

Regulations
76/759/EEC, ECE-R6, StVZO (FMVSS/CUR) § 54.
Group 1 turn-signal indicators (front-mounted), Group 2 (rear-mounted), and Group 5 (side-mounted) are specified for dual-track vehicles (Group 2 motorcycles).

The indicator lamp may be in any color desired. The flash frequency is defined as 90 ± 30 cycles per minute.

Front turn signals
Two amber-colored lamps are stipulated, and an indicator lamp must be fitted.

Side turn signals
Two amber-colored lamps are stipulated.

Side-marker and clearance lamps

Purpose
Side-marker and clearance lamps alert other traffic to the presence of large vehicles.

Regulations
76/758/EEC, ECE-R7, StVZO (FMVSS/CUR) §§ 51 and 53. Vehicles and trailers wider than 1,600 mm must be equipped with (forward-facing) side-marker lamps. Vehicles wider than 2,100 mm (such as trucks) must also be equipped with clearance lamps visible from the front.

Side-marker lamps
Two white-light side-marker lamps are specified.

Clearance lamps
Two white-colored clearance lamps are specified for installation at the front. Positions: As close as possible to the side extremities of the vehicle and at the greatest practicable height.

Front parking lamps

Purpose
Parking lamps are intended to ensure that stationary vehicles are visible for other road users. They must be suitable for operation without the need to switch on the headlamps or any other lighting equipment. The parking-lamp function is usually assumed by the side-marker lamps.

Regulations
77/540 EEC, ECE-R77, StVZO (FMVSS/CUR) § 51.

Vehicles may be equipped with two parking lamps at front and rear, or one parking lamp on each side. White light is prescribed to the front.

They must be suitable for operation without the need to switch on the headlamps or any other lighting equipment. The parking-lamp function is usually assumed by the side-marker lamps.

Daytime running lamps

Purpose
The light from special daytime-running lamps (or the low beam from the headlamps) is intended to make a vehicle more clearly recognizable during the day.

Regulations
ECE-R87.
In Europe, ECE-R87 governs the installation of daytime-running lamps.

The use of these lamps, or the headlamp low beam, is mandatory in Norway, Sweden, Finland, and Denmark. Poland and Hungary stipulate that the low-beam headlamps should be on for daytime driving.

Rear lighting system: Components

Various lamps and projection devices are installed at the rear of the vehicle; and their respective functions are described below. This section also describes the European regulatory framework applicable to both OEM units and aftermarket equipment intended for mounting or installation on the vehicle.

Backup (reversing) lamps
Purpose
These lamps are intended to illuminate the area to the rear of the vehicle when the vehicle is reversing.

Regulations
77/539/EEC, ECE-R23, StVZO (FMVSS/CUR) § 52.
One or a maximum of two white-light lamps may be installed (Fig. 1).

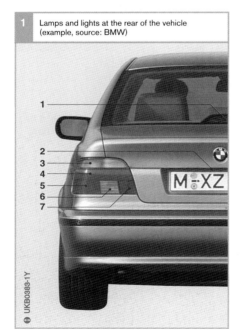

Fig. 1 Lamps and lights at the rear of the vehicle (example, source: BMW)

Fig. 1
1 High-mount stop lamp (CHMSL)
2 License-plate lamp
3 Turn-signal lamp
4 Parking lamp/tail lamp
5 Stop lamp
6 Backup lamp
7 Fog warning lamp

Bosch also has these lamps on offer for aftermarket requirements.
The switching circuit must be designed to ensure that the backup lamps operate only with reverse gear engaged and the ignition on.

Rear turn-signal lamps
Purpose
These lamps indicate intended changes in direction (turn-signal function) as well as potentially dangerous situations (hazard-warning-signal function). Design and location of the lamps must be such that their signals can always be clearly seen by other road users regardless of lighting and operating conditions.

Regulations
76/759/EEC, ECE-R6, StVZO § 54.
Group 2 (rear) flashers/turn signals are specified for dual-track vehicles (Fig. 1).
The dashboard-mounted monitoring lamp may be in any color desired.
The flash frequency is defined as 90 ± 30 cycles per minute.
Regulations call for two amber lamps.

Tail lamps and clearance lamps
Purpose
Tail lamps and clearance lamps are intended to provide following road users with an early warning of the vehicle's presence even when the brakes are not being applied.

Regulations
76/758/EEC, ECE-R7, StVZO §§ 51 and 53 (Fig. 1).
Rear-facing tail lamps are mandatory equipment on vehicles of all widths. Supplementary front and rear clearance lamps are also specified for vehicles wider than 2,100 mm (e.g., trucks).

Tail lamps
Two red-light tail lamps are mandatory equipment (Fig. 1).
The positions are the same as those prescribed for the rear turn-signal lamps.

When the tail and stop lamps are combined in a nested assembly, the luminous-intensity ratio for the individual functions must be at least 1:5. Tail lamps must operate together with the side-marker lamps.

Clearance lamps
Two red-light lamps visible from the rear are stipulated.
The clearance lamps must be positioned as far outward and as high as possible.

Rear parking lamps
Purpose
Parking lamps are intended to draw attention to a parked vehicle. They must be capable of operating when all other lamps are off. The parking-lamp function is usually assumed by the tail and side-marker lamps (Fig. 1).

Regulations
77/540/EEC, ECE-R77, StVZO § 51.
Two parking lamps at front and rear or one parking lamp on either side may be installed. Red is the color prescribed for the rear unit. Yellow (amber) is approved for use at the rear when the parking lamps are grouped with the side-mounted turn signals. Positioning is the same as prescribed for flashers and turn-signal lamps.

Stop (brake) lamps
Purpose
The purpose of the stop lamps is to alert following drivers to the fact that a vehicle is being braked.

Regulations
76/758/EEC, ECE-R7, StVZO § 53.

(Main) stop lamps
Two Category S1 or S2 stop lamps and a stop lamp Category S3 are mandatory for all passenger cars.
Two red-light stop lamps are prescribed as mandatory equipment on all vehicles.
When a nested design incorporating both the stop and tail lamps is used, the luminous-intensity ratio distinguishing the two functions must be a least 5:1.
The Category 3 high-mount stop lamp must not be nested with other lamps.

Supplementary high-mount stop lamps
Installation of a supplementary high-mount stop lamp situated at the vehicle's center (CHMSL, or Center High-Mount Stop Lamp) is mandatory for new vehicle models in Europe.
These center high-mount units must operate together with the conventional main stop lamps. Bosch also has aftermarket versions of this lamp available (Fig. 2).

Fog warning lamps
Purpose
The rear fog warning lamps are used to enable following drivers to detect unbraked vehicles when visibility is extremely restricted.

Regulations
77/538/EEC, ECE-R38, StVZO § 53 d.
The countries of the European Union prescribe two red-light fog warning lamps on all newly registered vehicles (Figs. 1 and 3). Bosch also has such fog lamps in its program for aftermarket installation.

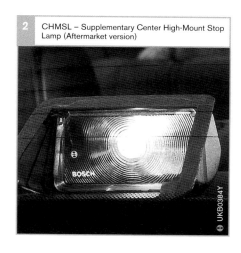

2 CHMSL – Supplementary Center High-Mount Stop Lamp (Aftermarket version)

The distance between the fog warning lamps and the stop lamps must be at least 100 mm (Fig. 3).

The visible illuminated area along the reference axis is not to exceed 140 cm². The electrical circuitry must ensure that the fog warning lamp operates only in conjunction with the low beam, high beam or front fog lamp. A provision must also be available for switching off fog warning lamps while the front fog lamps remain in operation.

Yellow is the prescribed color for the dashboard-mounted indicator lamp (green is also approved for vehicles initially registered prior to January 1981).

License-plate lamps
Purpose
The license-plate lamp is designed to make the vehicle's license plate visible to other road users.

Regulations
76/760/EEC, ECE-R4, StVZO § 60.
The lamp must ensure that the rear license plate is readable at a minimum distance of 25 m at night. Across the complete license-plate, luminance must be at least 2.5 cd/m². Measuring points are located on the license plate's surface between which the luminance gradient is not to exceed $2 \times B_{min}$/cm, whereby B_{min} is the lowest level of luminance measured at the measuring points.

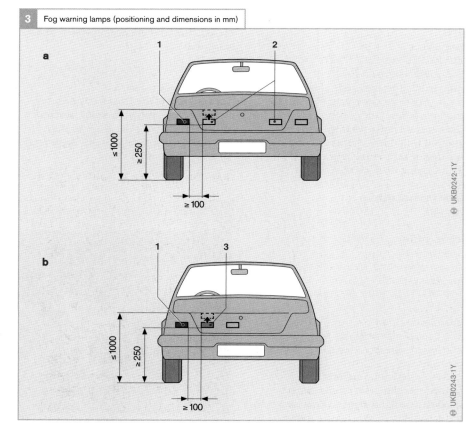

Fig. 3
a Rear of vehicle with two fog warning lamps
b Rear of vehicle with one fog warning lamp (positioning for lhd vehicle)

1 Stop lamp
2 Fog warning lamps (x2)
3 Fog warning lamp (x1)

Interior lighting system: Components

Inside the vehicle, the major emphasis for the control elements is on their operating and control efficiency, together with their providing the driver with adequate information on operating status while at the same time keeping driver distraction to an absolute minimum. These stipulations have priority over all other considerations. A well-lit instrument panel and the discrete illumination of specific function groups such as the radio or the navigation system is one of the prime prerequisites for relaxed and safe driving. The visual and audible signals passed to the driver must be graded in accordance with their urgency.

Interior lighting
Because there are no legal mandates specifying the character of interior lighting, vehicle manufacturers are free to equip their vehicles as they want. This results in a wide range of different lighting layouts.

Interior lamps
The interior lamp with switch settings for "on", "off" and "on with front doors open" has become the virtually universal standard. Supplementary rear ceiling lamps are also available which are switched on either by contact switches in the rear door pillars or by a dashboard-mounted switch.

Glove-compartment lighting
A contact activates the lamp(s) when the glove-compartment door is opened; there is no internal lighting when the door is closed.

Trunk lighting
Trunk lighting is now a standard passenger-car feature. The lamp is switched on and off by a contact switch operated by the trunk lid.

Instrument-panel illumination
Controls and display instruments in the dashboard and instrument cluster are illuminated to ensure that they are legible in the dark. The illumination for display instruments features a provision for automatic or manual adjustment, making it possible to adapt lighting intensity to actual conditions and avoid dazzling the driver with backglare. Indicator lamps in various colors are often used to monitor various operating conditions. Some of the colors are specified (such as blue for high beam, yellow for fog warning lamps, etc.). The identification symbols conform to a uniform symbol code valid throughout the ECE (refer to Fig. 1 for examples).

Controls and switches
In the interests of safety, official guidelines govern the design, installation and control/switching of all vehicular lighting equipment except the interior lamps. The object is to arrange the controls and switches to faciliate efficient operation with minimal driver distraction.

Illumination for controls and comfort and convenience equipment
The controls and comfort and convenience equipment (ventilation blower, heater and air

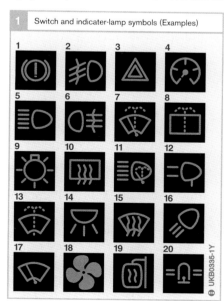

Fig. 1
1 Braking-system defect
2 Front fog lamp
3 Hazard-warning flashers
4 IInstrument-panel illumination
5 High beam
6 Fog warning lamps
7 Windshield wiper and washer
8 Rear-window washer
9 Main headlamp switch
10 Rear-window defroster
11 Headlamp wipe/wash
12 Spot lamp
13 Windshield washer system
14 Interior lighting
15 Windshield defroster
16 Floodlamp
17 Windshield wipers
18 Ventilation/heater fan
19 Heated mirror
20 Rotating beacon

conditioner, ashtray etc.) available to the vehicle's occupants must be illuminated either directly or indirectly. This is necessary so that they can be operated or reached in the dark, without extended searches that would distract the driver.

Switch illumination
Switch illumination fulfills two functions in the dark:
- It allows the driver to identify specific switches immediately (e.g., hazard warning flasher)
- Together with the ECE symbols (Fig. 1) it supports driver orientation

Frequent switching operations
Switches frequently used during normal driving must be designed to be within easy reach of the driver without it being necessary for him/her to take a hand off of the steering wheel. This stipulation is especially important for items such as the turn signals, the horn, the headlamp dimmer switch and the wipe/wash systems for windshield and headlamps. This is why in all vehicles these functions are united in combination stalks or switches on the steering column or within the steering wheel. There are still no standards specifying precise placement.

Occasional switching operations
Logical disposition of those switches that are used only rarely but still need to be accessible while the vehicle is being driven (headlamps, headlight leveling control, hazard-warning flashers, front fog lamps and rear fog warning lamps) makes a valuable contribution to active safety. The driver can locate these control elements "blindfolded" and can identify their function by touch alone, while remaining concentrated on the road and traffic.

Display elements
If operating conditions or equipment status are not indicated by illuminated switches, another option is to provide this information through indicator lamps or as a direct readout in a display panel.

Illuminated colored sectors and LEDs can furnish status reports (parking brake, lights, diesel preglow, etc.), while displays (liquid crystal) provide not only status messages but also the accompanying data (mileage, travel time, fuel consumption, remaining fuel, average speed, etc.).

Light-emitting-diode display
The **light-emitting diode**, or LED display is an active (self-luminous) device based on a

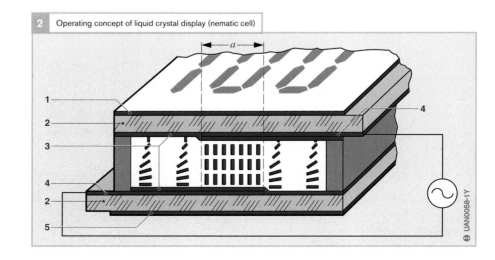

Fig. 2
1 Polarizer
2 Glass
3 Orientation and insulation layer
4 Electrodes
5 Polarizer (and reflector)

a Segment range

Fig. 2 Operating concept of liquid crystal display (nematic cell)

semiconductor element with a PN junction. During operation in the forward direction the charge carriers (free electrons and holes) recombine and release energy. Specific semiconductor substances then convert this energy into electromagnetic radiation with infrared wavelengths and in the visible portion of the spectral scale. Frequently used conductor materials are: gallium arsenide (infrared), gallium arsenide phosphide (red to yellow) and gallium phosphide (green).

Liquid-crystal display
The Liquid Crystal Display, or LCD is a passive display element. Visible contrast relies on supplementary illumination. The most widely used type of LCD is the twisted nematic, or TN cell (Fig. 2).

The liquid crystal medium is held between two glass plates. These glass plates are covered by a transparent electrically-conductive layer in the immediate vicinity of the display segment. Voltage can be applied to generate an electrical force field between the layers. An additional orientation layer rotates the light's plane of polarization passing through the cell. The cell's initial reaction to addition of polarizers applied at mutual right angles on the outside surfaces is to become translucent. Voltage application at the two opposed electrodes causes the liquid-crystal molecules to line up with the electrical field. This suppresses rotation of the polarization plane and induces opacity to form the actual display.

Separately controlled segment ranges can be used to portray numbers, letters and defined symbols. The only disadvantage is that a supplementary light source is needed to produce a readable display.

Light sources
Incandescent bulbs
Conventional systems rely on bulbs for illumination of the passive display elements. Color filters are used to modify the hue of the bulbs to meet operational and design requirements.

LED
Continuing advances in miniaturization and modular design techniques are causing the durability and installation advantages of the LED to appear in a new light. LEDs are available in red, green and yellow as well as blue.

Fluorescent lamps and fluorescent foils
Recent advances in the configuration of fluorescent lamps and foils permit, for instance, bright and extremely consistent backlighting for displays.

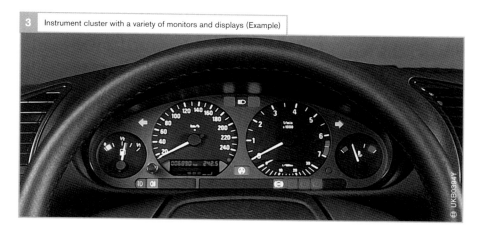

3 Instrument cluster with a variety of monitors and displays (Example)

Instrument clusters

Design

Microcontroller technology and the ongoing networking of motor vehicles have in the meantime transformed instrument clusters from precision-mechanical instruments to electronically dominated devices. A typical instrument cluster (LED-illuminated, with TN-type conductive-rubber-contacted segment LCDs, see Fig.) is a very flat component (electronics, flat stepping motors) and virtually all the components (mainly SMT) are directly contacted on a printed-circuit board.

Operating concept

While the basic functions are the same in most instrument clusters (see typical block diagram), the partitioning of the function blocks into (partly application-specific) microcontrollers, ASICs and standard peripherals sometimes differs significantly (product range, display scope, display types).

Electronic instrument clusters indicate measured variables with high accuracy thanks to stepping-motor technology, and also take over "intelligent" functions such as engine-speed-dependent oil-pressure warning, prioritized fault indication in matrix displays, or service-interval indicator. Even online diagnostic functions are standard and take up a significant part of the program memory.

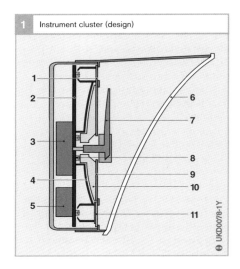

Fig. 1
1 Tell-tale lamp
2 Printed-circuit board
3 Stepping motor
4 Reflector
5 Plug
6 View cover
7 Needle
8 LED
9 Dial face
10 Optical waveguide
11 LCD

Because instrument clusters are standard features on all vehicle types and all the bus systems merge here anyway, the former are becoming increasingly established as gateways, i.e. bridges between the different bus systems in the motor vehicle (e.g. engine CAN, body CAN and diagnostic bus).

Measuring instruments

In Europe, manufacturers still tend towards the classical instrument with mechanical needle and dial face. Here, initially the compact, electronically triggered moving-magnet quotient measuring instrument replaced

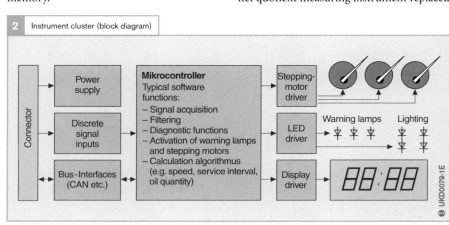

the bulky eddy-current speedometer. Geared stepping motors with very low overall depth and still higher precision are currently used. Thanks to a compact magnetic circuit and (mostly) 2-stage gearing with a power output of only approx. 100 mW, these motors permit swift and highly accurate needle positioning.

Lighting

Instrument clusters were originally lit by *frontlighting technology* in the form of *electric bulbs. Backlighting technology* has in the meantime gained acceptance on account of its attractive appearance. Bulbs have been replaced by long-lasting *light-emitting diodes* (LEDs, see P. 143). LEDs are also suitable for warning lamps and for the backlighting of scales, displays and (via plastic optical waveguides) needles (refer to table headed "Overview of lighting sources").

Highly efficient InAlGaP-technology LEDs are meanwhile available in the colors yellow, orange and red. The newer InGaN technology is currently producing significant efficiency improvements in the colors green, blue and white. Here the color white is obtained through the combination of a blue LED chip with an orange-emitting luminescent material (yttrium-aluminum granulate).

However, new technologies are also being used for special configurations:

- *CCFLs* (Cold Cathode Fluorescent Lamps): mainly for "black screen" instruments, which appear black when they are deactivated. The combination of a tinted view cover (e.g. 25% transmission) with these very bright lamps (high luminance, high voltage) produces a brilliant appearance with outstanding contrast. Since color LCDs have very low transmission (typically approx. 6%) it is imperative that CCFLs are used to backlight them in order to obtain good contrast even in daylight.
- *EL (Electroluminescent film):* AC is applied to this flat film in order for it to light up. It features very uniform light distribution, and has only recently become suitable for automotive applications. It offers extensive freedom of design for color combinations and/or for superimposing dial areas on display surfaces.

1 Overview of lighting sources

Lighting source	Possible colors	Typical Data [1]	Technically suitable for	Conventional ICs	Black-screen instruments	Service life [2] in h	Activation
Bulb	White (every color possible with filter)	2 lm/W, 65 mA, 14 V	Dial face / Needle / Display	+ / ○ / +	− / − / ○	$B_3 \approx 4{,}500$	No special activation required
SMD-LED luminescent diode	Red, orange, yellow (InAlGaP)	8 lm/W, 25 mA, 2 V	Dial face / Needle / Display	○ / + / +	− / + / −	$B_3 \geqslant 10{,}000$	Series resistors or control required
	Blue, green (InGaN), white (with converter)	2...5 lm/W, 20 mA, 3.6 V	Dial face / Needle / Display	○ / ○ / ○	− / ○ / −	$B_3 > 10{,}000$	
EL film electroluminescent	Blue, violet, yellow, green, orange, white	2 lm/W, 100 V~, 400 Hz	Dial face / Needle / Display	+ / − / −	− / − / −	approx. 10,000	High voltage required
CCFL cold-cathode lamp	white (every color possible depending on luminescent material)	25 lm/W, 2 kV~, 50...100 kHz	Dial face / Needle / Display	+ / − / +	+ / ○ / +	$B_3 \approx 10{,}000$	High voltage required

[1] Efficiency in lm/W (lumen per Watt), current in mA, voltage in V or kV, activation frequency in kHz.
[2] B_3 Point with 3% failure probability. Suitability: + preferred, ○ conditional, − no application.

Table 1

Display types

TN-LCD

With its high status of development, TN-LCD technology ("Twisted Nematic Liquid-Crystal Display") is the most commonly used form of display. The term stems from the twisted arrangement of the elongated liquid-crystal molecules between the locating glass plates with transparent electrodes. A layer of this type forms a "light valve", which blocks or passes polarized light depending on whether voltage is applied to it or not.
It can be used in the temperature range of −40 °C...+85 °C. The switching times are relatively long at low temperatures on account of the high viscosity of the liquid-crystal material.

TN LCDs can be operated in positive contrast (dark characters on a light background) or negative contrast (light characters on a dark background). Positive-contrast cells are suitable for front- and backlighting while negative-contrast cells can only be read with satisfactory reading contrast when strongly lit from the rear. TN technology is suitable not only for smaller display modules but also for larger display areas in modular or even full-size LCD instrument clusters.

Graphics displays for instrument clusters

Dot-matrix displays with graphics capabilities are needed to display infinitely variable information. They are activated by line scanning and therefore require multiplex characteristics. Under the conditions prevailing in a motor vehicle, conventional TN LCDs can today produce multiplex rates of up to 1:4 with good contrast and up to 1:8 with moderate contrast. Other LCD display technologies are needed to achieve higher multiplex rates. STN and DSTN technologies are now being used for modules with moderate resolution. DSTN technology can be implemented to provide monochrome or color displays.

STN LCD and DSTN LCD

The molecule structure of an STN LCD (**S**uper **T**wisted **N**ematic Liquid-Crystal Display) display is more heavily twisted inside the cell than in a conventional TN display. *STN LCDs* permit only monochrome displays; usually in blue-yellow contrast. Neutral color can be obtained by applying "retarder film", but this is not effective throughout the entire temperature range encountered in the vehicle. *DSTN LCDs* (**D**ouble-layer *STN* Liquid-Crystal Display) feature considerably improved characteristics, which permit neutral black-and-white reproduction over wide temperature ranges with negative and positive contrast. Color is created by backlighting with colored LEDs. Multicolor reproduction is created by incorporating red, green and blue thin-film color filters on one of the two glass substrates. Under automotive conditions, shades of gray are only possible to a very limited extent, the result of which is that the range of colors is limited to black, white, the primary colors red, green and blue and their secondary colors yellow, cyan and magenta.

AMLCD

The task of the visually sophisticated and rapidly changing display of complex information in the area of the instrument cluster and the center console with high-resolution liquid-crystal monitors with video capabilities can only effectively be performed by an *AMLCD* (Active-Matrix Liquid-Crystal Display).

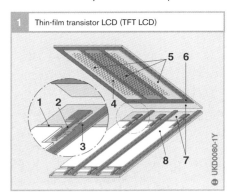

Thin-film transistor LCD (TFT LCD)

Fig. 1
1. Row circuit
2. Thin-film transistor
3. Column circuit
4. Front-plane electrode
5. Color layers
6. Black matrix
7. Glass substrate
8. Pixel electrode

The best developed and mostly widely used are the TFT LCDs (Thin-Film Transistor LCDs) addressed with thin-film transistors. Display monitors with diagonals of 4″...7″ in the center-console area and an extended temperature range (–25 °C...+85 °C) are available for motor vehicles. Formats of 10″...14″ with a still wider temperature range (–40 °C... +95 °C) are planned for programmable instrument clusters.

TFT LCDs consist of the "active" glass substrate and the opposing plate with the color-filter structures. The active substrate accommodates the pixel electrodes made from tin-indium oxide, the metallic row and column circuits and the semiconductor structures. At each intersecting point of the row and column circuits, there is a field-effect transistor which is etched in several masking steps from a previously applied sequence of layers. A capacitor is likewise generated at each pixel. The opposite glass plate accommodates the color filters and a "black-matrix" structure, which improves the contrast of the display. These structures are applied to the glass in a sequence of photolithographic processes. A continuous counter-electrode is applied on top of them for all the pixels. The color filters are applied either in the form of continuous strips (good reproduction of graphics information) or as mosaic filters (especially suitable for video pictures).

Head up Display (HUD)

Conventional instrument clusters have a viewing distance of 0.8...1.2 m. In order to read information in the area of the instrument cluster, the driver must adjust his vision from infinity (observing the road ahead) to the short viewing distance for the instrument. This process of adjustment usually takes 0.3...0.5 s. Older drivers find this process strenuous and in some cases, depending on their constitution, even impossible. HUD, a technology involving projection, can eliminate this problem. Its optical system generates a virtual image at such a viewing distance that the human eye can remain adjusted to infinity. This distance begins at approx. 2 m, and the driver can read the information with very little distraction, and without having to divert his eyes from the road to the instrument cluster.

Design

The HUD features an activated display for generating the image, a lighting facility, an optical imaging system and a "combiner", with which the image is reflected into the driver's eyes. The untreated windshield can also take the place of the combiner. Active displays (light emitters) such as CRTs or VFDs and passive displays (light modulators) are used for HUD. PDLC or DSTN displays are suitable but they do need very bright lighting because they absorb approx. 70 % of the transmitted light on account of the required polarizers.

Indication of HUD information

The virtual image should not cover the road ahead so that the driver is not distracted from the traffic or road conditions. It is therefore displayed in a region with a low road or traffic-information content. In order to prevent the driver from being overwhelmed with stimuli in his primary field of vision, the HUD should not be overloaded with information, and is therefore not a substitute for the conventional instrument cluster. It is however particularly well-suited to displaying safety-related information such as warnings, safety distance and route directions.

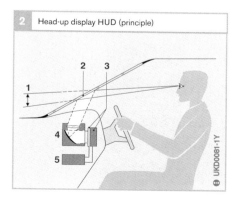

Fig. 2
1 Virtual image
2 Reflection in windshield
3 LCD and lighting (or CRT, VFD)
4 Optical system
5 Electronics

Special-purpose lamps

Special-purpose lighting equipment for motor vehicles makes a substantial contribution to highway security and occupational safety. Emergency identification lamps, floodlamps, and spot lamps alert other road users to potentially critical traffic situations and also make it possible to carry out vital and pressing jobs and operations in the dark.

Compliance with the regulations contained in the Highway Traffic Code (StVO) and the StVZO (FMVSS/CUR) is absolutely essential.

Emergency identification lamps

Emergency identification lamps must project what appears to be an intermittent, flashing beam of light around a 360° radius at a flashing frequency of between 2 and 5 Hz. *Blue emergency flashers* are approved for use on officially designated vehicles such as police cars, fire trucks and ambulances (Fig. 1). *Yellow flashers* are intended to warn of potential hazards (e.g., construction sites) and dangerous loads on trucks (e.g., overwidth or overlength cargo).

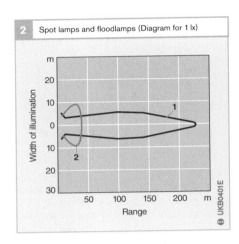

Spot lamps and floodlamps (Diagram for 1 lx)

Police vehicle with blue emergency identification lamp

A basic distinction is made between *mechanically-actuated rotating emergency beacons and electronically-controlled high-intensity discharge (HID) flashers*. Mechanical emergency beacons rely for their flashing effect on a reflector with belt or worm drive rotating around a bulb. HID flashers, on the other hand, use a high-intensity discharge tube to generate signal images with periodic gaseous-discharges.

Because HID devices remain fully operational under severe operating conditions, and resist extreme cold and contamination (no moving mechanical components), they are ideal for heavy-duty applications. The only critical element is the electronic control circuitry, which must also be capable of withstanding severe conditions (it must be impervious to vibration, condensation and water spray; and must have a high level of corrosion resistance).

Minimum luminous intensities are specified for emergency identification lamps: 20 cd for blue and 40 cd for yellow lights in a plane parallel to the road surface. In their light beams, blue emergency flashers must achieve at least 10 cd at ±4° and yellow lights a minimum of 20 cd at ±8°.

Floodlamps

Floodlamps provide consistent illumination for stationary or mobile job sites by provid-

Fig. 2
1 Spot lamp: Stretched beam of light over extended range (12 V/100 W)
2 Floodlamp: Large-area of light in the near range 12 V/155 W)

ing even illumination of large areas. Robust designs make these units particularly suitable for installation on heavy commercial vehicles, as well as for a wide variety of applications (construction, search and rescue, agriculture and forestry, shipping, etc, Figs. 2 and 3).

The use of floodlamps with the vehicle underway is approved only when vehicle travel is a part of the actual work process; examples are night-time road work and accident-recovery vehicles.

Spot lamps

Spot lamps project a concentrated, high-intensity beam in applications where recognition of objects at extended ranges is of prime importance (rescue operations, police and fire department activities, technical support, Figs. 2 and 4). Above all, spot lamps with transitionless focus are particularly effective in maintaining exact beam alignment at any range extending to approximately 225 m, allowing optimal exploitation of the unit's lighting power.

3 Illumination of construction sites or mobile working activities using a floodlamp

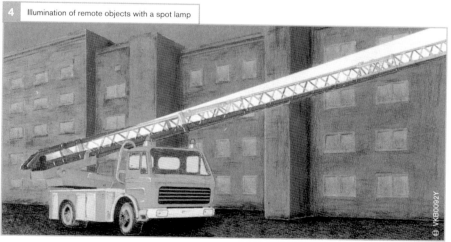
4 Illumination of remote objects with a spot lamp

Wiper and washer systems

It is the job of the vehicle's wipe/wash systems to always guarantee sufficient all-round visibility from inside the vehicle as dictated by legal regulations, irrespective of driving conditions.

Automotive wipe/wash systems are subdivided into the following categories:
- Windshield wiper systems
- Rear-window wiper systems
- Headlamp wiper systems
- Headlamp washer systems
- Combined washer and wiper systems

Windshield cleaning

Assignments and requirements
Dirty, wet windshields impair the driver's visibility. This fact makes windshield-wiper systems of paramount importance for safety on the roads (Fig. 1).
   A passenger car's wiper system must still be fully operational even after having performed 1.5 million wiping operations (the wiper blade must successfully complete 500,000). The figures for the wiper system equate to an area as large as about 200 football fields. Truck wiper systems must complete 3 million wiping operations.

This results in the following demands being made on a wiper system:
- The wiper and washer systems must completely free the windshield (and in some cases, the rear window) of rain, snow, and dirt (of mineral, organic, and biological form).
- In order to ensure that there is adequate visibility for registering traffic signs, and clearly seeing the road shoulder and traffic lights, the wipe pattern and therefore the wiped area must have a specific size as stipulated by legal regulations.
- Wipe quality must ensure that scattered light and the associated glare effects for oncoming traffic are avoided completely.
- The wiper system must be practically noiseless and must operate efficiently over very long periods at temperatures as high as +80° and as low as -30° C. Intermittent wiping must be possible or this function must be coupled to a rain sensor.
- The wiper system must be adequately corrosion-proof, and it must successfully complete a stall test.

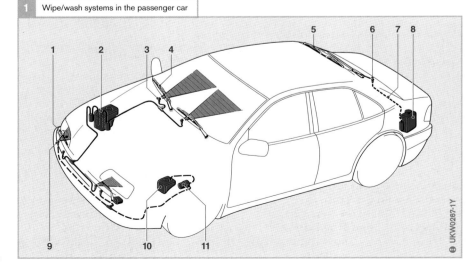

1 Wipe/wash systems in the passenger car

Fig. 1
1 Headlamp wipe/wash system
2 Pump with water reservoir (front)
3 Nozzle (windshield)
4 Windshield wiper
5 Rear-window wiper
6 Nozzle (rear window)
7 Rear-window washer system
8 Pump with water reservoir (rear)
9 High-pressure washer system (headlamps)
10 Water reservoir
11 High-pressure pump

Wiper and washer systems   Windshield cleaning   279

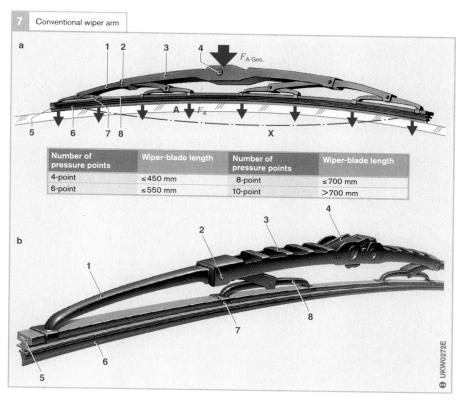

Fig. 7
a Wiper blade: Side view, under load (no load: Shape X)
b Wiper blade: Opposite side, under load

1 Intermediate bracket
2 Joint
3 Center bracket, perforated
4 Center rivet and adapter
5 Spring strip
6 Rubber wiper element (wiper rubber)
7 Claws
8 Claw brackets

$F_{A Ges}$ Total force on the center-bracket joint
F_A Force at the pressure points A

Number of pressure points	Wiper-blade length	Number of pressure points	Wiper-blade length
4-point	≤ 450 mm	8-point	≤ 700 mm
6-point	≤ 550 mm	10-point	> 700 mm

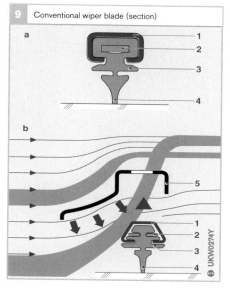

Fig. 8
Hook the wiper arm (1) in the "Quick-Clip" adapter (3) already installed in the wiper blade (2). Pull briefly, the adapter clicks into place and the new wiper blade is ready to go.

Fig. 9
a Basic version
b Spoiler version

1 Claw bracket
2 Spring strip(s)
3 Rubber wiper element (wiper rubber)
4 Wiper-element lip
5 Spoiler

Jointless wiper blade
(Aerotwin wiper blade)

The jointless wiper blade (Aerotwin wiper blade, Fig. 10) represents the state-of-the-art in wiper technology. The application pressure is no longer distributed by the claw brackets and the joints as on the conventional wiper-blade bracket. The jointless version uses two special spring strips (leaf springs) which have been specially formed to adapt to the vehicle's windshield. They ensure that the wiper-element lip is forced against the windshield with even pressure, and in doing so they reduce lip wear and improve wiping quality. The fact that the bracket system has been dispensed with means that joint wear is a thing of the past. A further advantage lies in the condiderably reduced installation height, less weight, and less noise (among other things, less wind noise).

The outside surface of the wiper blade is in the form of a spoilers (Figs. 10 and 11), and without any further measures being needed permits operation at very high speeds. This wiper blade's elastic material is also an ideal protection against injury in case of an accident involving a pedestrian.

The appropriate connection to the wiper arm not only ensures that the wiper blade is firmly held during operation, but also means that wiper-blade replacement is a simple matter.

Rubber wiper element (wiper rubber)

The wiper rubber is the most important element of the wiper system. The wiper rubber's micro double edge is only 0.01...0.015 mm thick and moves over the surface of the windshield. In the case of conventional wiper blades, pressure is applied to the wiper rubber through the bracket-system claws, and it is strengthened by spring strips. The Aero wiper blade only uses pre-formed spring strips. Considering the coefficients of dry friction in the range of 0.8...2.5 (depending upon air humidity) and coefficients of wet friction in the range of 0.6...0.1 (depending upon the friction speed), the correct pairing of the lip profile and the rubber's technical characteristics mean that the wiper-

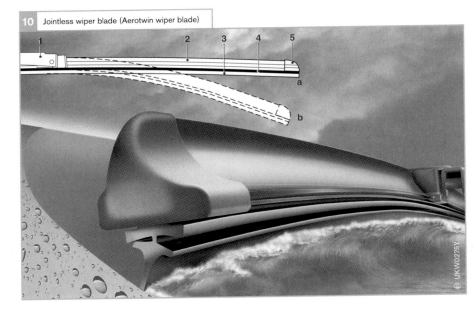

Fig. 10 Jointless wiper blade (Aerotwin wiper blade)

Fig. 10
a Loaded wiper blade seen from the side
b Contour of the non-loaded wiper blade

1 Wiper arm
2 Spoiler
3 Wiper-blade element (wiper rubber)
4 Spring strip
5 End clip

the size of the wiped area (Fig. 6b). The system also detects such hindrances as the heaps of snow which accumulate at the wiper-blade reversal point. The control prevents system blockage and resulting damage by automatically reducing the size of the wiped area. Lowering the motor speed just before the reversal points leads to quieter running.

There are also advantages inherent in the combination of a *rain sensor* with the electronic control of the rotational speed. The drive-motor speed can then be varied infinitely as a function of the amount of rain on the windshield

The *extended parking position* is another supplementary function. Here, the wiper disappears below hood level when the wipers are switched off (Fig. 6b, item 2). Parking the wiper blades in this manner improves the vehicle's dynamics, reduces the wind noise, and ensures that the driver's field of vision is not impaired. At the same time, the risk of injury for pedestrians and cyclists is reduced should they be involved in an accident with the vehicle.

The electronically controlled drive motor is also suitable for powering two-motor wiper systems in which each wiper arm has its own

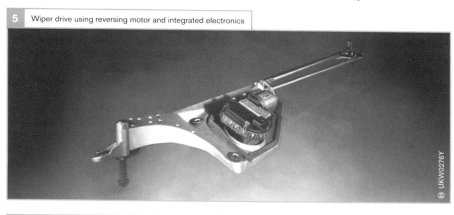

5 Wiper drive using reversing motor and integrated electronics

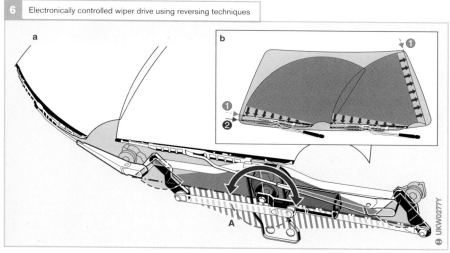

6 Electronically controlled wiper drive using reversing techniques

Fig. 6
a Electronic wiping-angle control (reduces the installation space A)
b Extended wiping area

1 Upper/lower reversing point
2 "Hidden" (extended) parking position

motor. The electronic control ensures that the sequence of motions is coordinated. This system has no connecting rods and linkages, and compared to conventional systems thus needs far less space and is much lighter.

Centrifugal pumps and water reservoirs

Washer systems are imperative for efficient cleaning of windshields and headlamps. These system use electrically powered centrifugal pumps of very simple design (Figs. 7 and 8) to force a pointed jet of water (with cleaning additive) through 2…4 nozzles and onto the windshield or headlamp lens. Fig. 9 shows a typical pump characteristic curve.

Normally, the reservoirs for cleaning water contain between 1.5 and 2 l. If one and the same reservoir is used for windshield and headlamp cleaning, it can have a capacity of up to 7 l. It is possible to have a separate reservoir for the rear-window wiper.

Very often, the washer system and the wiper system are coupled electronically so that when a knob is pressed water is first sprayed onto windshield or headlamp lens before the respective wiper system goes into operation for a number of cycles.

Fig. 7
1 Intake fitting
2 Impeller
3 Pump housing
4 Pressure fitting
5 DC motor

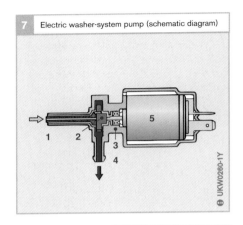
Fig. 7 Electric washer-system pump (schematic diagram)

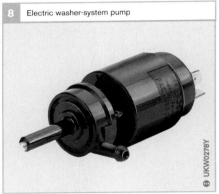
Fig. 8 Electric washer-system pump

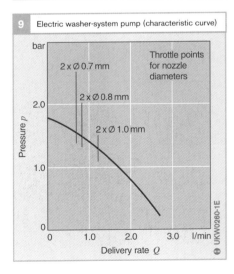
Fig. 9 Electric washer-system pump (characteristic curve)

A brief history of windshield cleaning

The windshield-wiping problem cropped up as soon as the first vehicles were equipped with a windshield to protect the driver against road dust and wind. And this inevitably then led to the necessity for freeing the windshield of rain, snow, and dirt.

The first device to be used for this job though, was not the wiper blade as we know it today but rather a straight slat. Prince Heinrich, the younger brother of the German Kaiser Wilhelm, was a passionate automobilist and in 1908 applied for a patent for his hand-operated wipe slat. In order to clean the windshield the driver simply pulled the wiper slat up and down and wiped off the rain, but every time he did so he was distracted from the job of driving and could only steer with one hand. And the pneumatically powered windshield wipers installed in America as from 1916, had the marked disadvantage that their wiping speed dropped the faster the engine turned or the higher the vehicle's road speed.

In 1926, Bosch was the first on the market with a system which used electric motors and wiped across the windshield at constant speed. The system cost 36 Reichsmarks, and its electric motor moved a rubber-covered wiper arm across the windshield at a speed which was independent of the engine. Drive was through a worm and gear transmission, and the system operated with 30 wiping cycles per minute. A mere 16 years later, there were already 32 variants of this system.

The major advances though came after World War II along with the emergence of mass motoring. In 2001, Bosch was able to look back on 75 years of development in windshield cleaning.

Wipe/wash milestones

1908	Patent awarded for wipe-off slat
1926	Bosch introduces its first wiper
1950's	The first wiper with bracket appeared (for curved windshields)
1959	Windshield washers
1960's	Trapezoidal-shaped rubber wiper elements using micro double edge, streamlined wipers
1971	Intermittent-wiper switch
1973	Torsionally stable, rivetless wiper blades
1975	Rear-window wipers
1982	Heatable double nozzle for windshield washer systems
1984	Winter wiper blades
1985	"Tandem" double wiper blades
1986	Spoiler wiper blades
1987	Electronic wiper-blade application-pressure control
1989	Rain sensor
1990's	Twin two-component technology
1995	Wiper arm with elongation control
2000	Jointless Aerotwin wiper blade (1st generation)
2001	Aerotwin (2nd generation)

Prince Heinrich and his wipe-off slat

Jointless Aerotwin wiper blade

Automotive microelectronics

Microelectronics have revolutionised motor vehicle technology. Initially, mechanical components were replaced by electronic devices in order to make systems more reliable. This was the case with the contact breaker points in the conventional coil ignition system, for example. Gradually, however, more and more new vehicle systems were introduced which simply would not have been possible without the use of electronics. The impetus for these new developments was provided by increasingly demanding requirements placed on the exhaust-emission characteristics of the internal-combustion engine (e.g. emission-control systems), on comfort and convenience (e.g. climate control and navigation systems) and on safety (e.g. antilock braking system [ABS] and airbags).

Overview

Cars and commercial vehicles equipped with the latest available technical improvements are absolutely brimming with electronic systems. Those systems can be subdivided into the following areas of application:
- Engine and drivetrain
- Safety
- Comfort and convenience
- Communication and multimedia

Fig. 1 provides an overview of the electronic systems that can be found on modern motor vehicles. Many of the systems referred to are now standard equipment on all new vehicles. By contrast, there are others that remain (as yet) the preserve of the most expensive luxury models.

Electronic systems can be subdivided into the following functional areas:
- Sensors and setpoint generators
- Control units (ECUs)
- Actuators
- ECU communication links (networks) and
- Electronic diagnosis

Sensors and setpoint generators
Sensors detect operating conditions (e.g. engine speed, wheel speed, temperature). They convert physical variables into electrical signals. Setpoint generators (e.g. controls operated by the driver) specify desired settings.

Control units (ECUs)
Control units process the information received from the sensors and setpoint generators using specific mathematical calculation sequences (control algorithms). They control the actuators by means of electrical output signals. Control units also form the interface with other systems and the vehicle diagnostics.

Actuators
Actuators convert the electrical output signals from the control unit into physical variables. Examples of actuators are:
- Gasoline-engine fuel injectors
- Diesel-engine fuel injectors
- Electric motors (e.g. for driving the power-window regulator mechanism or as throttle-valve positioner on vehicles with ETC (Electronic Throttle Control))
- Fans

Networks
As the number of electronic systems in vehicles grows, so does the amount of interconnecting wiring needed. The total length of the wiring in the wiring harness of a midrange car is now roughly 1.6 km on average and incorporates up to 300 connectors with a total of around 2000 connector pins.

Networking of the various systems reduces the overall length of the wiring required. A shared data bus consisting of only two wires (e.g. CAN bus) carries data that is read by all bus users. Depending on the specific requirements, a vehicle may also have separate busses for engine and drivetrain, comfort and convenience systems, and communication systems.

Another advantage of the bus system is that sensor signals need only be analysed by a single control unit. For example, the instrument cluster can use the individual wheel-speed signals delivered by the ABS control unit to calculate the vehicle's road speed. This complex calculation even takes account of cornering differences and spinning wheels. The road-speed signal is transferred on the CAN bus to all other users (e.g. the ABS control unit which uses it for controlling brake application, the engine-management ECU which uses it as the basis for cruise control, or the car radio for the speed-related volume adjustment).

Electronic diagnosis

Electronic diagnosis functions on the control unit continuously monitor the operation of the system and its components. Any faults that occur (e.g. short circuits in the wiring, sensor failure) are stored in the control unit's fault memory. Those stored faults can then be read out in the course of service using a dedicated system tester that is connected to the control unit's diagnosis interface. The system tester can also be used to scan sensor signals and send commands to actuators to perform specific operations. Electronic diagnosis thus enables faults to be located more quickly and easily in the course of vehicle service.

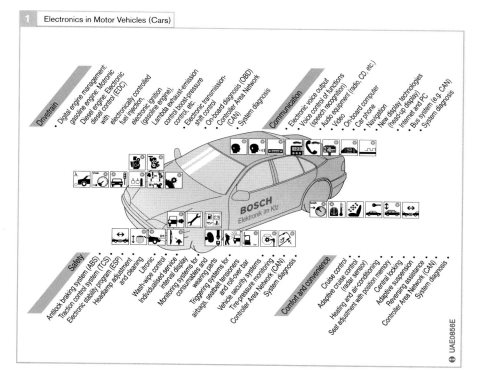

1　Electronics in Motor Vehicles (Cars)

Demands on electronic systems

Electronic systems in motor vehicles are exposed to extreme stresses (e.g. due to extreme temperature variations, unusual climatic conditions, poor road surfaces and the effect of corrosive substances). These are some of the requirements they must meet in order to be able to function reliably and without faults over long periods:

- Resistance to temperatures ranging from −40 °C ... 125 °C
- EMC (electromagnetic compatibility): immunity to external interference (e.g. mobile phone signals) and no emission of electromagnetic radiation likely to cause interference on other equipment
- Resistance to shocks and vibration
- Resistance to water and damp
- Resistance to corrosive fluids (e.g. oils and salt-water spray)
- Light weight
- Economical production costs and
- Secure and trouble-free mounting

History of development

The amount of electronic equipment in motor vehicles is continually increasing. Fig. 2 provides an overview of the growth of electronic equipment expressed as a proportion of vehicle cost.

Because of their cost, electronic systems were initially reserved for vehicles at the luxury end of the market. This explains why in 1980 electronic equipment accounted for only half a percent of vehicle cost. From that time onwards, and particularly in the 1990s, that proportion grew rapidly as the price of electronic equipment continually dropped with the result that more and more systems could be fitted to mid-range and even small cars.

Gradually, more and more electronic systems were fitted to motor vehicles (Table 1). And the trend continues.

The new science of mechatronics deals with the interaction between mechanical, electronic and data processing devices.

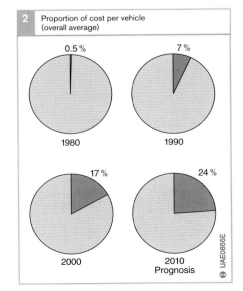

2 Proportion of cost per vehicle (overall average)
1980: 0.5 %
1990: 7 %
2000: 17 %
2010 Prognosis: 24 %

	Historical development of electronic systems in motor vehicles (examples)
1958	DC generator with variode
1962	3-phase alternator with variode
1965	Transistorised ignition
1967	D-Jetronic gasoline-injection system (pressure-controlled)
1978	Antilock braking system (ABS)
1979	Motronic (combined ignition and fuel-injection system)
1982	Electronic ignition system
1982	Knock control
1986	Electronic diesel control (EDC)
1986	Electronic throttle control (ETC)
1987	Traction control system (TCS)
1989	Electronic transmission-shift control ("stand-alone" system Tiptronic)
1989	CAN (Controller Area Network)
1989	Vehicle navigation system (Travelpilot)
1991	Litronic
1994	ME-Motronic (integrated ETC)
1997	Electronic stability program (ESP)
2000	MED-Motronic (gasoline direct injection)
2000	Adaptive Cruise Control

Table 1

Miniaturisation

Transistorised ignition was first used on gasoline engines in 1965. It did away with the negative effect on ignition timing accuracy of erosion caused by electrical arcing between the contact-breaker points. The transistor had gained a foothold in the motor vehicle, heralding the start of the electronic age. But is wasn't until electronic componentry was miniaturised that the decisive step was taken in making electronic systems in automobiles capable of the levels of performance that are taken for granted today. Enormous advances in miniaturisation were made in the area of semiconductor components in particular, making it possible to integrate more and more functions within a component that occupied only a tiny amount of space. Every ECU contains microcontrollers that combine millions of transistor functions on semiconductor chips that take up only a few square millimetres.

It has also been possible to substantially reduce the dimensions of power components such as output stages for controlling actuators. For example, multiple ignition output stages are now combined in a single component. This means that an external ignition output stage is no longer required. It is now integrated in the engine-management ECU. Consequently, the external ignition output stage previously used can be dispensed with. The associated reduction in the number of components also improves the reliability of the system.

The size of discrete components (resistors, capacitors) has similarly been significantly reduced. SMDs (**S**urface **M**ounted **D**evices) are soldered or bonded to the circuit board without wire connections.

In spite of the continual growth in the number and complexity of the functions performed, the miniaturisation of electronic equipment has meant that the size of the ECUs continues to shrink (Fig. 3).

Memory capacity

Whereas a memory capacity of 4 kilobytes was adequate for the modest requirements of, for instance, a management system for a gasoline engine in the late 1970s, 10 years later the figure had reached 30 kilobytes. The incorporation of more and more functions in the engine-management ECU led to an explosion in the demand for memory capacity. By the year 2000, the required capacity had reached 500 KB. Other automotive electronic systems have followed a similar pattern of development. And there is no foreseeable end to this trend.

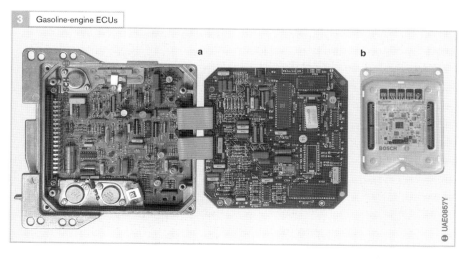

Fig. 3
a 1979 Jetronic ECU with 290 components and a weight of 1.14 kg
b 1996 Motronic hybrid ECU with 82 components and a weight of 0.25 kg

Basic principles of semiconductor technology

Semiconductors occupy a unique position between conductors and non-conductors. The electrical conductivity of semiconductors is dependent on pressure, temperature, intensity of incident light or the proportion of foreign atoms in the semiconductor material. Those properties are the basis for semiconductor technology.

Terminology

Electronics
According to the definition of the IEC (International Electrotechnical Commission), electronics is the branch of science and technology that deals with the study and utilisation of the physical phenomena in gases, solids and vacuums that are related to the flow of electricity.

Microelectronics
According to DIN 41 857, microelectronics is a branch of technology that deals with the conception, design, technology, manufacture and use of highly miniaturised electronic circuits.

It is evident from that definition that microelectronics is concerned with miniature circuits made up of multiple individual components rather than miniaturised discrete components (i.e. components with clearly definable individual functions). Only integrated film and semiconductor circuits and composite microcircuits (hybrid circuits) fall into that category.

Electrical conductivity

The specific suitability of different materials for conduction of electricity is determined by the number and mobility of free charge carriers that they contain. The electrical conductivity of solids at room temperature can vary by 24 powers of ten between different materials.

Solids are subdivided into three classes of material according to their electrical conductivity (Table 1).

Table 1. Classes of material based on conductivity (with examples)

Conductors (Metals)	Non-conductors (Insulators)	Semiconductors
Silver	Teflon	Silicon
Copper	Quartz glass	Germanium
Aluminum	Aluminum oxide	Gallium arsenide

All solids contain around 10^{22} atoms per cubic centimetre which are held together by electrical forces.

Conductors (Metals)
In metals, the number of free charge carriers is very large (one or two free electrons per atom). Their level of mobility is moderate. The electrical conductivity of metals is high. In good conductors it can be as much as 10^6 siemens/cm.

Non-conductors (Insulators)
In insulators, the number of free charge carriers is practically zero and consequently the electrical conductivity virtually non-existent. The conductivity of good insulators is of the order of 10^{-18} siemens/cm.

Semiconductors
The electrical conductivity of semiconductors is somewhere between that of conductors and insulators. Under normal conditions they also have a very small number of free charge carriers, but that number can be substantially increased by the application of energy from an external source. Their conductivity, therefore – in contrast to that of metals

and insulators – is heavily dependent on:
- Pressure (affects the mobility of charge carriers)
- Temperature (affects the number and mobility of charge carriers)
- Exposure to light (affects the number of charge carriers) and
- Added impurities (affects the number and type of charge carriers)

The sensitivity of semiconductors to pressure, temperature and light makes them suitable for use as sensors.

The ability to accurately modify and localise the conductivity of semiconductors by the controlled introduction of impurities that affect electrical properties (doping) is the basis of semiconductor technology. The electrical conductivity that can reliably be brought about in silicon by doping ranges from 10^4 to 10^{-2} siemens/cm.

As silicon is by far the most important semiconductor material, the explanations that follow will restrict themselves exclusively to that material. When solid, silicon consists of a crystal lattice in which each silicon atom is linked to four equally spaced adjacent atoms. Every silicon atom has four outer electrons (Fig. 1). Each pair of adjacent atoms is linked together by two shared electrons. In such a perfect crystal lattice, therefore, there are no free charge carriers, which means that the silicon is a non-conductor. This condition is changed fundamentally by the introduction of suitable impurities (doping), or energy from an external source.

n-type doping
The addition of foreign atoms with five outer electrons (e.g. phosphorus) introduces free electrons because only four are required to bind each atom within the silicon crystal lattice. Each phosphorus atom introduced therefore supplies one free, negatively charged electron. The silicon becomes negatively conductive (Fig. 2). It is then referred to as n-type silicon.

p-type doping
The addition of foreign atoms with three outer electrons (e.g. boron) creates electron gaps. The boron atom is one electron short of the number required to properly bind it within the silicon crystal lattice. The result-

1 Structure of a perfect, pure-silicon crystal lattice with four outer electrons (valance electrons •)

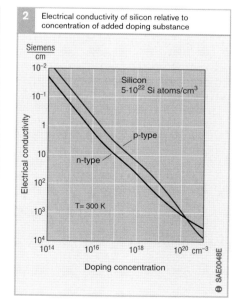

2 Electrical conductivity of silicon relative to concentration of added doping substance

ing gaps are referred to as "holes". In silicon, those holes are mobile, and within an electrical field they move in the opposite direction to electrons. Holes therefore act like free positive charge carriers. Each boron atom introduced therefore supplies one free, positively charged hole. The silicon becomes positively conductive (Fig. 2) and is therefore referred to as p-type silicon.

The conductivity of n-type silicon is around 10 times as high as that of p-type silicon assuming other factors are equal (e.g. number of foreign atoms, temperature, pressure) because electrons can move about 10 times as fast as holes.

Intrinsic conductivity

The application of heat or light can generate free charge carriers even in undoped silicon. They consist of electron-hole pairs and make the semiconductor intrinsically conductive. Such conductivity is generally low compared with that produced by doping. The number of electron-hole pairs increases exponentially with rising temperature and ultimately erases the electrical differences between areas of p-type and n-type silicon created by doping. Consequently there are maximum limits for the operating temperatures of semiconductor components, as shown in the following table:

Material	Max. operating temperature
Germanium	90...100 °C
Silicon	150...200 °C
Gallium arsenide	300...350 °C

An n-type semiconductor always has some holes and a p-type semiconductor always has some free electrons. Such minority charge carriers are fundamental to the way in which almost all semiconductor components work (refer to section entitled "Electronic Components").

p-n junction

The boundary between the p-type and n-type regions of the same semiconductor crystal is called the p-n junction. Its properties are fundamental to almost all semiconductor components.

p-n junction without external voltage

In the p-type region there are a large number of holes (○) and few free electrons. In the n-type region, by contrast, there are extremely few holes and a large number of free electrons (●) (Fig. 3). Due to the concentration differentials, the mobile charge carriers in each region diffuse into the other region in each case (diffusion currents). As a result, the p-type region is negatively charged and the

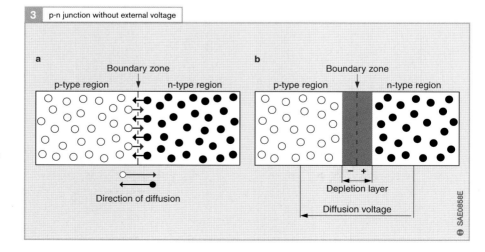

Fig. 3
a   Diffusion of mobile charge carriers
b   Formation of depletion layer (space-charge region)
○   Holes
●   Electrons

n-type region positively charged. A potential difference (diffusion voltage) is thus created between the p-type region and the n-type region which counteracts migration of the charge carriers. This brings the equalisation of holes and electrons to a halt. As a result, a region that is deficient in mobile charge carriers, and therefore has poor electrical conductivity, is created at the p-n junction. It is referred to as the depletion layer or space-charge region. Due to the diffusion voltage, the depletion layer has as strong electric field.

p-n junction with external voltage
If an external voltage is applied to a p-n junction, it produces the effects explained below (Fig. 4).

Reverse bias
If the negative terminal is connected to the p-type region and the positive terminal to the n-type region, the width of the space-charge region increases. Consequently, current flow is largely inhibited apart from a very small residual current (reverse current) produced by minority charge carriers.

Forward bias
If the positive terminal is connected to the p-type region and the negative terminal to the n-type region, the depletion layer is broken down. When the diffusion voltage is exceeded, the charge carriers "flood" the p-n junction and a large current flows in forward direction.

Breakdown voltage
The breakdown voltage is the reverse-direction voltage above which a small increase in voltage brings about a steep rise in the reverse current.

The cause of this effect is the release of bound electrons from the crystal lattice in the space-charge region due to the high field strength (Zener breakdown) or due to surges of accelerated electrons. The accelerated electrons strike other electrons, breaking them free of their bonds and starting an avalanche-like increase in the number of charge carriers ("avalanche breakdown" or "first breakdown"). Both effects are reversible. The nature of the cause and the level of the breakdown voltage are dependent on the doping concentration profile.

A second breakdown occurs if there is localised heating of a semiconductor component caused by current constriction so that the area concerned becomes more conductive. This results in a self-accelerating increase in current and leads to the destruction of the semiconductor component.

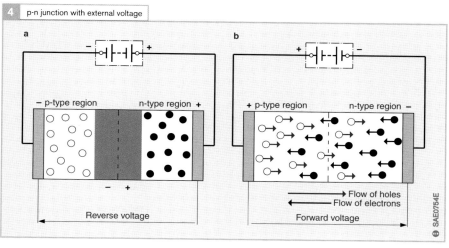

Fig. 4
a Reverse bias
b Forward bias

Electronic components

Electronic devices are made up a large number of components that can be subdivided into two main categories: passive components and semiconductor components (active components); the latter can be further subdivided into subcategories.

Passive components

Ohmic resistors, capacitors and inductances are classed as passive components.

Ohmic resistors
Ohmic resistors are generally made of materials with conductive properties similar to metals e.g. carbon (carbon-film resistors) or special metal alloys (metal-film resistors). They are constructed so as to reduce to the absolute minimum the effect of voltage, current and temperature on their electrical resistance. The conductor has a very small cross-section relative to its length, this being achieved either by the application of thin films to insulators or by winding wire into a coil.

In electronic circuits, resistors limit the current flow I or produce a voltage U proportional to the current. The resulting energy loss is converted into heat. An extreme example of this is a light bulb, in which an ultra-fine tungsten filament is heated to such a degree that it starts to glow.

The resistance in a circuit is referred to by the symbol R and its unit of measurement is the ohm (Ω).

Capacitors
The most simple type of capacitor consists of two parallel metal plates that are not in contact with one another. The area and separation of the plates as well as the medium separating them (dielectric) determine the quantity of charge carriers that can be stored by the capacitor (its capacitance). Using a vacuum (or air) as the dielectric offers the lowest capacitance. The capacitance can be substantially increased by the use of other insulating materials. The "amplification factor" is referred to as the relative permittivity or dielectric constant ε_r. Since, in many cases, very large plates would be required for the frequently used capacitance levels, capacitors are normally made by winding long strips into a coil (wound capacitors) or packing a large numbers of small plates together (multi-layer capacitors). The electrolytic capacitor uses a thin layer of oxide as the dielectric. This method makes it possible to manufacture small capacitors with large capacitances.

When direct current is applied to a capacitor, it stores up charge until the limit of its capacitance is reached; at that point current can no longer flow. Thus – for a certain period at least – the capacitor stores electrical energy that is then available if the external power supply fails. This effect is utilised in the circuit for triggering the airbag, for instance. Even if the wires connecting the airbag triggering unit to the vehicle's battery are severed in a serious accident, the capacitor still holds sufficient electrical energy to ensure that the airbag is deployed.

In alternating-current circuits, a capacitor has a similar effect to a resistor. Its resistance is dependent on the capacitance and the frequency of the alternating current. As the frequency decreases, the resistance increases. If the frequency is reduced to its lowest limit, i.e. zero (direct current), the resistance is equal to infinity and, consequently, no current flow is possible. This relationship is utilised, for example, by frequency filters in speaker systems in order to filter out the high-frequency sounds from the lower-frequency ones (high-pass).

The capacitance in a circuit is referred to by the symbol C and its unit of measurement is the farad (F).

Inductances

When an electric current flows through a coil, a magnetic field is created. The strength of the magnetic field depends on the strength of the current, the number of windings in the coil and the properties of the coil core (generally a ferrite or iron core). Inductance is the characteristic of a coil which indicates the amount of magnetic energy the coil can store for a given current.

If the current or the magnetic flux changes, a voltage is induced in the coil which counteracts the generation of the magnetic field. Once the magnetic field of a coil carrying direct current has reached its maximum strength, current flow is no longer restricted. Thus inductance does not represent a lasting hindrance to the flow of direct current.

In an alternating-current circuit, because of the constant generation and collapse of the magnetic field (and the energy contained within it), an inductance acts as a frequency-dependent resistor, the characteristics of which, however, are exactly the opposite of those of a capacitor. In this case, the higher the frequency, the greater is the resistance. This means that in frequency filters such as those referred to above in the description of capacitors, inductors can be used to filter out the low-frequency sounds from the high-frequency ones (low-pass).

The inductance in a circuit is referred to by the symbol L and its unit of measurement is the henry (H).

Semiconductor components

Semiconductor components are active components that are generally subdivided into four categories (Table 1). These categories are further subdivided according to the way in which the components are manufactured and their function. The first two "classical" categories are discrete semiconductor components and monolithic integrated circuits (see Table 1 for examples).

Discrete semiconductor components are self-contained, individually distinguishable components (discrete being derived from "discretus", the past participle of the Latin verb "discernere" meaning "to distinguish"). According to the currently accepted definition, semiconductor components with fewer than 100 transistor functions are referred to as discrete.

Monolithic integrated circuits
ICs (**I**ntegrated **C**ircuits) are active components which incorporate more than 100 individual functions on a single chip (mono-

Table 1. Semiconductor components (Examples)

Discrete active components
– Diodes
– Semiconductor resistors
– Transistors
– Thyristors
Integrated circuits (ICs)
– Analog circuits
– Digital circuits
– Mixed-signal circuits
Opto-electronic components
– Photoresistor
– Photodiode
– Photovoltaic cell
– Laser diode
– Phototransistor
– Charge-coupled device
Micromechanical sensors (Examples)
– Pressure sensors
– Acceleration sensors
– Yaw rate/angle sensors
– Flow sensors
– Temperature sensors
– Position/angle sensors (Hall-effect sensors)
– Gas sensors

lithic literally meaning "made from a single stone" from the Greek "monolithos" meaning "single stone").

Optoelectronic components form the third category. They are so distinctly different from the classical active components in terms of their method of production and usage, that classification as a separate category makes sense.

Micromechanical sensors using MST (Micro-System Technology) or MEMS (Micro ElectroMechanical Systems) have more recently come into being as a fourth category.

In this case too, the methods of production and the type of use differ substantially from conventional active components.

Diodes

Diodes are semiconductor components with a p-n junction and two connections – one to the p-type region and one to the n-type region ("diode" means "two ways"). A diode utilises the characteristics of the p-n junction. The pattern of doping impurity concentration within the crystal determines the specific characteristics of diodes.

Diodes designed for a forward current of more than 1 A are referred to as power diodes.

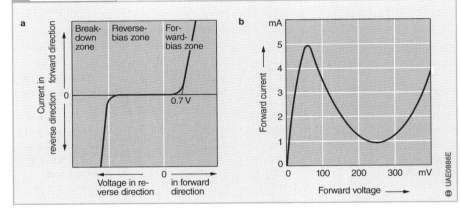

Fig. 1
a Characteristic of diodes, e.g. rectifier diode, Zener diode and Schottky diode
b Section of tunnel-diode characteristic

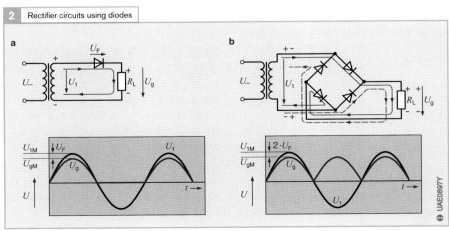

Fig. 2
a Half-wave rectifier
b Bridge rectifier

U_F Diffusion voltage (approx. 0.7 V)
$U_\sim$ AC voltage
U_1 Transformed AC voltage
U_g Rectified voltage
R_L Load resistance
U_{1M} Amplitude of U_1
U_{gM} Amplitude of U_g
t Time

Rectifier diode
A rectifier diode allows current to pass in one direction (positive terminal connected to p-type region) but not in the other (positive terminal connected to n-type region). It acts like a flow control valve and is therefore the ideal component for rectifying alternating current (Figs. 1a and 2).

The current flowing in reverse direction (reverse current) is much smaller than the forward current (approx. 1/100 the strength). It increases rapidly with rising temperature.

Amongst other applications, rectifier diodes are used in automotive alternators to convert the alternating current into direct current. Because of the high ambient temperatures to which the alternator is subjected, the reverse current is a critical factor and has to be taken into account in the design of the diodes used.

Rectifiers for high reverse voltages
In order to obtain a high reverse voltage in rectifiers, at least one region of the rectifier must have low conductivity. However, that means a high resistance in forward direction and consequently high power loss and excessive heating.

Creating a very lightly doped i-type region between the heavily doped p and n-type regions produces a p-i-n rectifier which, despite having a high reverse voltage, has a low forward resistance (conductivity modulation). The i-type region acts like intrinsically conductive silicon.

Uses: all rectification applications involving high voltages.

Zener diode
The Zener diode is a semiconductor diode in which the reverse current rises abruptly upwards of a certain voltage as a result of Zener breakdown and/or avalanche effects. Even though large numbers of these diodes depend for their function on avalanche effects rather than Zener breakdown, they are still referred to as Zener diodes.

Zener diodes are rated for different breakdown voltages depending on their particular application. They are designed for continuous operation at the breakdown voltage. They are used chiefly for voltage limitation and for generating a voltage reference, e.g. in direct-current power supply units.

Variable-capacitance diode
The space-charge region at the p-n junction acts like a capacitor; the semiconductor material depleted of charge carriers acts as the dielectric. An increase in the applied voltage increases the width of the depletion layer and reduces the capacitance; reducing the voltage increases the capacitance. Capacitance diodes are used mainly for resonant circuit tuning and frequency multiplication (e.g. in tuners).

Schottky diodes
The Schottky diode is a semiconductor diode with a metal-semiconductor junction. Because electrons transfer more easily from the n-type silicon to the metal film than in the opposite direction, an electron-depleted boundary layer known as the Schottky barrier is created in the semiconductor. Transmission of charge is performed exclusively by electrons; as a result, extremely high switching speeds are achieved because no minority storage effects occur.

Schottky diodes are suitable for use as fast switches and microwave rectifiers.

Tunnel diode
The tunnel diode (or Esaki diode) is a semiconductor diode with a very heavily doped p-n junction in which the tunnel effect, which is explainable only by quantum mechanics, occurs to such a degree that when operated in forward direction a negative differential conductivity occurs over a certain range of the current/voltage characteristic (Fig. 1b).

Tunnel diodes are used as oscillators in the gigahertz (GHz) range and for low-noise amplifiers.

Semiconductor resistors

In contrast with ohmic resistors, these components are voltage, current and temperature-dependent. They generally consist of polycrystalline semiconductor materials. The effects that occur in semiconductor resistors are based in part on depletion-layer properties that appear at the crystallite boundaries.

Varistors
The resistance of a varistor decreases as voltage increases. The polarity of the voltage makes no difference. A varistor consists of polycrystalline ZnO or SiC powder mixed with a binder, compressed and sintered.
Uses: e.g. voltage stabilisers, surge protectors for semiconductor circuits.

NTC resistors (thermistors)
NTC (**N**egative **T**emperature **C**oefficient) resistors (thermistors or thermal resistors) are, as the name suggests, resistors that have a marked negative temperature coefficient. Their electrical resistance decreases as the temperature increases and consequently, NTC resistors conduct electricity better at high temperatures than at low temperatures.
NTC resistors are made from polycrystalline metal oxides such as Fe_2O_3, $ZnTiO_4$ or $MgCr_2O_4$ by a process of compression and sintering. Their temperature coefficients can be as much as $-6\%/K$.
Uses: e.g. temperature sensors.

PTC resistors (thermistors)
PTC (**P**ositive **T**emperature **C**oefficient) resistors are resistors that have a positive temperature coefficient (of the order of a few ‰/K). Their electrical resistance increases as the temperature increases and consequently, PTC resistors conduct electricity better at low temperatures than at high temperatures. Most metals are thermal resistors with very low temperature coefficients.
The term PTC resistors refers to resistors made of semiconductor materials. PTC resistors made of ferro-electrical ceramic material (e.g. polycrystalline barium titanate) have a relatively narrow temperature range and a very high positive temperature coefficient (+6 to +60 ‰/K). PTC resistors made of silicon have wide temperature range and a virtually constant positive temperature coefficient (approx. +0.8 ‰/K).
Uses: e.g. liquid level sensors, heating regulators.

Magnetoresistors
A magnetoresistor is a magnetically controllable semiconductor resistor. Its resistance increases as the magnetic flux density B increases (Fig. 3).
The indium antimonide film (approx. 25 μm thick) contains minute needles of nickel antimonide with very high electrical conductivity. The flux density prevents the charge carriers from following a direct path. They move at an angle from one needle to the next. Within the metallic needles, differing charge-carrier densities are immediately equalised. As the flux density B increases, the current paths become more and more angled and the distance travelled by the charge carriers therefore greater. As a result, the resistance of the magnetoresistor increases.
Uses: e.g. magnetic-field sensors, controllable resistors.

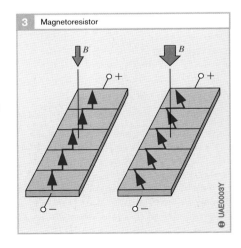

Fig. 3
Current path under the influence of a low (left) and a high (right) magnetic flux density B

Transistors

Transistors can be used control a large current with a small control current. Consequently, these semiconductor components can be used as power amplifiers or switches.

Transistors have three or more connections. Figure 4 illustrates the classification of the transistor family. "Transistor" is a contraction of the term "**trans**fer re**sistor**".

Bipolar transistors

Bipolar transistors consist of three semiconductor regions of differing conductivity arranged to form a p-n-p or an n-p-n configuration. These regions (and their connections) are called the emitter (E), the base (B) and the collector (C).

Depending on their applications, transistors are categorised as low-signal transistors (up to 1 watt power loss), power transistors, switching transistors, low-frequency transistors, high-frequency transistors, microwave transistors, phototransistors, etc.

Such transistors are called bipolar because they make use of charge carriers of both polarities (holes and electrons). In an n-p-n transistor, positive charge carriers (holes) in the base current control roughly 100 times as many negative charge carriers (electrons) flowing between the emitter and the collector. This corresponds to a current amplification factor of around 100. Bipolar transistors are thus controlled by means of the base current.

Method of operation of a bipolar transistor (with reference to an n-p-n transistor): The emitter-base junction (EB) is polarised for forward bias (Fig. 5 overleaf). This means that electrons migrate into the base region. The base-collector junction (BC) is polarised for reverse bias. This creates a space-charge region with a powerful electrical field.

Discernible coupling (transistor effect) occurs if the two p-n junctions are very close to one another (less than 10 µm apart in silicon). In that case the electrons crossing the EB junction pass through the base to the collector. As soon as they come within the

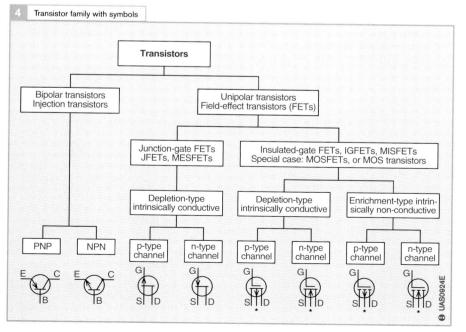

Fig. 4
B Base
E Emitter
C Collector
S Source
D Drain
G Gate
* Bulk connection (connection designation not usual)

Fig. 4 Transistor family with symbols

range of the electrical field at the BC junction, they are accelerated into the collector region and travel onwards as the collector current. The concentration differential within the base thus remains and, therefore, so does the impetus for continued electron migration from the emitter to the collector (Fig. 5).

In conventional transistors, 99% or more of electrons originating from the emitter pass into the BC space-charge region and form the collector current. The few that are lost have found their way into the electron gaps in the p-type base while passing through it. If this effect were left unchecked, the base would become negatively charged and the resulting repulsion forces would completely prevent the continued flow of electrons within an extremely short space of time (50 ns). Such a build-up of negative charge can be partially or entirely counteracted in a transistor by a low base current of positive charge carriers (holes). Small changes in the base current bring about large variations in the emitter-collector current.

With moderate base currents, the n-p-n transistor acts as a bipolar, current-controlled, amplifying semiconductor component. If the base current alternates abruptly between very low and very high levels, the transistor acts as a switch.

Field-effect transistors (FETs)
In a field-effect transistor (Fig. 6a), the current flowing from the source (S) to the drain (D) through a conductive channel is controlled by an electric field. That field is generated by a voltage applied via a control electrode known as the gate (G). The use of an electric field to control the current flow explains the origin of the term "field-effect transistor" (FET). In contrast with bipolar transistors, field-effect transistors use only one type of charge carrier (either electrons or holes) and are therefore also referred to as unipolar transistors. Field-effect transistors are subdivided into
- Junction-gate field-effect transistors (junction FETs, JFETs)
- Insulated-gate field-effect transistors, in particular metal-oxide semiconductor (MOS) field-effect transistors (MOSFETs)

MOSFETs are particularly suited to use in highly integrated circuits. Power FETs have superseded bipolar transistors for most applications.

Method of operation of a junction-gate FET (with reference to a JFET with an n-type channel, Fig. 6a):
DC voltage is applied to the ends of an n-type crystal. Electrons flow from the source, S, to the drain, D. Two p-type regions at the sides forming the gate (G), and the negative voltage applied to them, determine the width of the channel. If the negative gate voltage is increased, the space-charge regions extend further into the channel and constrict the current path. The voltage at the control electrode, G, thus controls the current between the source, S, and the drain, D. The FET only requires charge carriers of a single polarity in order to function. The current is controlled virtually without power consumption. The junction FET is thus a unipolar, voltage-controlled component.

Method of operation of a MOS FET (with reference to a MOS FET with p-type enrichment): while no voltage is present at

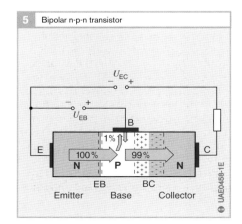

Fig. 5
N n-type silicon
P p-type silicon
E Emitter
B Base
C Collector

the gate electrode, no current flows between the source, S, and the drain, D; the p-n junctions are reverse biased (Fig. 6 b). A negative voltage applied to the gate, G, forces the electrons in the n-type region below the electrode into the centre of the crystal and draws holes – which are always present as minority charge carriers even in n-type silicon – to the surface. A narrow p-type layer is created below the surface, in effect a p-type channel. Current, which is comprised only of holes, can now flow between the two p-type regions (source and drain).

As the gate voltage acts through an insulating oxide layer, no current flows in the control circuit and control therefore requires no power. The MOS FET is thus a unipolar, voltage-controlled component.

PMOS, NMOS and CMOS transistors
In addition to the **p**-type channel **MOS** transistor, or PMOS transistor, there is the NMOS transistor in which the p and n-type regions are reversed in comparison with the PMOS transistor. Due to the higher mobility of electrons, NMOS transistors are faster than PMOS transistors which, for physical reasons, are easier to manufacture and were therefore available first.

If PMOS and NMOS transistors are created in pairs on the same silicon chip, they are referred to as CMOS transistors (complementary **MOS** transistors, Fig. 6c). Their particular advantages are very low power loss, high interference immunity, TTL-compatibility (transistor-transistor logic in electrical circuits), low power-supply voltage and suitability for analog signal processing.

The low power consumption, particularly for digital circuits, is due to the fact that current only has to flow when a digital cell changes (e.g. switches from "0" to "1"). Retention of the information requires only that a voltage is applied. Since, with this configuration, the PMOS and NMOS transistors are connected to one another, the leakage current is minimal. More than 80 % of ICs are now manufactured using CMOS technology.

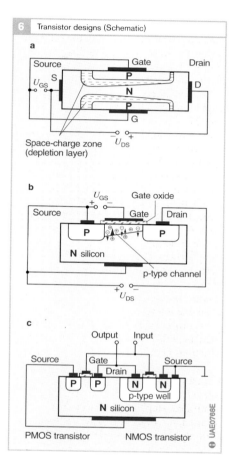

Fig. 6
a Depletion layer, field-effect transistor
b PMOS transistor
c CMOS transistor pair

N n-type silicon
P p-type silicon
S Source
G Gate
D Drain

U_{GS} Voltage between G and S
U_{DS} Voltage between D and S

Thyristors
Thyristors are semiconductor components with at least three p-n junctions (one of which may also be replaced by a suitable metal-semiconductor contact) that can be switched from a reverse-bias condition to a forward-bias condition (or vice versa). The term thyristor is used as the generic term for all types of component which conform to that definition. It is a contraction of the two words **thyra**tron (gas-filled tube triode) and re**sistor**.

Uses in power electronics: speed and frequency control, rectification and conversion, switching.

Monolithic integrated circuits (ICs)

Monolithic integration
The planar process is a method of manufacturing semiconductors which makes it possible to produce all components of a circuit (resistors, capacitors, diodes, transistors) and the conductive connections between them on a single silicon chip in a single production process. It involves the creation of multiple p and n-type layers in a multi-stage doping process using a pre-doped single-crystal silicon disc or wafer. An integrated circuit IC (Integrated Circuit, see section entitled "Manufacture of Semiconductor Components and Circuits") does not contain any "separate" (discrete) components but rather switching elements or functional elements.

Degree of integration
The degree of integration is defined as the number of functional elements, transistors or gates on a single chip. The following categories are defined based on the degree of integration (and chip surface area):
- SSI (small-scale integration): up to around 1,000 elements per chip, average chip area 3 mm² (varies considerably depending on level of power loss)
- MSI (medium-scale integration): up to around 10,000 elements per chip, average chip area 8 mm²
- LSI (large-scale integration): up to 100,000 elements per chip, average chip area 20 mm²
- VLSI (very-large-scale integration): more than 100,000 elements per chip, average chip area 30 mm²

The number of functional elements on VLSI chips is constantly increasing. Microprocessors can now have up to 10 million transistor functions per chip. Memory modules (DRAM) can even contain several hundred million transistors. The number of such functional elements is subject to an exponential increase over time. This fact was first recognised by Gordon Moore

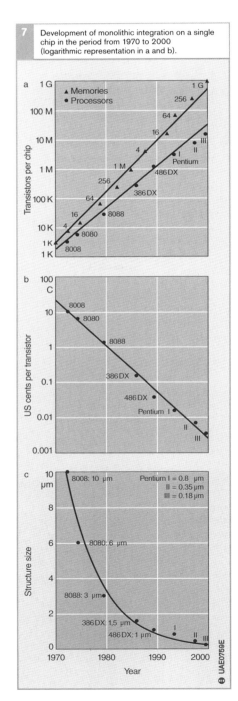

Fig. 7
a Increase in number of transistors per chip in memories and processors (Gordon Moore's Law)
b Fall in cost of transistors
c Decreasing size of structural units of new processors

(co-founder of the company Intel) and is expressed by "Gordon Moore's Law" (Fig. 7a):
Elements per chip = $2^{(year - 1956) \cdot 2/3}$

This law states that the number of functions per chip will double every 18 months. Since the structural units in which the chips are manufactured are becoming smaller at the same time as the number of functional elements is growing, the size of the ships alters very little and the costs per transistor are similarly dropping exponentially (Fig. 7b). Thus the structural units of new processors have shrunk over a period of 28 years from 10 μm in 1972 to 0.18 μm in the year 2000 (Fig. 7c).

The degrees of integration LSI and VLSI demand methods such as CAD (computer-aided design). VLSI circuits in particular can only be created with the help of highly sophisticated programs which can convert entire function blocks into corresponding circuit subdivisions. Such program systems are called HDL (Hardware Description Language) or VHDL (Visual HDL).

IC classification
There are various systems for classifying ICs, although there are invariably mixed categories within each system.
- *Classification by method of production:*
  Bipolar/Unipolar (MOS);
  mixed category: e.g. BiCMOS, BCD
- *Classification by function:*
  Analog/Digital;
  mixed category: e.g. mixed-signal IC
- *Classification by application:*
  Standard IC/ASIC
  mixed category: e.g. ASSP

ASIC (**a**pplication-**s**pecific **IC**): developed and produced specifically for a particular application and exclusively for a particular customer.
ASSP (**a**pplication-**s**pecific **s**tandard **p**roduct): developed and produced specifically for a particular application for a particular customer but is also sold to others for the same type of application.

Apart from a few exceptions, ICs are now always based on MOS or combined technologies, and for that reason bipolar ICs are not described in any further detail at this point.

Analog circuits
Analog circuits are required wherever electronic equipment has to communicate with the outside world – which is always analog. This applies in particular to the areas of "preparation" and "processing" of input and output signals. Examples of analog-input signal generators are microphones (telephone) and sensors. Examples of analog-output signal receivers are speakers or actuators.

Figure 8 illustrates the stages of development of an analog IC from conception to installation in an ECU.

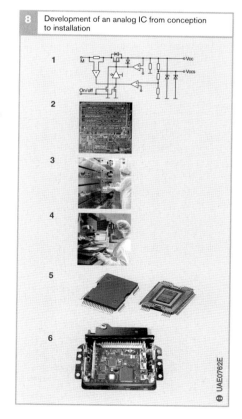

Fig. 8
1 Development and simulation
2 Layout
3 Production
4 Testing, analysis and release
5 Packing, final testing
6 Installation in an ECU

Basic structures: stabilised-voltage sources, stabilised-current sources, differential amplifier stages, coupling circuits, potential shifting circuits, output stages.
Application-based ICs: operational amplifiers, voltage regulators, comparators, timers, transducers, interface circuits.
Specialised ICs: voltage references, broadband amplifiers, analog multipliers, function generators, phase-locking circuits, analog filters and switches.

Digital circuits
Digital circuits (Fig. 9) are used wherever large volumes of data have to be processed within a short space of time. High-performance microcontrollers in particular can perform several hundred million computational operations per second. They are thus able to utilise the input data to provide the required, highly accurate response at the output of the particular device. Similarly, large volumes of data can be transmitted via a cable link in coded form, e.g. several thousand telephone calls can be transmitted simultaneously via a fibre-optic cable without cross-talk.

Since inputs and outputs for communication with the real world always have to be analog, analog-digital converters (ADCs) are used at the inputs of a digital circuit, and digital-analog converters (DACs) at the outputs.

Digital systems make use of a whole series of recurring basic circuits and variations of them. The range extends from simple gates to memories, microprocessors and microcontrollers.

Digital modules can only be connected up to form a complete system if power supply voltages, logic signal levels, switching speeds and signal transmission times are compatible. This requirement is met within a family of circuits. Since, apart from a few specialised applications mainly in military systems, CMOS logic circuits are the only type now used, such compatibility is no longer a problem.

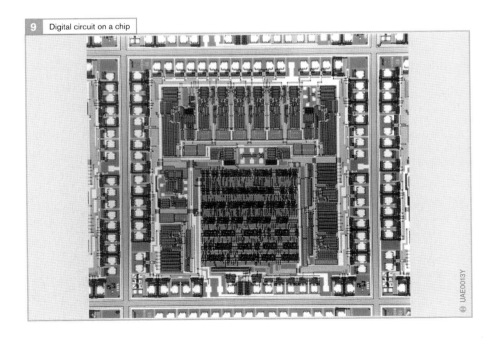

9 Digital circuit on a chip

Mixed-signal ICs

ICs in which both analog and digital circuit elements are combined are called mixed-signal ICs. They are an essential requirement for the combination of entire electronic systems on a single piece of silicon (SoC: System-on-a-Chip).

With ever-increasing degrees of integration, this type of circuit is becoming more and more attractive and the number of applications is continually increasing. It is now conceivable, for example, to integrate the entire electronic circuitry for a mobile phone or a complete Internet-access system on a single chip. Where only low levels of electrical power are required, such as in the examples quoted, both the analog and the digital sections are manufactured using CMOS technology.

Mixed-signal ICs for applications involving higher electrical power levels, such as are frequently encountered in automotive systems, are typically hybrids of various technologies, such as BCD systems (where B stands for bipolar, C for CMOS and D for DMOS).

Figure 10 shows how the functions, "sensing", "analysing" and "acting" are integrated in a mixed-signal IC. In such circuits, the bipolar section is used for converting the analog inputs. The CMOS section of the IC performs the logical processing operations using digital technology. The DMOS section enables high analog output performance.

Elements of digital circuits

Digital circuits assign voltage signals one of the two signal levels "0" (low voltage level) or "1" (high voltage level) and process those signals digitally. As digital circuits make up the greater part of all ICs and our day-to-day lives are no longer conceivable without them, a few typical circuit elements are described below.

Gates

Gates are logical-operation circuits with two or more inputs. The input signals are logically linked to one another so that their various combinations determine the output signals according to a defined logic (e.g. AND gate).

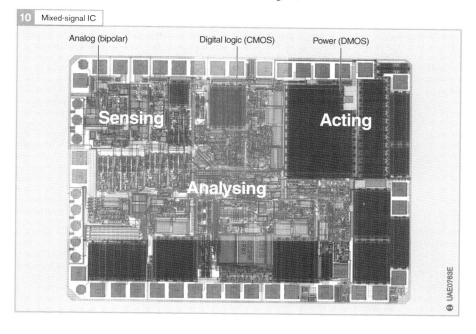

10 Mixed-signal IC

Inverters
Inverters are digital circuits that convert the digital input signal "0" into the output signal "1" and vice versa.

Bus
The bus links up the individual elements of a digital circuit; this relates both to connections within an integrated circuit and the interconnection of separate digital ICs.

A bus can connect up a large number of separate modules with different functions but with electrically identical interfaces. A serial bus (e.g. for connecting up serial EEPROMs) transmits the data along a single lead. A parallel bus (e.g. address bus, data bus, control bus) is a bundle of parallel lines. The number of separate items of information that can be simultaneously transmitted (equal to the number of bus lines) is, together with the data transmission speed, a measure of the performance capacity of the data bus. The most common types of bus are 8-bit, 16-bit and 32-bit.

The dimensioning of the data bus is determined by the capacity of the CPU (central processing unit). In order for a system to achieve its maximum performance capability (i.e. maximum processing speed) the CPU and data bus should have the same capacity, i.e. an 8-bit CPU should use an 8-bit bus, a 16-bit CPU a 16-bit bus, etc.

In addition, the bus should be able to transmit the data as quickly as the CPU can process it. This is often not the case with external busses (e.g. CPU clock speed $f_{CC} = 400$ MHz and bus clock speed $f_{CB} = 133$ MHz).

Only one bus node at a time can write data to the bus. According to the bus type, this addresses the data in such a way that it can only be read by the addressee, or else it switches all other bus nodes off before sending the information and then on again afterwards.

Clock-pulse generator (CPG)
A clock-pulse generator ensures that all operations on the microcomputer are synchronised with a defined timing pattern. The clock-pulse generator must be matched to the required speed of computing operations.

Input/output (I/O) unit
The I/O unit handles data exchange with the outside world. The input signals are scanned at the required frequency. The output signals are transmitted at the speed required by, and in the optimum order for, the application concerned or are held in temporary storage until called for.

Microcomputers
The microcomputer consists of the following interacting components (Fig. 11):
- A *microprocessor* as the CPU (central processing unit). For its part, the microprocessor consists of the control unit and the arithmetic and logic unit. The arithmetic and logic unit performs arithmetical and logical operations, as the name suggests, while the control unit carries out the instructions from the program memory.
- *Input and output units* (I/O units) which handle the exchange of data with peripheral devices. Peripheral devices include input and output devices and external data storage media.
- A *program memory* in which the operating program (user program) is permanently stored (ROM, PROM or EPROM).
- *Data memories* for holding the data being processed at any particular time. This data changes continually (RAM).
- The *bus system* which links up the individual components of the microcomputer.
- A *clock-pulse generator (oscillator)* which ensures that all operations on the microcomputer are synchronised with a defined timing pattern.
- *Logic circuits* which are modules with specialised tasks such as program interrupts. They are integrated in individual I/O units.

Electronic components | Semiconductor components | 309

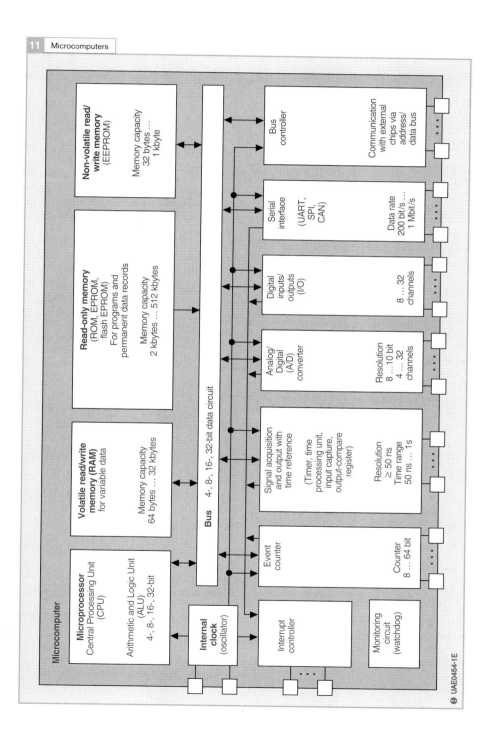

11 Microcomputers

The chief components of a microcomputer are generally separate modules connected to one another on a printed-circuit board. For simpler tasks such as are more and more frequently required for Internet access on wireless communication devices, single-chip computers are becoming increasingly common. They integrate the functions detailed above on a single silicon chip (System-on-a-Chip). The relatively small amount of RAM that can be accommodated on a chip at reasonable cost limits the performance capabilities of such highly integrated systems.

Microprocessor
A microprocessor is a central processing unit (CPU) in the form of an integrated circuit on a chip. The concept of the microprocessor avoids individualisation despite the high degree of integration, and enables adaptation to the multiplicity of practical demands by reliance on programming. A microprocessor is not capable of functioning on its own; it is always part of a microcomputer. At present, there are 16-bit, 32-bit and 64-bit microprocessors. Processors are subdivided into two main categories:
- PCs use CISC (Complex Instruction-Set Computer) processors. These are extremely versatile and permit unrestricted programming.
- Workstations normally use RISC (Reduced Instruction-Set Computer) processors. These perform specific tasks such as are frequently required on workstations much faster, but are distinctly slower for all other tasks.

Microcontrollers
The microcontroller is a component that incorporates the following elements on a chip:
- CPU
- Random-access memory (RAM)
- Peripheral modules (input/output, interrupt, timer, serial interfaces) and
- An optional read-only memory (ROM)

With these integrated components, the microcontroller is capable of stand-alone operation. For that reason, it is also referred to as a single-chip microcomputer.

The microcontroller is used to control self-regulating systems such as an engine-management system.

At present, there is a choice of 4-bit, 8-bit, 16-bit or 32-bit microcontrollers for the various types of application. Depending on application, they may also have expansion modules connected to them (e.g. additional memory for data and program code).

The program that is run by the CPU is permanently fixed in the read-only memory and is not changed for different applications. This distinguishes the single-chip microcontroller from the PC.

Transputer
A transputer is a special type of microprocessor that is particularly suited to the construction of parallel computer networks. In addition to the usual components of a microprocessor, it has communication and process-handling hardware on a chip.

Programming
The only command form capable of direct interpretation by a microprocessor is a bit pattern, i.e. the binary representation of a number. Since, however, this form of instruction is not easy to work with for a programmer, and is therefore susceptible to errors, easily memorable abbreviations (mnemonics) are used. These are automatically translated by an assembler program into bit patterns (machine code) that can be understood by the microprocessor. Microcomputers for single-purpose applications are programmed in the assembler language specific to the processor.

For more complex systems and programs, "high-level programming languages" such as C are needed, as otherwise it would be impossible to keep extensive programs manageable and free of errors. Such languages require sophisticated translation programs (compilers) which convert the text of the high-level language into a form that can be processed by the processor.

Semiconductor memories

Applications
Memories are used to store large volumes of
- Digital signals representing data (I/O data, statuses, intermediate results involving frequent and rapid reading and writing)
- Program code (usually permanently stored) and
- Constants (permanently stored)

Storage involves
- Recording (writing)
- Permanent retention (actual storage) and
- Location and retrieval (reading) of information

Memories make use of physical effects that make two different statuses clearly distinguishable and easy to generate and identify. The advantage of semiconductor memories lies in their technological compatibility with the modules used in other parts of a computer, and thereby in new opportunities for functional integration.

Classification of semiconductor memories
Digital signals are stored by utilising the status alternatives conducting/not conducting or charged/uncharged. The most important terms are explained below according to their standardised definitions, where applicable, or their most common usage (see Fig. 12 for overview).

Random-access memory (RAM)
Random-access memory or RAM is a short-term memory that allows direct access to any storage location. Information can be read/written from/to the memory any number of times. However, it must exist in binary form, i.e. encoded as a series of "yes or no" statuses (logical "1" or logical "0"). Such a "yes or no" unit of information is called a bit (**bi**nary digi**t**). Random-access memories, like read-only memories, are organised on a bit or word basis depending on the particular application. A "word" is a group of bits that can be processed as a single unit. The word length is equal to the number of bits processed as a single unit. On microcomputers, word lengths of 4, 8, 16, 32 and 64 bits are common. 8 bits make up one byte, i.e. 1 byte = 8 bits.

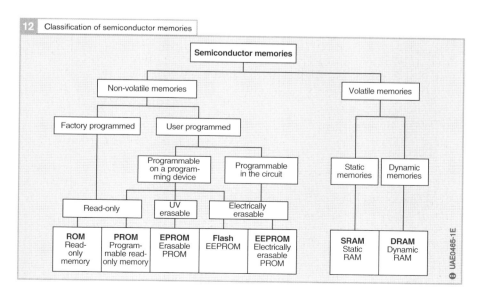

12 Classification of semiconductor memories

Memories can be organised on the basis of a variety of word lengths. The way the memory is organised is generally indicated by a multiplication sign, e.g. an 8 M × 8 RAM or 8-megabyte RAM
≈ 8 million bytes
= 64 million bits
= 64 Mbit

That means that the memory has eight data inputs and eight data outputs at which the eight bits of one of the eight million stored data words are simultaneously present. Its word length is thus 8 bits. Since all memory specifications are based on the binary system, the exact number of bits is as follows:
64 Mbit
= 8 Mbyte × 8 bit
= 2^{23} × 8 bit
= 2^{26} bit
= 67,108,864 bits

Static RAM (SRAM)
Static RAMs use bistable switching elements as the data storage cells. Their function is similar to that of a flip-flop, a simple circuit with two transistors, of which either the one (logical "1") or the other (logical "0") conducts at any one time. In SRAM, the information remains stored until the storage cell concerned is addressed and overwritten, or the power supply is switched off.

Static RAMs are currently available in sizes up to 16 Mbit. Although considerably more expensive than dynamic RAMs, because they can be written into and read significantly faster they are used as high-speed access computing memories (cache memory) for the CPUs in computers. Typical read-write times are currently less than 30 ns.

Dynamic RAM (DRAM)
The information in a DRAM is stored as an electrical charge in the gate capacitor of a CMOS transistor which either passes (logical "1") or does not pass (logical "0") a current as a result. As such capacitors are susceptible to leakage, the charge is gradually lost. In order that the information (charge) is retained, the memory has to be refreshed at regular intervals (every few ms).

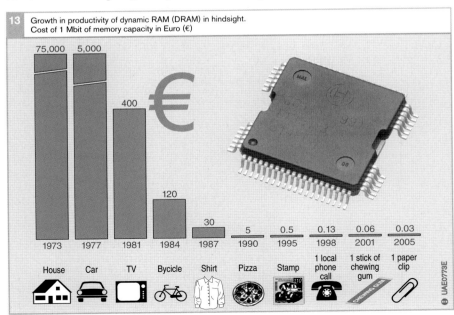

13   Growth in productivity of dynamic RAM (DRAM) in hindsight. Cost of 1 Mbit of memory capacity in Euro (€)

DRAMs are currently available commercially in sizes up to 256 Mbit (on one chip), and in laboratory conditions up to 1 Gbit (gigabit = 109 bit). Since every storage cell consists of a transistor and a capacitor, this requires more than a billion transistors on a single chip.

DRAMs are volatile memories. This means that they lose their information when power is switched off.

DRAMs are very widely used nowadays as system memories for all types of computer. The fall in price for 1 Mbit of memory over the last few decades has been so dramatic that ample memory can now be fitted in any PC without substantially affecting its price (Fig. 13).

Read-only memory
Read-only **m**emory (ROM) is permanent-storage memory that allows any memory location to be accessed directly but – as the name indicates – allows the information only to be read and not altered.

A ROM is a non-volatile memory, i.e. the information it contains is retained even when the power supply is switched off. It is usually used to store program codes (control programs) and fixed data (function tables, encoding rules, character generators, engine characteristic-data maps) that need to be retrievable at any time. The information may be indelibly entered in the memory either by the manufacturer (in one of the last stages of production) or by the user by appropriate programming of specially prepared memories (PROMs or programmable **ROMs**).

Erasable ROM
There are also ROMs whose contents can be erased and reprogrammed as outlined below.

EPROM (Erasable **PROM**)
This type of erasable read-only memory can have its contents completely wiped by irradiation with UV light and can then be reprogrammed. Such reprogramming is only possible with special equipment at relatively high cost.

EEPROM (Electrical **EPROM**)
EEPROM is also referred to as E^{2}PROM (E squared PROM). This type of erasable read-only memory can be electrically wiped and reprogrammed. The wiping and reprogramming operation can be performed either in a separate unit or in situ. Every storage cell of an EEPROM be individually overwritten. For that reason, this type of memory module can also be used as non-volatile data memory (e.g. for status information in engine-management systems).

Flash EEPROM
Yet another variation of EPROM and EEPROM is the flash EEPROM. In this case, electrical flash pulses are used to erase specific storage areas or the entire contents of the memory. The erased areas can subsequently be reprogrammed.

The flash memory can be reprogrammed on a programming station. However, the advantage of flash EEPROM is that is can also be reprogrammed while still inside the sealed control unit. When this is done, the memory area which contains the programming routines must not be erased, or alternatively, the programming routines must be transferred to the RAM before the memory is wiped. The microcontroller then works with the RAM as the program memory. Flash EEPROM is used wherever relatively large volumes of data need to be stored but also have to be changed from time to time (e.g. in mobile phones, digital cameras and as program memories in electronic control units in motor vehicles).

Opto-electronic components

Photoresistor

A photoresistor is a resistor whose resistance decreases when exposed to light (Fig. 14). Light (photons) generates free charge carriers in the semiconductor as soon as the energy of a photon is equal to the energy holding a charge carrier within the lattice (internal photoelectric effect).

Photoresistors generally consist of polycrystalline semiconductor materials in which the effect is particularly marked such as CdS, CdSe, PbS, PbSe, CdTe, ZnO, Se, InSb, InAs, Ge or Si.

A photoresistor is not equally sensitive to all wavelengths. Its maximum sensitivity lies within a narrow wavelength band that is specific to the material used.

Photoresistors are mainly suitable for use in cameras as light meters.

Photodiode

A photodiode is a semiconductor diode which utilises the depletion-layer photoelectric effect. Reverse voltage is applied to the p-n junction. Incident light releases electrons from the crystal lattice. As a result, additional free electrons and holes are produced. They increase the reverse current (photoelectric current) in proportion to the intensity of the light (Fig. 15).

Photodiodes can be used in light meters, photoelectric beams and in positioning and remote control applications that use infrared light.

Photovoltaic cell

Like the photodiode, the photovoltaic cell releases charge carriers when exposed to light; no external voltage is applied to the p-n junction, however. If the electrons and holes reach the p-n junction, they are separated by the internal field of the space-charge region. A photoelectric voltage is generated which produces a photoelectric current in the external circuit. Light energy is thus converted into electrical energy.

Photovoltaic cells are used to measure light intensity and to generate electricity with solar panels.

Light-emitting diode

The light-emitting diode or LED emits light when a forward current is passed through it. In other words, it operates in the opposite way to the photodiode effect – free electrons and holes are recombined. The energy released is emitted in the form of light.

The semiconductor materials most commonly used are gallium compounds such as gallium arsenide (GaAs) and gallium phosphide (GaPh). The wavelength of the light

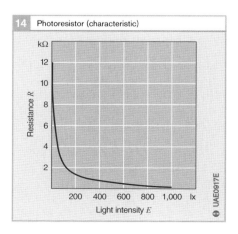

14 Photoresistor (characteristic)

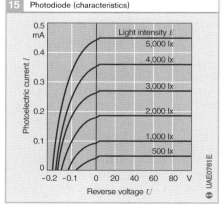

15 Photodiode (characteristics)

emitted by an LED depends on the semiconductor material used. There is now a choice of colours available (infrared, red, yellow, green, blue).

LEDs are used for numerical and alphabetical displays as well as for warning lamps. Since it has been possible more recently to substantially increase the amount of light generated, LEDs are now also used in motor vehicles to illuminate the instrument panel or as a third (high-level) brake light. They have the advantage of a relatively high light yield and, as a result of their very long service life, do not have to be replaced.

Laser diode

With the appropriate mechanical/optical design, LEDs can also be made to emit laser light, i.e. a parallel beam of monochromatic (of a single wavelength) and coherent (the waves are in phase) light. Such properties are required for the transmission of data by fiber-optic cables. Laser diodes were therefore a necessary requirement for the high data flow rates such as occur in computer networks with access to the Internet or the transmission of television signals by cable.

Phototransistor

A similar effect to that used by the photodiode also occurs in the phototransistor. Incident light striking the base alters the conductivity between the emitter and the collector in such a way that a current proportional to the intensity of the light is able to flow through the transistor. This type of component can thus be used as a switch for photoelectric beams, for instance.

There are also phototransistors that can be operated the opposite way around so that they emit light. The advantage compared to the LED is that the phototransistor can be switched on and off very quickly. It is therefore particularly suited for use in the manufacture of flat screens. There is a transistor for every screen dot (determined by the screen raster in a cathode ray tube). As the areas required are generally substantially larger than the silicon slices available, and such slices would, in any case, be much too expensive, flat screens are manufactured using thin-film techniques (vapour deposition through a screen). The advantages of flat screens over cathode-ray tubes are their slimmer dimensions, lower power consumption and the absence of flicker.

Charge-coupled device (CCD)

CCDs are integrated circuits arranged as a charge-coupled array. They are used to record images in video and digital cameras as well as in scanners. Thousands or even millions of individual elements laid out in a matrix pattern (array) are created on a chip (Fig. 16). Digital cameras can record up to 3.5 million pixels with a colour resolution of 16 million colours. CCDs are not only capable of converting the visual information into electrical signals, they can also store the data until it can be transferred to an external storage medium. To do this, the output register (5) is scanned at a high rate.

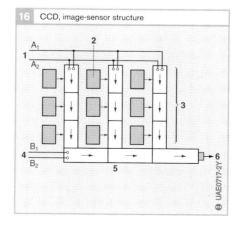

Fig. 16
1 Column clock pulse A_1/A_2
2 Photosensors
3 CCD array
4 Row clock pulse B_1/B_2
5 Output register
6 Video output

Micromechanical sensors

It has been possible for a number of years now to manufacture some sensor components using silicon technology. Since a large number of sensors can be produced simultaneously on a single silicon wafer in similar fashion to electronic components, the unit costs are substantially lower than in the production of conventional sensors. In addition, there is also the possibility of having the sensor signals electronically processed directly on the same chip.

Examples of micromechanical sensors:
- Magnetic-field sensors measure an external magnetic field by purely electrical means.
- Acceleration and pressure sensors measure the relevant variables by the physical deformation of specific areas of the silicon.
- Temperature sensors utilise the change in conductivity of the semiconductor material according to temperature.
- Chemical sensors make use of the effect whereby certain chemicals (gases, liquids) alter the electrical conductivity of some semiconductor materials in a very specific way.

Some of the more important examples of sensors and their design are described below.

Hall-effect sensor for detecting magnetic fields

An electrical potential U_H is generated at the edges of a thin plate through which a current is flowing if a magnetic field is acting perpendicular to the current flow I_V (Fig. 17). It is described by the formula

$$U_H = R_H \cdot I_S \cdot B/d$$

where
U_H Hall-effect voltage
R_H Hall-effect constant
I_V Supply current
B Flux density of the magnetic field
d Thickness of the plate

In metals, the Hall-effect constant is very small: $R_H \approx 10^{-9}$ m³/As. In semiconductors, R_H is five orders of magnitude greater, e.g.
- Indium antimonide (InSb)
  $R_H \approx 2.4 \cdot 10^{-4}$ m³/As
- Indium arsenide (InAs)
  $R_H \approx 1.2 \cdot 10^{-4}$ m³/As

Hall-effect sensors can be used in motor vehicles as position sensors (e.g. for sensing camshaft position). The advantage of this measurement principle is that position can be detected even when the component concerned is not moving. By comparison, inductive speed sensors only produce an electrical signal above a certain minimum rotational speed.

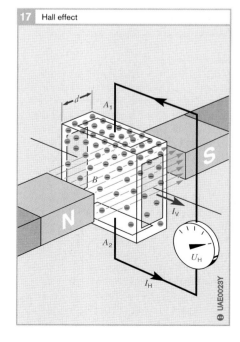

Fig. 17
A_1, A_2 Measurement points for the Hall-effect voltage
B Flux density of the magnetic field
I_V Supply current
I_H Hall-effect current
U_H Hall-effect voltage
d Thickness of the plate

Pressure sensor

Figure 18 shows the structure of a micromechanical pressure sensor. The silicon chip (2) is attached to a glass base (4) through which there is a hole which acts as the pressure channel for the pressure p that is to be measured. At the point where the measured pressure acts on the silicon chip, it has been hollowed out on the underside. Consequently, the chip is in effect a thin diaphragm (1) at that point, and deforms under the action of pressure.

Attached to the top of the diaphragm there are resistors which change their resistance in response to physical deformation (piezoresistive effect). By measuring the resistance, the pressure can be calculated with the aid of the integrated bridge circuit.

Pressure sensors are used in motor vehicles for such tasks as measuring the intake manifold pressure and the atmospheric pressure.

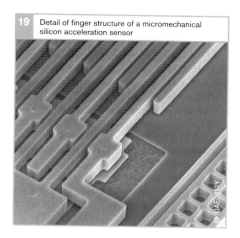

Detail of finger structure of a micromechanical silicon acceleration sensor

Acceleration sensor

Finger-like structures are etched into the surface of a silicon ship (Fig. 19). Every other "finger" is rigidly attached to the chip, while the alternate fingers are only attached at one end so that they can oscillate freely at the other. These fingers are connected up as a multi-layer capacitor. If the sensor is accelerated in a particular direction, the distance between the fixed and the movable fingers changes, thus altering the capacitance. That change can be measured as an electrical signal from which the acceleration can then be calculated.

Acceleration sensors are used in motor vehicles wherever changes in the vehicle's dynamic behaviour need to be detected.

They are employed, for example, to trigger the airbags in the event of an accident (abrupt deceleration).

The Electronic Stability Program (ESP) uses a yaw rate sensor to detect the rate of rotation of the vehicle around its own axis and counteracts such yawing motion at an early stage by controlled operation of the brakes.

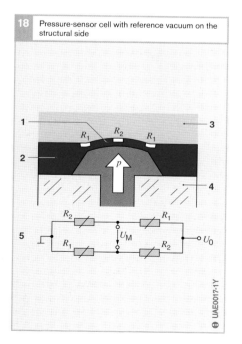

Pressure-sensor cell with reference vacuum on the structural side

Fig. 18
1 Diaphragm
2 Silicon chip
3 Reference vacuum
4 Glass base (Pyrex)
5 Bridge circuit

U_0 Supply voltage
U_M Measurement voltage

Strain resistors R_1 (compressed) and R_2 (stretched)

Microcontrollers

Microcontrollers are single-chip computers that are used for event-driven control systems or for controlling technical processes. Modern-day life is no longer conceivable without them and they are to be found wherever control systems for electrically operated devices are required.

Microcontroller developments

Five billion microcontrollers were produced in 1999, that is roughly equivalent to one for every person on the Earth. Its triumphant march into every aspect of our day-to-day lives is clearly demonstrated by the number of microcontrollers produced in the last ten years (Fig. 1).

A microcontroller consists of the integration of a CPU (central processing unit) with memories and peripheral circuits on a single chip. These are referred to as "embedded systems". Frequently such systems incorporate a microprocessor that was originally developed for computer applications and subsequently adapted and modified so as to be marketed as part of an embedded system.

Nevertheless, there are also many microcontroller architectures that were developed from the outset as controllers. They include the M-Core™, the TriCore™ and the SH7000, 8051 and C166 families. Such microcontrollers are characterised by rapid data exchange with their environment, low power consumption, high data density and low production costs.

What then is the actual origin of microcontrollers? Even before the transistor was invented, most of the devices in which microcontrollers are now used were already in existence – cars, telephones, radios, televisions and household appliances. The difference was that for most applications the control systems were mechanical and only in very rare and very costly exceptions were functions controlled by electron tubes and relays. In addition to their high cost, the problem of such control systems was that their reliability was heavily influenced by manufacturing tolerances, environmental conditions and wear.

In the 1960s, Intel, followed shortly afterwards by Motorola, developed the first microprocessors – initially for use in cash registers. Following the development of integrated circuits and the first microprocessors, the inexorable progress of digital technology began in earnest in the 1970s. For the first time, it was now possible to eliminate a large number of interference factors.

The end of the 1970s saw the dawn of the age of the single-chip microcomputer. Prominent representatives of that period were the Intel 8048 and Motorola 6800 microcontrollers, the structure of which is still the basis for many of today's microcontrollers. In the succeeding period, advances in high-integration technology allowed more and more peripheral circuits to be incorporated in the chip. These were the be-

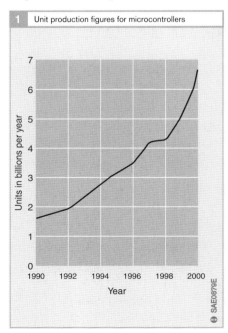

1　Unit production figures for microcontrollers

ginnings of the microcontroller as we know it today.

In parallel with these developments, scientists realised that single-chip computers could be used to perform control tasks in motor vehicles. The first applications involving high data volumes were the management systems for internal-combustion engines, followed later on by antilock braking systems (ABS) and entertainment systems.

At the end of the 1980s, the term "embedded control system" was introduced to distinguish them from infinitely programmable systems such as the PC. Embedded systems are designed exclusively for a specific application so that the entire program together with the vehicle-specific data in a ROM or EPROM are a permanent part of the product that cannot be altered by the user.

Most microcontrollers sold today have an 8-bit CPU. Although the numbers of more powerful 16 and 32-bit controllers are continually growing, simple 4-bit designs are still used in large quantities (Fig. 2).

Microcontrollers are produced in a variety of designs for a vast range of applications. The various types differ by virtue of the choice of integrated memory and the peripheral modules offered. The range of integrated memories available encompasses RAM, ROM, EEPROM and, more recently, flash EPROM.

To make it worthwhile to develop a stand-alone microcontroller design, the production figures should run to at least 1 million units a year. The unit price will then be between 0.5 and 15 Euro.

Applications that hold the promise of such production volumes include control systems for automotive applications, communications devices, computer peripherals, domestic appliances and entertainment systems. In addition, microcontrollers are also popular in industrial electronics, automation systems and testing and measuring equipment, although recourse is generally made to standard products.

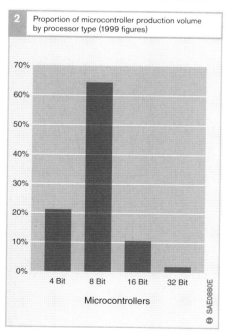

2 Proportion of microcontroller production volume by processor type (1999 figures)

Microcontroller components

The microcontroller is a programmable electronic module that contains all the necessary components for a microcomputer system. It consists of the
- *CPU*
- The *memories* for instructions (program memory) and data (data memory) and
- The *peripheral modules*

Those components exchange data and monitoring information via bus links (Fig. 1).

CPU

The CPU (**c**entral **p**rocessing **u**nit; also referred to as the microcontroller core) is the programmable unit for addressing and manipulating data and for controlling the timing and logical sequence of a program.

Memory

The memory is used to store data and program instructions. The memory for variable data is a **r**andom-**a**ccess **m**emory (RAM). The memory used for program instructions and unvarying data is a read-only memory (e.g. ROM or PROM). In addition, microcontrollers have a small register memory integrated in the CPU for rapid access (cache memory).

Peripherals

The peripherals are used for the input and output of data originating from or destined for external systems. The peripherals are programmable to a limited degree in order that their functions are adaptable to the requirements of the application.

Typical peripheral modules digitise analog external signals or convert internal digital signals into analog signals for output (analog-digital and digital-analog converters). Counters and timers count external pulses and time intervals between events. Communication interfaces are used for the exchange

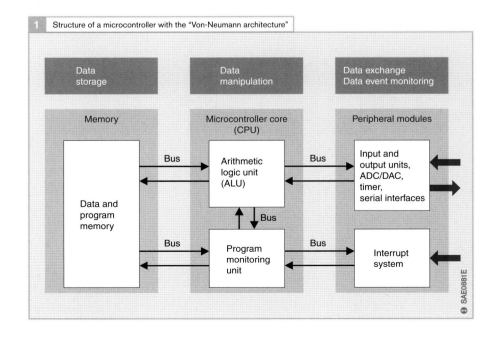

Figure 1: Structure of a microcontroller with the "Von-Neumann architecture"

of data with other modules via standardised bus links (e.g. CAN bus).

There are, of course, many other functions that can be integrated in the microcontroller depending on the requirements of the user concerned.

Main operations

The three blocks illustrated in Figure 1 enable the four main operations of the microcontroller, i.e.
- Data manipulation (data processing)
- Data storage
- Data exchange with external systems (data movement) and
- Data event monitoring (control mechanism)

These functions enable the microcontroller to be used to transfer, store and manipulate data (both in the memory and externally). The sections which follow describe the various modules of the microcontroller that make those operations possible.

Design and operating concept

CPU (Central Processing Unit)

The CPU processes the data received from external sources via the peripherals and monitors the data flow. In the CPU there is a small memory (register) in which operands, results and addresses are stored. Figure 2 shows the basic structure of a CPU, which may also be extended by additional components in order to increase the processing speed.

Programming model

The "programming model" of a microcontroller refers to the sum total of all registers that are available to, i.e. are "visible" to, the programmer. In principle, there does not have to be any "visible" register in the CPU. But in that case, every alteration to the program would necessitate modification of the hardware, which would be very involved, expensive and time-consuming. Rarely altered configurations are therefore set by means of bits in special control registers. The control

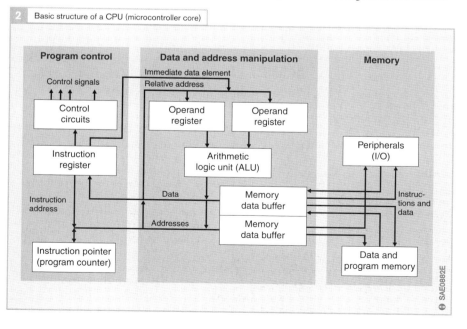

2 Basic structure of a CPU (microcontroller core)

registers are thus quasi-static extensions to the instructions. The interrupt control register, for example, specifies which interrupts are allowed and which are barred. Other control registers define the function of the ALU (arithmetic logic unit) or the peripheral modules.

Various operations can alter the operating status of the microcontroller. If, for example, a signal is received from an external source (interrupt), this generally causes the program sequence to divert to a defined memory address. While the interrupt service routine at that location is being processed, only signals with a higher priority can interrupt that procedure. All other interrupt requests are stored and only processed once the interrupt service routine in progress has been completed.

The status information that accrues in the process could be temporarily stored in the instruction memory. However, this would result in very long instructions. For that reason, special registers that store the status of the CPU are integrated in it in addition to the control registers. These status registers include, among others, the program status register, the interrupt status register, the multiplier status word, etc.

In order to reduce the number of loading and storage operations by the microcontroller, the register file architecture incorporates several accumulators (special computation registers in the CPU). These enable interim results or important variables to be held in the CPU. This increases the maximum clock speed and reduces power consumption.

Operand memories
Depending on the instruction-set architecture, there are various possible ways of providing the operands (data to be linked) involved in mathematical calculations or logical operations before and after the computing operation.

Memory-memory architecture
The memory-memory architecture uses the general random-access memory (RAM) to provide the operands. When doing so, it encodes the memory addresses of the operands and the result of a mathematical operation (function that links the operands) explicitly in the instruction. In that way, for example, two numbers that are both stored in the memory can be added together with a single instruction. The result can then be written back to the memory immediately. The term "memory-memory architecture" is derived from the storage location of the operands.

Accumulator architecture
With the accumulator architecture, the CPU has an integral memory cell that is permanently defined as both the source and the destination of every mathematical operation. This memory cell is called the accumulator. Only the address of the second operand is encoded in the instruction. Before every mathematical operation, the first operand must be copied from the memory to the accumulator by a load command. Following the operation, the result is copied back from the accumulator to the memory.

Memory-register architecture
In a memory-register architecture, there is a whole series of special memory cells called registers integrated in the CPU. Both operands are explicitly encoded in the instruction. However, only one of the two operands can be addressed directly in the memory using the memory address. The second operand and the result are addressed in a register. As in the case of the accumulator architecture, one of the operands must be copied from the memory to a register before the mathematical operation is performed. Following the operation, the result must then be written back to the memory. If, however, the number of registers is large enough, interim results can be held in registers rather than being continually copied backwards and forwards. The term "mem-

ory-register architecture" is derived from the location of the operands.

Register-register architecture
This last architecture category – also called "load-store architecture" – addresses both operands of an operation explicitly in the registers. Before any mathematical operation, therefore, both operands first have to be loaded into a register. The result is then copied back to the memory.

Operand addresses
Another important distinguishing feature among instruction-set architectures is the possible number of implicitly and explicitly encoded addresses. The operation $C = A + B$ requires three addresses:
- The address of operand A
- The address of operand B and
- The address of the result operand C

Instruction-set architectures which allow free choice of these three addresses (possibility of encoding three addresses) are referred to as "non-destructive instruction-set architectures". But as three addresses normally occupy too many bits in the instruction code, many architectures use an implicit method of addressing.

With implicit addressing, one of the addresses of the two source operands is also used as the destination address. Thus, the address of one of the source operands is used to store the result of the operation, and that means that the operand in question is overwritten, i.e. destroyed. That "destruction" has led to the adoption of the term "destructive instruction-set architecture".

Requirements placed on instructions
Instruction-set architectures differ from one another not only by virtue of their operand memories and operand addresses, but also on the basis of the length of their instructions. The following two very different requirements must be met in this connection:

Number of bits available for encoding
Different operations require very widely varying numbers of bits for encoding instructions of different levels of complexity. For example, the null operation NOP (No Operation) requires no operators (no addresses) nor any additional bits for defining the precise function. The operation MAC (Multiply & Accumulate: $A = A + (B \times C)$), on the other hand, requires three operators and additional bits for encoding the function, such as the behaviour in the event of a mathematical overflow (saturation addition), identifying the data format (Q format), etc.

Decoding complexity
Instructions of differing lengths are difficult to decode. First of all, the length of the current instruction has to be established. Then there is a check as to whether the instruction has been completely read. Both operations generally require several clock cycles and limit the possible processing speed to a substantial degree.

Length of instructions
There are currently three different concepts for the length of instructions.

- *Fixed-length instruction set:* every instruction is the same length, e.g. always 16 bits.
- *Multiple instruction sets:* it is possible to alternate between two different fixed-length instruction sets, e.g. 16 bits and 32 bits. The first bit of each instruction then indicates the format.
- *Multiple instruction length:* the length of the instructions varies according to complexity, e.g. the NOP command can have a length of 16 bits and the intersegment jump a length of 32 bits.

Instruction set
The complete instruction set of a microcontroller enables it to execute any expression of a higher-level programming language (e.g. C) by a sequence of instructions. The instructions are classified as follows:

- *Data processing operations:* mathematical, logical and conversion instructions.
- *Control instructions:* jumps, comparisons, etc.
- *Input/output instructions:* instructions for inputting and outputting data.
- *Memory instructions:* instructions for reading and writing data from/to the memory.

The bits available in an instruction for addressing the operands/result are limited. Embedded applications, however, require ever larger programs and data volumes, which leads to larger and larger numbers of addresses. As a result, the addresses become longer and longer and encoding them becomes more and more complex. There are a number of different methods for solving this problem.

Instruction execution

Execution of instructions on modern microcontrollers
The various phases of instruction execution are illustrated in Figure 3.
- *Fetch-instruction phase (1):* before an instruction can be executed, it has to be fetched from the program memory to the CPU. This involves first calculating its address, which is required in order to load the instruction into the CPU.
- *Decode-instruction phase (2):* the instruction transferred to the CPU is analysed. The relevant function is then initiated.
- *Fetch-operand phase (3):* the operands are fetched from the registers or memories to the executing unit (e.g. ALU).
- *Execute phase (data operations) (4):* once the operands have been transferred to the executing unit, the data operation specified by the instruction is executed.
- *Store-operand phase (5):* the result of the data operation is written back to a register or memory in this phase (which is therefore also called the "write-back phase").

Execution of instructions on older microcontrollers
On older microcontrollers, instructions are processed purely sequentially (CISC architecture: Complex Instruction-Set Computer). This requires a complete clock cycle for each phase:
- Phase 1: calculating the next instruction address and reading the instruction
- Phase 2: decoding the instruction

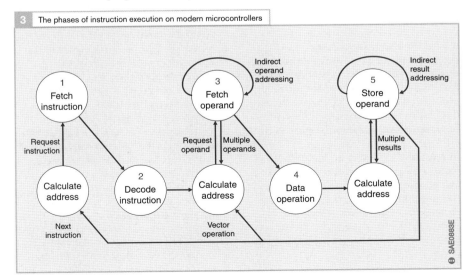

Fig. 3
The numbers indicate the five phases of execution of an instruction

- Phase 3: calculating the operand address and reading the operand
- Phase 4: executing the data operation
- Phase 5: calculating the address and storing the result
- Phase 6: additional cycles for calculating indirect addresses

Sequential execution is necessary with this process because certain units of the CPU are used a number of times in different phases. The arithmetic logic unit (ALU) in particular is used to calculate all addresses (instruction, operand, result, indirect address) and to execute the data operation. That means that the ALU is in use in every phase.

Parallel execution of instructions
By adding extra address calculation units (address ALUs), the various operations can be performed simultaneously. Thus, while one instruction is being decoded, the next is being addressed and loaded (pipelining). This is the way in which RISC (**Reduced Instruction Set Computer**) architectures generally operate. A RISC architecture is distinguished by the following characteristics:
- Reduced number of instructions
- Large number of general-purpose registers and
- Optimised pipeline

In a microcontroller with a pipeline architecture, the execution of multiple consecutive instructions is handled simultaneously. Rather like on a conveyor belt, the various instructions are at differing stages of execution and it is not possible to reduce the execution time of individual instructions.

In order to ensure that one instruction is completed in each cycle, all stages of a pipeline must operate simultaneously and independently of one another. Once the CPU pipeline is full, one instruction is completed in each clock cycle.

Microcontroller memories

It is the job of memories to store both data (constants and variables) and the instructions of a program and to make them available when required.

There are a number of different ways in which memories are implemented. This range of alternatives is necessary because there is no single implementation which suits all requirements. In order to satisfy the different requirements, microcontrollers usually have several different memory blocks.

The features by which memories are distinguished include the implementation location, the method of addressing data, the access method and volatility/non-volatility.

Implementation location
- External to the microcontroller (external memory)
- Integrated as separate module on the microcontroller (internal memory) or
- Integrated as a submodule in a peripheral module on the microcontroller (e.g. timer) or in the CPU (register)

Data addressing
- Free access (reading or writing) to all data by allocation of a fixed address (RAM: random-access memory).
- Direct addressing: specification of the memory address by an external source (e.g. the programmer).
- Indirect addressing: the microcontroller calculates the address itself according to external events.
- Content-related access to data by addressing based on the data content. For example, address the word whose first bit is "one" (CAM: content addressable memory, or associative memory). This type of memory is used in CAN communication modules, for example.

- Sequential access to data with the aid of a hard-wired buffer structure (e.g. FIFO: first in, first out; LIFO: last in, first out). These memories are primarily used for synchronising separate processor elements (e.g. microcontroller for antilock braking system [ABS] with two CPUs).

Access
- Read-only memory (ROM)
- Random-access memory (RAM) or
- Erasable read-only memory (EPROM)

Volatility/Non-volatility
- Volatile: data is lost when the power supply is switched off or
- Non-volatile: data is retained when the power supply is switched off

A RAM that has a back-up power supply from a battery can act as a non-volatile memory.

Microcontroller peripheral modules

In addition to the CPU and the memory, the peripheral modules (input/output modules or I/O modules) are essential components of microcontrollers. They make it possible to acquire external signals in a variety of ways and to manipulate the controlled variables by means of outgoing signals. Peripheral modules represent an "intelligent" link between the microcontroller and its environment. Each module has an internal connection to a system bus and external connections to sensors and actuators.

Figure 4 shows a schematic diagram of the structure of a peripheral module. Its tasks can be subdivided into the following:
- Communication with the internal system bus
- Communication with the environment
- Data storage
- Monitoring and timing and
- Fault detection

Peripheral modules: Addressing
Isolated peripherals (isolated I/O)
There are two separate address areas for memory and peripherals. Programming of the peripherals is strictly limited in this case as only special instructions can be used for peripheral modules.

Memory-mapped peripherals (memory-mapped I/O)
Peripherals and memory modules share a common address area. The advantage of this is that the large number of instructions for addressing memories can also be used for peripheral modules. However, this uses up valuable address space which can be disadvantageous with 4-bit and 8-bit microcontrollers.

Microcontrollers with 16-bit or 32-bit data width now almost exclusively use memory-mapped peripheral architectures.

silicon crystal, giving it n-type conducting properties.

Manufacturers of semiconductor components buy their raw material in the form of wafers that have been pre-doped according to precisely defined specifications.

Doping by diffusion
At high temperatures, the doping material can diffuse into the silicon crystal. The effect is achieved by creating a specific concentration level of the impurity atoms at the surface of the wafer. The difference in concentration causes the doping atoms to diffuse into the silicon wafer.

The process involves placing between 50 and 200 wafers in a kiln together and exposing them to boron or phosphorus compound vapours at temperatures of around 1,000 °C. Boron produces p-type regions while the effect of phosphorus is to create n-type regions. Surface concentration levels, temperature, and time, are the parameters that determine the penetration depth of the doping material.

Doping by ion implantation
In this process, the atoms of a gaseous doping material are first of all ionised in a vacuum and then accelerated by a high voltage (up to 300 kV) so as to be "catapulted" into the semiconductor. This allows particularly precise control of concentration level and localisation of the implanted material. Embedding of the implanted atoms and restoration of the crystal lattice requires subsequent thermal treatment.

Epitaxy
This doping method creates a monocrystalline doped semiconductor layer a few micrometres thick on a monocrystalline substrate. If gaseous silicon tetrachloride and hydrogen are passed over silicon wafers heated to around 1,200 °C in a quartz tube, the vapour breaks down and silicon is deposited at a rate of about 1 µm/min, form-

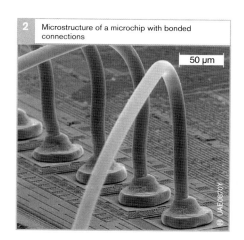

2 Microstructure of a microchip with bonded connections

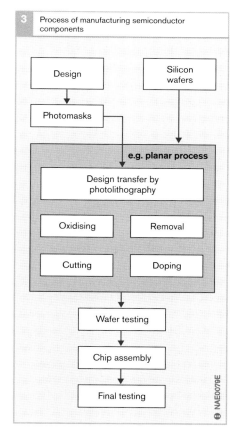

3 Process of manufacturing semiconductor components

ing a monocrystalline layer. If a defined quantity of doping impurity is added to the gas flow, an epitaxial layer is created whose electrical conductivity and conduction type differs fundamentally and abruptly from that of the substrate.

Photolithography and the planar process

In the *photolithography* process, the pattern of the component design is transferred to the wafer by using metal screens. When the design has been created, the data for producing the screen is saved onto a storage medium (e.g. magnetic tape or CD). It is then used to control a photographic exposure device that transfers the design pattern to photographic plates. This pattern is subsequently reduced by optical means to the actual size for the application and copied onto metal screens on which it is repeated as many times as will fit onto the size wafer being used.

As this method can be used for structures that are many times smaller than the wavelength of the light used, it will continue to be used in the future. The size of the smallest achievable structures depends on the wavelength of the light source. Using UV lasers, photolithographic methods have already produced structures as small as 0.08 µm in the laboratory (by comparison, a human hair has a diameter of 40...60 µm).

Even smaller structures can be produced using other methods such as X-ray or electron-beam lithography. However, those methods are substantially more expensive as they can only "expose" *one* IC at a time on the wafer. For that reason, they are only used in special cases.

Silicon wafers are easily oxidised by oxygen or water vapour. The oxide layer thus created prevents penetration by the doping atoms in the doping process. In the *planar process*, holes are created in the oxide layer so that when the wafer is doped, localised areas of n-type and p-type material are created.

The wafer is coated with a special lacquer and then exposed to light while covered with a metal screen. Following developing, the areas of the lacquer coating previously covered by the screen and the oxide layer below it can be removed by etching. The position, size and shape of the holes thus created match precisely the specified design. In the subsequent doping process in the diffusion furnace (Fig. 4) or by ion implantation, impurities which have a known electrical effect such as boron or phosphorus pass through those holes in the oxide layer into the silicon, thereby creating n-type or p-type regions in the desired locations. Afterwards, the oxide layer is removed again and the wafer is ready for the next stage in the manufacturing process.

The photolithographic process and the doping process are repeated as many times as required to produce the desired number of layers of different conductivity on the semiconductor component. In the case of complex integrated circuits, this may involve as many as 20 or more separate manufacturing stages. In order to provide the electrical connections between the resulting functional elements, the wafers are coated with aluminum or copper and then the metal conductor track patterns are formed. This process too may involve the formation of multiple layers of metal, one above the other.

Completion of the wafer production process is followed by electrical testing (preliminary testing) of the individual chips on the wafer. Chips that do not meet the specifications are marked with coloured dots. The wafers are then cut up into individual chips using a diamond cutting tool (Fig. 5). The functional chips are subsequently placed in metal or plastic casings and fitted with external connections (Fig. 6). After being hermetically sealed or encased in plastic, they go through a final testing stage.

4 Wafers about to be placed in a diffusion furnace

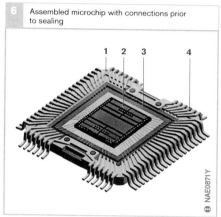

6 Assembled microchip with connections prior to sealing

Fig. 6
1 Base
2 Chip
3 Bonded connection wire
4 Connection pin

5 Cutting the silicon wafer into individual chips

Micromechanics

Micromechanics is the name given to the production of mechanical components from semiconductors (generally made of silicon) utilising semiconductor technologies. In addition to its semiconducting properties, the mechanical characteristics of silicon are made use of as well. This makes it possible to create sensors with the most minute of dimensions. The following techniques are used:

Bulk micromechanics

The silicon wafer material is removed up to the full thickness of the wafer by anisotropic (alkaline) etching with or without electrochemical resist. The material is etched away from the reverse side of the silicon layer (Fig. 1, item 2) in those areas where it is not protected by the etching screen (1). This method can be used to create very small diaphragms (a) with typical thicknesses of between 5 and 50 µm, holes (b) and bars and ridges (c), e.g. for pressure or acceleration sensors.

Surface micromechanics

The base material is a silicon wafer on the surface of which miniature mechanical structures are formed (Fig. 2). First of all, a "sacrificial layer" is applied and shaped (A) using semiconductor production processes (e.g. etching). This is deposition-coated with a layer of polysilicon approximately 10 µm thick (B) which is then shaped by vertical etching using a lacquer screen (C). In the final stage, the sacrificial oxide layer below the polysilicon layer is removed using gaseous hydrogen fluoride (D). In this way, structures such as flexible electrodes (Fig. 3) for acceleration sensors can be created.

Wafer bonding

Anodic and seal glass bonding are methods used to join wafers together by the action of electricity and heat or heat and pressure in order, for instance, to hermetically seal a reference vacuum or to protect sensitive structures by placing a cap over them.

1 Structures producible using bulk micromechanics

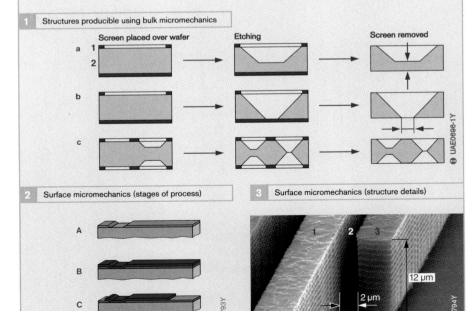

Fig. 1
- a Production of a diaphragm
- b Production of a hole
- c Production of bars and ridges

1 Etching screen
2 Silicon

2 Surface micromechanics (stages of process)

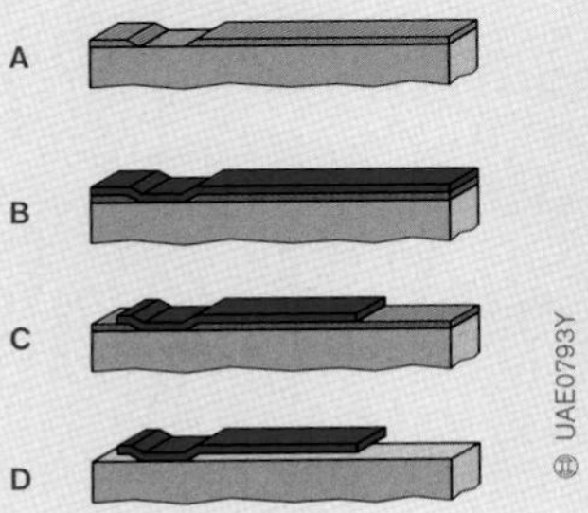

Fig. 2
- A Deposition and shaping of the sacrificial layer
- B Deposition of the polysilicon
- C Shaping the polysilicon
- D Removing the sacrificial layer

3 Surface micromechanics (structure details)

Fig. 3
1 Fixed electrode
2 Gap
3 Oscillating electrode

Conventional printed-circuit boards

The printed-circuit board has become an electronic component in its own right. It has to have precisely defined electrical and mechanical properties. In a motor vehicle, for example it may be required to withstand temperatures ranging from −40...+145 °C. The standards required in respect of EMC (electromagnetic compatibility), maximum current capacity and complexity are becoming more and more demanding. And at the same time, the circuit boards are continually expected to become smaller and cheaper – in spite of shorter product life cycles.

The basic material – the baseboard – is glass fibre. It can be rigid or semi-flexible. The conductor tracks on the surface are made from a thin layer of copper 12...70 μm thick (copper base). To protect it against corrosion, the surface of the copper is protected by a coating of tin alloy, gold or an organic surface sealant, depending on the type of application.

Designs

Depending on the complexity of the circuit, printed-circuit boards may be made up of a number of layers (Fig. 1). Circuit boards used in motor vehicles consist of between two and eight layers.

Single-sided circuit board
The pattern of conductor tracks and solder eyes is only on one side of the baseboard (a).

Double-sided, non-interconnected circuit board
There are patterns of conductor tracks on both sides of the circuit board. However, those two circuits are not connected to each other (b).

Double-sided, interconnected circuit board
The conductor track patterns on each side are interconnected by copper linings inside the holes (c).

Multilayer circuit board
In addition to the two outer layers (in this case the 1st and 4th layers), there are additional conductor layers on the inside of the baseboard (internal layers). Those conductor layers may be electrically connected with one another. To that end, the relevant conductor layers are connected to the copper linings inside the holes (d).

1 Types of printed circuit board

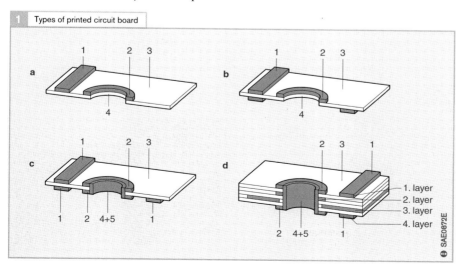

Fig. 1
Schematic diagram: conductor tracks are shown as raised surfaces

a   One-sided
b   Double-sided, non-interconnected
c   Double-sided, interconnected
d   Multilayer

1   Conductor track
2   Solder eye
3   Baseboard
4   Hole
5   Copper lining

Production process

Pattern plating and panel plating have become the established methods of manufacturing printed-circuit boards (Fig. 2).

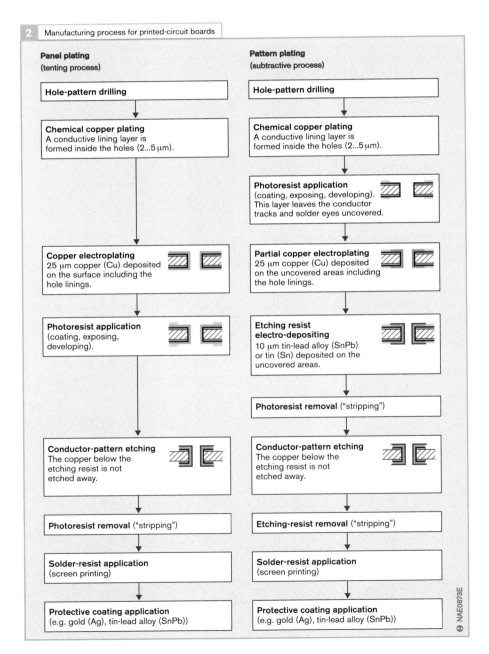

Fig. 2
The base material is coated on both sides with a copper film

Subsequent processing

Once the circuit board has been produced, it has to be fitted with the circuit components. These subsequent processing stages also demand the highest quality standards so as to ensure reliable operation of the finished product in the vehicle. There are two methods of mounting the components on the board – through-fitting and surface mounting.

Through-fitting method
With the through-fitting method, the component connections are passed through the holes in the circuit board and then soldered (Fig. 3a).

Surface-mounting method
The surface-mounting method SMT uses special electronic components whose connections lie flat against the surface of the circuit board (Fig. 3b). Such components are referred to as SMDs (surface mounted devices). Another advantage – in addition to the greater component density achievable – is that the components can be fitted to the circuit board fully automatically. The surface mounting method is therefore becoming more and more widely adopted. The component fitting machines used can achieve work rates of more than 60,000 components an hour, increasing productivity as a result.

Production sequence
There are various production systems that are used in the subsequent processing of circuit boards and they are classified according to the spatial arrangement of the production equipment and the work areas. The categories distinguished are job-shop, pool, flow-shop and flow production.

The process described below is an example of flow production (Fig. 4). According to the standard DIN 33 415, flow production involves "a process of operations organised according to the flow principle in a rigid sequence, aligned to a particular spatial arrangement and tied to a cyclic timing pattern". Furthermore, the flow principle involves "arrangement of the individual work stations according to the order in which the operations are performed". This method is also referred to as line production.

Production of the electronic circuits of an ECU involves the following stages:

Basic materials
The basic materials consist of printed-circuit boards that have not yet been fitted with circuit components. Usually, one or more circuit boards are arranged on a "panel" of standardised dimensions. After being fitted with components, the circuit boards are cut out of the panel.

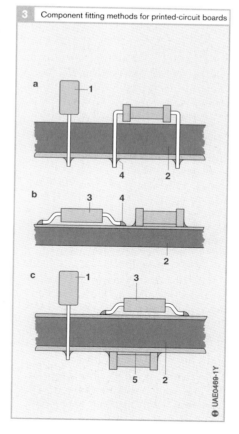

Fig. 3
a Through-fitting method
b Surface-mounting method
c Mixed mounting

1 Wired component
2 Circuit board
3 SMD
4 Soldering point
5 Cement

Solder-paste application
The first stage is the application of solder paste for the SMDs using a screen printing process. The solder paste is a mixture of metal powder, flux and other organic additives. It is pressed through the spaces in the printing screen onto the panel with the aid of a "scraper".

SMD fitting (reflow side)
An automatic component-insertion machine presses the SMDs into the solder paste applied to the panel.

Reflow oven
The panel passes through a reflow oven on a conveyor belt. The heat of the oven melts the solder paste on the panel. This electrically and mechanically connects the components to the pcb.

Position checking
The camera of a vision system examines the soldered connections (position checking system). Depending on the findings of the examination, the panel may be diverted to a repair station or automatically moved on to the next stage of production.

Turning the circuit board
In this stage, the circuit board is rotated through 180 degrees so that the "underside" is facing upwards and can be worked on. The way in which components are fitted to the underside depends on the nature of the second soldering stage.

If the *reflow/reflow method* is used, the process described above is repeated.

If the alternative *reflow/wave method* is used, since the soldering method applied differs to that from the top side, the SMDs are fitted differently and a cementing station is required.

Cementing station (underside)
At the cementing station, cement spots are first of all applied in those positions on the "underside" where SMDs are to be placed. This is done using pipettes or a screen, similar to the procedure for applying the solder paste. The cement holds the SMDs in position until they are soldered.

SMD fitting (underside)
At succeeding stations along the assembly line, the circuit components are placed on the cement spots. On this side of the circuit board, only components that are suitable for attachment by wave soldering can be fitted.

Cement hardening
In order to prevent the components falling off during the soldering process, the cement is hardened in an oven.

Component checking
A visual component checking system checks that all components are present and that they are in the correct positions on the underside of the circuit board. Non-compliant products are indicated and diverted to a repair station.

Turning the circuit board
The circuit board is once again rotated through 180 degrees, returning it to its original orientation.

Wired-component fitting
This stage involves fitting components that have leads attached (e.g. large coils and connectors) into ready-made holes (solder eyes) in the panel using the through-fitting method. They are subsequently soldered on the reverse side at the same time as the SMDs during the wave-soldering stage. The wired-component fitting stage completes the component fitting operations.

Wave soldering
The wave soldering stage solders the connections of all SMDs and "wired components" on the underside. The wave soldering process involves the following three stages:
1. Flux is applied to the underside of the panel.

2. The panel then passes through a pre-heating zone so that the components are not damaged by the sudden increase in temperature in the subsequent stage.
3. The panel passes over a wave of liquid solder created by a jet. As it does so, the solder is deposited on the soldering points (pads) on the panel. The solder resist on the circuit board prevents the solder attaching itself in the wrong places.

Position checking
Another camera checks the finished soldered connections. If faults are found, the panel is passed to a repair station.

In-circuit testing (ICT)
The purpose of the in-circuit test is to check that the electrical circuit is working properly. The testing equipment is connected via a testing adaptor and checks the components for correct function and electrical readings.

Cutting
The individual circuit boards are now cut out of the panel using computer-controlled machines.

Final assembly
In the final-assembly stage, the circuit board is fitted into the ECU casing.

Temperature testing
The finished ECU is subjected to very high temperatures in order to test its ability to withstand extreme conditions. The tests simulate operation of the ECU in real conditions. In this way, component or soldering faults can be detected.

Final testing
Before the ECU leaves the production line, it is subjected to a final test to make sure it performs the functions required of it in actual use.

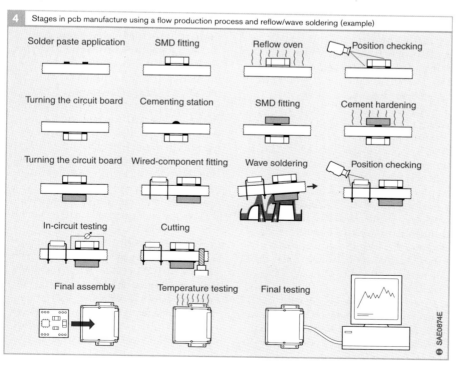

4 Stages in pcb manufacture using a flow production process and reflow/wave soldering (example)

Film and hybrid circuits

Film circuits
In film integrated circuits, passive circuit elements (usually conductor tracks, insulation and resistors, but also capacitors and inductors) are applied to a base (substrate) in layers. The advantages of these circuits are:
- Fine structures (up to approx. 10 µm) with high element density and
- Good high-frequency characteristics

Those advantages are counterbalanced by relatively high production costs.
The thickness of the films originally gave rise to the terms "thin-film circuit" and "thick-film circuit". Classification is now based on differences in the method of production.

Thin-film circuits
With thin-film circuits, the films are applied to glass or ceramic substrates using vacuum-coating processes.

Thick-film circuits
For thick-film circuits, the preferred method of manufacture involves application of the films to ceramic substrates by screen printing processes followed by firing.

Ceramic multilayer substrates
Ceramic multilayer substrates are made up of unfired ceramic foils onto which conductor tracks are applied by screen printing methods. Several such foils are then combined to form a multilayer laminate which is fired at 850...1,600 °C so that it becomes a solid ceramic body with integral conductor channels. A typical hybrid substrate consists of four or five layers. Particularly high wiring densities can be achieved with LTCC (low-temperature cofired ceramic) line-line substrates.

To make the electrical connections between layers, holes are punched in the individual films and filled with metal paste. Those holes are referred to as so-called "vias". By using suitable material systems, resistors and capacitors can also be integrated. The wiring densities of these circuits is considerably greater than with thick-film circuits.

Hybrid circuits
Hybrid circuits are integrated film circuits with additional discrete components such as capacitors and integrated semiconductor circuits (ICs) that are attached by soldering or cementing. The use of unpacked semiconductor chips, to which the connections are made by bonding, or SMDs makes high component densities possible. Extremely small hybrid ECUs (microhybrids) can be made by using ceramic multilayer substrates. The advantages of these circuits are:
- High permissible installation temperatures due to their good heat dissipation properties
- Compactness of design combined with good shock resistance and
- Good resistance to mediums

Hybrid circuits are therefore particularly suited to use in telecommunications systems and for automotive applications where they can be found in ABS, traction control, ESP, transmission control and engine-management ECUs (engine-mounted).

Figure 1 shows the basic stages in the production of a hybrid-circuit substrate. The tapes are punched with holes for the vias for each separate wiring layer and the holes are then filled with silver paste (Fig. 2a). Screen-printing stations print the conductor tracks onto the film. The various layers are aligned with each other, laminated and then fired at 890 °C. A specially controlled sintering process keeps the ceramic firing to within tolerance limits of roughly 0.03 %. This is important for the packing density. The circuit's resistors are printed on the reverse side of the substrate and fired (Fig. 2b).
For bonding on the top side, the surfaces are finished using a plating process adapted to the LTCC. The spacing of the microcontroller contacts (bond land grid on the

substrate) ranges from 450 to 260 μm. The component connections are bonded using 32-μm gold wire and 200-μm aluminum wire.

Alongside the electrical vias, there are also thermal vias with a diameter of 300 μm for optimum cooling of ICs with high power losses. The effective thermal conductivity of the substrate can thus be increased from approx. 3 W/mK to 20 W/mK.

All components are cemented with conductive cement. There are two methods used for final assembly of the finished hybrid.

Method 1: the finished hybrid is cemented to the steel panel of the casing using heat-conducting cement and connected to the glass feed-through for the connector using bonded 200-μm aluminum wire. The casing is then hermetically sealed.

Method 2: the finished hybrid is cemented to the aluminum casing using heat-conducting cement and connected to the plastic-encased connector pins using bonded gold or 300-μm aluminum wire. Before the cover is cemented in place, a gel is applied to protect the circuit.

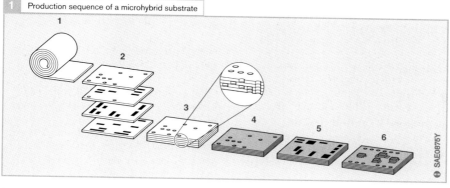

Fig. 1
1 Unfired glass-ceramic film
2 Punching of holes, filling with conductor paste and printing of conductor tracks
3 Aligning and stacking (laminating)
4 Sintering
5 Printing of resistors (reverse side), firing and plating of the bonding pads (top side)
6 Fitting of components and wire bonding

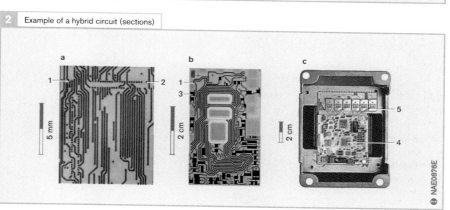

Fig. 2
a Inner layer
b Reverse side with resistors
c Top side inside ECU

1 Conductor track
2 Via
3 Resistor
4 Microcontroller
5 Bonded wire

Glossary and tables for automotive microelectronics

This glossary provides a quick guide to the most important automotive microelectronics terms.

An arrow pointing to a term printed in italics (e.g. → *semiconductor*) indicates that this italicised term is also explained in this glossary.

A

Analog circuit → *Electronic circuit* with which analog signals are processed – as opposed to → *digital circuit*. Analog signals are infinitely variable within a specific range. Examples: battery voltage, speedometer display.

Avalanche breakdown: Sudden increase in the number of → *charge carriers* resulting from the release of bound electrons from the crystal lattice of a → *semiconductor* in the → *space-charge zone* caused by a high-strength electric field (→ *avalanche effect*).

Avalanche effect: see → *avalanche breakdown*

B

Binary code: Information encoded in the form of binary numbers and therefore consisting of a sequence of the digits 0 and 1.

Bipolar transistor: → *Transistor* with three regions of differing conductivity (→ *p-type semiconductor* and → *n-type semiconductor*). It is constructed either as a p-n-p or n-p-n transistor. The terminals are called emitter (E), base (B) and collector (C).

Bit: Contraction of **binary digit**. A single digit in → *binary code*

Breakdown voltage: reverse voltage at a → *p-n junction* above which a small increase in voltage brings about a large increase in the → *reverse current*.

Bus: Connection linking the elements of a → *digital circuit* and consisting of a single lead (serial bus) or multiple leads (parallel bus). A bus can connect up a large number of separate elements with different functions but with electrically identical interfaces. A parallel bus transmits all bits of a data "word" simultaneously (one lead for each bit). A serial bus transmits the bits sequentially (one after the other).

Byte: Coherent group ("word") of eight → *bits*.

C

Capacitance: Property of a capacitor which indicates how much electrical charge – and therefore electrical energy – it can store when a specific voltage is applied.

CCD: Abbreviation for → *charge-coupled device*.

Central processing unit (CPU): Main operating unit of a → *microcontroller*.

Charge-coupled device (CCD): Silicon image sensor in which incident light striking a transparent electrode releases → *charge carriers* in proportion to the light intensity and exposure time. The charge carriers collect in a Si-SiO$_2$ boundary layer ("potential well").

Charge carrier: Carrier of electric charge, usually of a single elementary unit of charge. Free electrons and anions are examples of negative charge carriers. → *Lattice defects* (holes), cations and protons are examples of positive charge carriers.

Chip: Basic building block of microelectronic circuits. Base material is a monocrystalline block of semiconductor material, primarily silicon.

Where does the term "electronics" come from?

This term really originates from the ancient Greeks. They used the word electron for "amber" whose forces of attraction for wool and similar materials had already been described by Thales von Milet 2,500 years ago.

The term "electronics" originates directly from the word "electrons". The electrons, and therefore electronics as such, are extremely fast due to their very small mass and their electrical charge.

The mass of an electron has as little effect on a gram of any given substance as a 5 gram weight has on the total mass of our earth.

Incidentally, the word "electronics" is a product of the 20th century. There is no evidence available as to when the word was used for the first time. Sir John Ambrose Fleming, one of the inventors of the electron tube could have used it around 1902.

The first "Electronic Engineer" though goes back to the 19th century. He was listed in the 1888 Edition of a form of "Who's Who", published during the reign of Queen Victoria. The official title was "Kelly's Handbook of Titled, Landed and Official Classes". The Electronic Engineer is to be found under the heading "Royal Warrant Holders", that is the list of persons who had been awarded a Royal Warrant.

And what was this Electronic Engineer's job? He was responsible for the correct function and cleanliness of the gas lamps at court. And why did he have such a splendid title? Because he knew that "Electrons" in ancient Greece stood for glitter, shine, and sparkle.

Source:
"Basic Electronic Terms" ("Grundbegriffe der Elektronik") – Bosch publication (reprint from the "Bosch-Zünder" (Bosch Company Newspaper)).

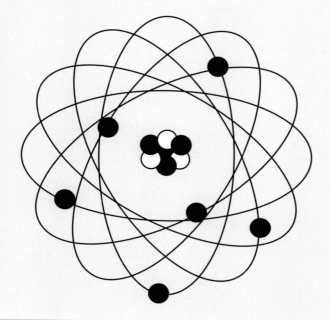

Automotive sensors

Today's state-of-the-art vehicle is equipped with a large number of sensors. These can be regarded as the vehicle's "sensory organs", and from their physical or chemical inputs they generate the electrical output signals needed by the vehicle's ECUs for implementing the closed and open-loop control functions used in its engine-management systems, and in its safety, comfort, and convenience systems.

Basics

Terms and definitions
The terms sensor, probe and pickup are synonymous. This manual uses the term "sensor". Taking into account disturbances Y_i, the sensor converts a physical or chemical (usually non-electrical) input quantity Φ into an electrical output quantity E. This often takes place with the help of non-electrical intermediate stages. The electrical sensor outputs are not only in the form of current and voltage alone, but are also available as current or voltage amplitudes, frequency, phases, pulse durations, and cycles or periods of an electrical oscillation, or as the electrical parameters "Resistance", "Capacitance", and "Inductance" (Figs. 1 and 2).

A sensor can be defined using the following equation:
1. Sensor output signal
$$E = f(\Phi, Y_1, Y_2, ..) \quad (1)$$
2. Required measured variable
$$\Phi = g(E, Y_1, Y_2, ..) \quad (2)$$

If functions f or g are known, these equations represent a "sensor model" with which the required measured variable can also be derived mathematically and practically without error using the output signal E and the disturbance Y_i (refer to "Intelligent sensors").

There are no specific rulings on whether the sensor can contain part of the signal-processing or not.

Applications in the vehicle
As part of the vehicle's periphery, the sensors and actuators form the vehicle's interface to its complex drive, braking, chassis, and bodywork functions, as well as to the vehicle guidance and navigation functions and the (usually digital) ECUs which operate as the processing units. As a rule, a matching circuit (refer to "Signal processing") adapts the sensor signals to the standard form required by the ECUs (measuring chain, measured-value acquisition, Fig. 3).

These matching circuits are tailor-made for specific sensors and are adapted to the particular vehicle. They are available in integrated design and in a wide variety of versions. They are a highly essential and worthwhile complementary device for the sensors described below, but due to lack of space are not gone into in detail. It would be impossible to use sensors in practice without these matching circuits. To be precise, definition of the sensor's measuring quality applies to the sensor and the matching circuit.

The vehicle can be regarded as a highly complex process, or control loop, which can be influenced by the sensor information from other processing units (ECU), as well as from the driver using his/her controls. Display units keep the driver infomed about the status and the process as a whole. Fig. 4 provides an overview of the abundance of electronic vehicle systems which are already on the market. Undoubtedly, this number will increase immensely in the years to come.

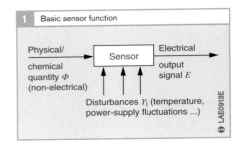

1 Basic sensor function

Automotive sensors Basics

Classification
Automotive sensors can be divided into three categories:

Assignment and application
Here, sensors can be allocated to two different groups:
- Functional sensors mainly used for open and closed-loop control assignments
- Sensors for safety and safeguarding (theft-deterrent) assignments and
- Sensors for vehicle monitoring (On-Board Diagnosis (**OBD**), fuel-consumption and wear parameters) and for driver/passenger information

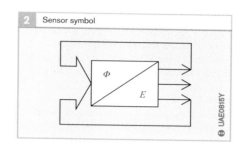

2 Sensor symbol

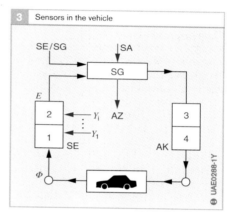

3 Sensors in the vehicle

Fig. 3
1 Measuring sensor
2 Matching circuit
3 Driver
4 Actuators
AK Actuator
AZ Display
SA Switch
SE Sensor(s)
SG ECU
Φ Physical quantity
E Electrical quantity
$Y_{1...}$ Disturbances

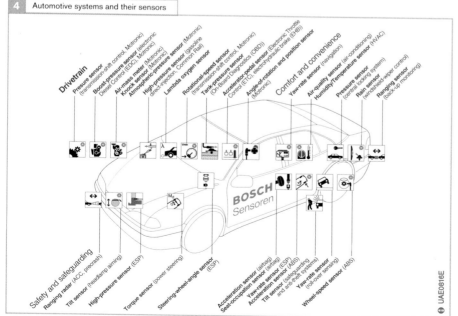

4 Automotive systems and their sensors

Curve types

In order to perform their various monitoring and closed and open-loop control assignments, sensors must feature a variety of different characteristic curves (Fig. 5):

Continuous, linear curves
Such curves are used mainly for control assignments covering a wide measuring range.
Linear curves are also distinguished by uncomplicated testing and calibration.

Continuous, non-linear curves
Such curves are often used for the closed-loop control of a measured variable across a very restricted measuring range (e.g. exhaust-gas control to $\lambda = 1$, vehicle spring-deflection level). When, for instance, the permissible deviation relative to the measured value is demanded throughout the complete measuring range (air-mass meter), curves which feature both pronounced non-linearity and a special shape (e.g. logarithmic) are at an advantage.

Discontinuous, two-step curves
Such two-step curves (possibly even featuring hysteresis) are used for limit-value monitoring in such cases where remedial measures are easy to apply when the limits are reached. If remedial measures are more difficult, then multiple-step curves can be used for an earlier warning.

Type of output signal

Sensors also differ with respect to their output signals (Fig. 6):

Output signals analog to:
- Current/voltage or a corresponding amplitude
- Frequency/period and
- Pulse duration/pulse duty factor

Discrete output signal:
- Two-step (binary coded)
- Multi-step, with irregular steps (analog coded) or
- Multi-step, with equidistant steps, that is with uniform spacing (analog or digital coded)

Furthermore, the sensors differ in their output signal being continuously available or only at discrete instants in time (continuous and discontinuous respectively). For instance, the signal is bound to be discontinuous if it is digital and outputted in bit-serial form.

Main requirements, trends

In contrast to the everyday universal-application sensors available on the market, automotive sensors are tailor-made to comply with the requirements of the vehicle's special electronic systems. The research and devel-

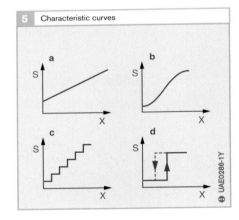

Fig. 5
S Output signal
X Measured variable
a Continuous linear
b Continuous non-linear
c Discontinuous multi-step
d Discontinuous two-step

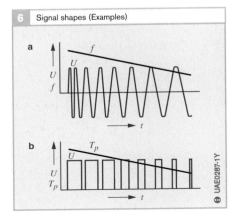

Fig. 6
a Output signal U Frequency f
b Output signal U Pulse duration T_p

opment departments are responsible for ensuring that they satisfy the five major demands as listed in Fig. 7. These requirements are also reflected in the most important trends in sensor engineering.

High reliability
In accordance with their assignments, automotive sensors are sub-divided into the following reliability classes, given in descending order of severity:
- Steering, brakes, passenger protection
- Engine/drivetrain, chassis/tires
- Comfort and convenience, OBD, information, and theft-deterrence

In automotive engineering, the specifications for the highest reliability class correspond to those for aviation and astronautics, and in some cases necessitate similar measures being taken.

Development trends:
Appropriate design measures guarantee built-in reliability. For instance, this necessitates the use of reliable, top-quality components and materials, coupled with rugged and well-proven techniques and engineering. Plug-in connections are a potential source of trouble, and to avoid them system integration takes place as far as possible. This is also the aim of "radio-scanned sensors" based on the antenna-coupled SAW[1]) elements which do without wiring completely. Safety considerations can dictate that redundant sensor systems are used. That is, sensor systems connected in parallel which perform identical measuring functions.

Low manufacturing costs
On board a modern-day, state-of-the-art vehicle, there can easily be as many as 60 to 70 sensors. Compared to other sectors of sensor application, this is a very large number and is only possible as long as low manufacturing costs are achieved. Typically, target costs are in the range between 2 and 50 DM (1 to 25 €), and are often 100 times lower than those of conventional sensors with the same performance, whereby when an innovative technology is introduced costs start at a high level, and then usually drop in the course of time.

Development trends
For the most part, sensor manufacture uses highly efficient automated production methods. For example, semiconductor sen-

[1]) SAW Surface Acoustic Wave

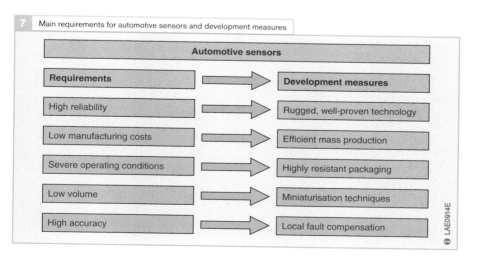

7　Main requirements for automotive sensors and development measures

sors are manufactured using "batch processing" in which there are typically 100 to 1,000 sensors on a single Si wafer.

On the other hand, such manufacturing equipment is only an economic proposition when correspondingly large numbers of sensors are produced. These quantities sometimes exceed an automotive-industry supplier's own in-house requirements, and can commonly be between 1 and 10 million per year. Here, the high numbers of sensors needed by the automobile industry played an unprecedented and revolutionary role, and set completely new standards.

Severe operating conditions
Sensors are installed at particularly exposed positions on the vehicle. Accordingly, they are subjected to particularly severe loading and must be able to withstand a wide variety of different stresses:
- Mechanical (vibration, shock)
- Climatic (temperature, dampness)
- Chemical (e.g. splashwater, saline fog, fuel, lube-oil, battery acid)
- Electromagnetic (irradiation, wire-conducted spurious pulses, excess voltages, polarity reversal)

Due to the inherent advantages involved, sensors are preferably installed directly at the measuring point. This tendency though has led to a considerable increase in the severity of the requirements made on the sensor.

Development trends:
Protective measures must be introduced to cope with the above loading. This necessitates a very high know-how level in the field of sensor "Packaging". Among other things, this includes:
- Passivation and connecting techniques
- Sealing and joining techniques
- EMC measures [2]
- Low-vibration installation
- Service-life, test, and simulation methods
- Use of highly resistant materials together with detailed knowledge of the loading to which the sensor will be subjected at the

particular installation point. Total competence in the selection and implementation of suitable protective measures is the decisive factor for sensor quality. Such measures often account for a far greater share of the overall sensor costs than the actual measuring element itself.

Fiber-optic sensors
In such sensors, the light flowing in the optical fiber (glass, plastic) can be modified as a function of the measured variable. Up to the point where the optical signal is converted back to an electrical signal, these sensors are regarded as being particularly immune to electromagnetic disturbances. Insofar as they are applied at all in the future, this will necessitate extensive development work on low-priced measuring elements and the accompanying technologies.

Low-volume design
On the one side the number of electronic systems in the vehicle continues to climb steadily. On the other, today's vehicles are becoming more and more compact. These facts, together with the need to retain the high level of passenger-compartment comfort forces development to concentrate on an extremely low-volume design. Furthermore, the increasing demand for further improvements

[2] EMC Electromagnetic Compatibility

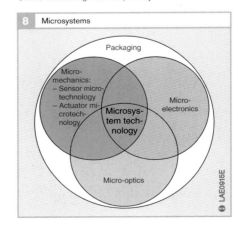

8 Microsystems

in fuel economy mean that minimization of the vehicle's weight is of prime importance.

Development trends

Widespread use is made of the familiar technologies applied in circuit engineering for the miniaturisation of electronic components (Fig. 8):
- Film and hybrid technologies (deformation-dependent resistors, thermistors, and magnetoresistors)
- Semiconductor techniques (Hall-effect and temperature sensors)
- Surface and bulk micromechanical techniques (silicon pressure and acceleration sensors, Fig. 9)
- *Microsystem technologies* (combinations of two and more microtechnologies such as microelectronics and micromechanics)

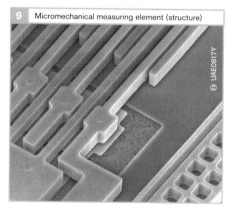

9 Micromechanical measuring element (structure)

Often, the indispensable mechanical part belonging to the function with which the sensor is associated is used to accomodate the sensor, and acts as its "housing". This combination of electronics and mechanics is known as *mechatronics* and is coming more and more to the forefront in the search for cost and space savings. In the foreseeable future, practically all systems will operate on this basis.

High accuracy

In comparison to the probes and sensors used for instance in the processing industry, with only a few exceptions (e.g. the air-mass meter) the demands on automotive-sensor accuracy are relatively modest. Generally, the permissible deviations are $\geq 1\%$ of the measuring-range final value. This applies in particular when considering the unavoidable effects of ageing. The permissible deviations are normally achieved by the application of complex techniques to compensate for manufacturing tolerances, and to balance the effective compensation measures used against interference. Particularly since the above-mentioned requirements have for the most part been satisfied, continually more demanding and sophisticated systems are imposing higher and higher demands in this sector.

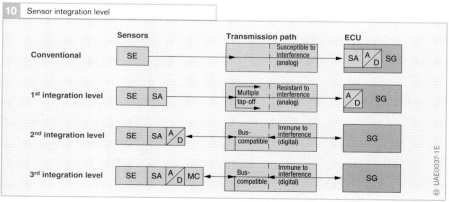

Fig. 10
SE Sensors
SA Signal conditioning (analog)
A/D Analog-digital converter
SG Electronic control unit (digital)
MC Microcomputer

Development trends
Initially, a tightening up of the tolerances in manufacture, and refinement of the calibration and compensation techniques help to guarantee a high level of accuracy. An important step forward here is the hybrid or monolithic integration of the sensor and signal electronics directly at the measuring point, up to complex digital circuits such as analog/digital converters and microcomputers (Fig. 10).

Such microsystems are also known as "intelligent sensors". They take full advantage of the sensor's inherent accuracy, and offer the following features:
- Reduce the ECU's working load
- Uniform, flexible, and BUS-compatible interface
- Sensors can be used for a number of different functions
- Due to local amplification and demodulation, it is possible to utilise low-output and high-frequency measuring effects
- The correction of sensor deviations at the measuring point, and the mutual calibration and compensation of both sensor and electronics, is simplified and improved by storing the individual correction information in a PROM

While simultaneously detecting and digitising disturbances, "intelligent sensors" can almost perfectly calculate the required measured variable by applying the mathematical sensor model given in the Paragraph "Terms, Definitions" (with Equations 1 and 2). Here, the item-specific model parameters (individual sensor samples are used in calculating a model auxiliary quantity) are defined in a preceding process which is equivalent to the calibration as previously performed, and stored in a PROM integrated with the sensor (Fig. 11, correction module).

In this manner, it is possible to considerably improve the sensor's static and dynamic characteristics (evaluation of the differential equation which defines the dynamic performance).

Local electronic circuitry (in other words directly at the measuring point), necessitates the use of *multi-sensor structures* which use a number of identical sensors, or a number of different sensors, to register a variety of highly complex facts and reduce these to their basic information content. This latter process can also take place locally. This applies in particular to *image sensors* which in future will play an every increasing role in registering the situation inside and outside the vehicle.

With a number of integrated *pressure sensors*, it is possible to not only increase the reliability of the measurement, but also to reduce the ageing drift (deviation due to ageing) by applying mean-value generation. If the individual sensor elements are designed for differing measuring ranges – and at the same time feature high overload capabilities (e.g. capacitive) – such a sensor can be used to considerably extend the high-acccuracy measuring range.

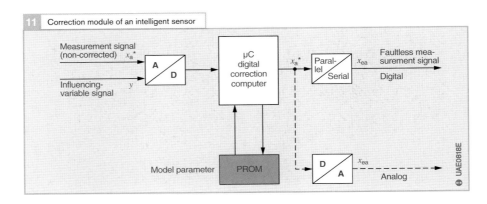

11 Correction module of an intelligent sensor

Miniaturization

Thanks to micromechanics it has become possible to locate sensor functions in the smallest possible space. Typically, the mechanical dimensions are in the micrometer range. Silicon, with its characteristics has proved to be a highly suitable material for the production of the very small, and often very intricate mechanical structures. With its elasticity and electrical properties, silicon is practically ideal for the production of sensors. Using processes derived from the field of semiconductor engineering, mechanical and electronic functions can be integrated with each other on a single chip or using other methods.

Bosch was the first to introduce a product with a micromechanical measuring element for automotive applications. This was an intake-pressure sensor for measuring load, and went into series production in 1994. Micromechanical acceleration and yaw-rate sensors are more recent developments in the field of miniaturisation, and are used in driving-safety systems for occupant protection and vehicle dynamics control (Electronic Stability Program ESP). The illustrations below show quite clearly just how small such components really are.

Micromechanical acceleration sensor

Electric circuit
Bonding wire   Sensor chip
Evaluation circuit

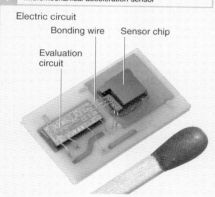

Comb-like structure compared to an insect's head
Suspension spring   Seismic mass with movable electrodes

200 µm   Fixed electrodes

Micromechanical yaw-rate sensor

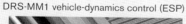

DRS-MM1 vehicle-dynamics control (ESP)

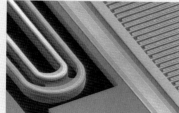

100 µm

DRS-MM2 roll-over sensing, navigation

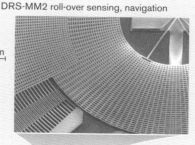

3.3 cm

Position sensors (travel/angle)

Characteristics

Position sensors register the most varied forms of travel and angular position, and are certainly the most common sensor in the vehicle. In this sector of applications, activities have long since been directed at changing over to *proximity* or *non-contacting* sensor principles. Such sensors are wear-free and thus have a longer service life as well as being more reliable. The costs involved though, often force vehicle manufacturers to retain the "wiper-type" sensor principle, and such sensors still perform efficiently enough at a number of points in the vehicle.

Position sensors are often referred to as so-called *"extensive sensors"*. Here, sensor size and measured quantity are always interrelated, although with regard to wave-propagation sensors, this only applies to a limited degree. In the classification method used here, sensor and measuring principles which only measure extremely minute shifts/movements (a few μm, for instance in the case of expansion) are allocated to other measured variables such as force, torque, and acceleration. Only those position sensors will be dealt with which are used for measuring larger distances ($\geq$1 mm) and angles ($\geq$1°).

Measured variables: Overview

In this sector there are a large number of applications in which position represents the actual measured variable. This is shown by the Table on the right.

1 Travel/angular position as the direct measured variable

Measured variable	Measuring range
Throttle-valve setting on the spark-ignition engine	90°
Accelerator-pedal/brake-pedal position	30°
Seat, headlamp, rear-view mirror position	
Control-rack travel and position for diesel in-line fuel-injection pumps	21 mm
Angular setting of the injected-fuel-quantity actuator on the diesel distributor pump	60°
Fuel level in the fuel tank	20...50 cm
Clutch-actuator travel	50 mm
Distance from vehicle to vehicle or between vehicle and obstacle	150 m
Steering-wheel angle	±2·360° (±2 revolutions)
Angle of inclination (tilt)	15°
Angle of vehicle travel	360°

In other cases, the measured position or angle represents a different measured variable (Table 2).

2 Travel/angular position as the indirect measured variable

Measured variable	Measuring range
Spring-deflection travel (headlamp range, vehicle inclination or tilt)	25 cm
Torsion angle (torque)	1...4°
Deflection of a sensor plate (throughflow)	30...90°
Deflection of a spring-mass system (acceleration)	0.5...1 mm

In practice, *"incremental sensor systems"* are also often referred to as angular-position (or angle-of-rotation) sensors, even when they are used for measuring rotational speed. Since the increments (steps with which a given quantity increases) which have to be measured with these sensors in order to measure the deflection angle must be counted with the correct preceding sign (in other words, added), these sensors are in reality not angular-position (or angle-of-rotation) sensors. Due to the danger of the counter being falsified due to spurious pulses, such angular-position measuring systems are only in limited use. Fixed, directly locatable reference marks only provide very little help in this dilemma. Another disadvantage of such angular-position measuring systems is the fact that the absolute position is lost when the power supply is switched off. Here, it is no use storing the final position in a non-volatile memory, since most angular positions can change mechanically. This also applies when power has been removed.

Measuring principles

Potentiometer-type sensors

For measuring purposes, the wiper-type potentiometer (Fig. 1) uses the correspondence between the length of a wire or film resistor (Cermet or conductive plastic) and its resistance. At present, this is the lowest-priced travel/angle sensor. Voltage is usually applied to the measurement track through low-resistance series resistors R_V (these can also used for calibration of zero point and curve-slope). The shape of the curve is influenced by shaping the measuring track (or only sections of it). Wiper connection is usually through a second contact conductor track with an identical surface applied over a low-resistance conductor track.

Wear and falsification of measured values can be kept to a minimum by keeping the electrical loading of the pick-off as low as possible ($I_A < 1$ mA) and by dust-proof encapsulation. One of the prerequisites for low wear is the optimal friction pairing between the wiper and the conductor track. To this end, wipers can be of "spoon" or "scraper-shape" design, and one or more can be mounted. Brush-shaped wipers are also in use.

A whole range of clear advantages are faced by a considerable number of serious disadvantages:

Advantages of the potentiometer-type sensors
- Simple design
- Very extensive measuring effect (measurement range ≅ supply voltage)
- No electronic circuitry required
- High level of interference immunity
- Broad temperature range (up to 250°C)
- High precision (better than 1 % of full range)
- Wide measuring range (almost 360° is possible)
- No difficulty with redundant design
- Calibration possible (Laser etc.)
- Flexible characteristic curve (variable conductor-track width)
- Flexible assembly (on curved as well as on flat surfaces)
- Wide range of manufacturers
- Samples can be supplied quickly

Disadvantages of potentiometer-type sensors
- Mechanical wear, abrasion
- Measuring errors due to abraded particles
- Problematic regarding operation in fluids
- Variation in contact resistance between wiper and measurement track
- Strong acceleration or vibration can result in wiper lift-off
- Testing is costly
- Limited possibilities of miniaturization
- Noise

Examples of potentiometer-type sensors
- Sensor-plate potentiometer (KE- and L-Jetronic)
- Throttle-valve angular-position sensor (M-Motronic)
- Accelerator-pedal sensor, accelerator-pedal module
- Fuel-level sensor

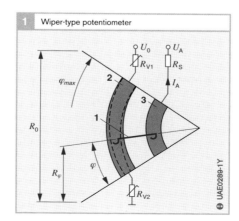

Fig. 1
1 Wiper
2 Resistance track
3 Contact conductor track
I_A Wiper current
U_0 Supply voltage
U_A Measurement voltage
R Resistance
φ_{max} Maximum angle of rotation
φ Measured angle

Magnetically inductive sensors

Of all the sensors using proximity and non-contacting principles for position measurement, the magnetic sensors have proved to be the most rugged and most insensitive to interference. This applies in particular to those principles relying on alternating current, in other words magnetically inductive principles. Compared with a micromechanical sensor though, the coil configuration needed here requires far more space. This means therefore that there is no favorable possibility of redundant (parallel measurement) design. Furthermore, coil contacting is less favorable from the costs and reliability viewpoint. Although there are a multitude of different principles in use for this form of sensor, only two have come to the forefront for automotive applications. Regarding their operating concepts, these are very similar to each other.

Eddy-current sensors

When an electrically conductive flat or curved (damping) disc (for instance Al or Cu) approaches a coil (usually ironless) to which high-frequency AC has been applied, it has an effect upon the coil's equivalent resistance and its inductance. This is the result of the eddy currents generated in the disc (otherwise known as a spoiler) due to the increasing magnetic coupling. The disc's position represents the measured travel (Fig. 2).

Although this principle functions satisfactorily in the kHz range, for detection of rapid movements it is recommended that a higher operating frequency in the MHz range is applied. This also uses less current. On the other hand, this generally means that the electronics must be in or on the sensor. In order to convert the measuring effect into an electrical output signal, either the damping-effect (equivalent resistance) or field-displacement (inductance) principles can be applied. In the first (damping effect) case, a variable-amplitude oscillator can be used, and in the second a variable-frequency oscillator or a constantly supplied inductive voltage divider (differential configuration).

There are many ways to adapt the eddy-current principle to the measuring assignment. It is just as suitable for the measurement of large travels and angles as it is for smaller quantities. Sensors applying this principle have low temperature sensitivity.

Short-circuiting-ring sensors

In contrast to eddy-current sensors, the coil of a short-circuiting-ring sensor has a straight or curved U or E-shaped soft-magnetic, (usually) laminated iron core (Fig. 3). The coil, or short-circuiting-ring ("spoiler"), is of highly conductive material (Cu or Al), and is located around one (or all) of the core limbs. Thanks to the iron core, such sensors have a far larger inductance than eddy-cur-

Fig. 2
1 Spoiler
2 Eddy currents
3 Air-core inductor
4 Variable-damping oscillator
5 Demodulator
s Measured travel
$A(s)$ Oscillator voltage
$U_A(s)$ Output voltage

Fig. 3
1 Short-circuiting ring
2 Soft-magnetic core
3 Coil
I Current
I_W Eddy current
$L(s)$ Inductance and
$\Phi(s)$ Magnetic flux for measured travel s

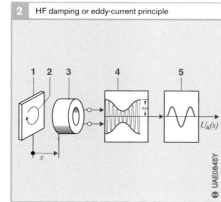

2 HF damping or eddy-current principle

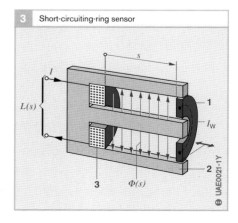

3 Short-circuiting-ring sensor

rent sensors. This means that they operate well at low frequencies, and do not necessarily require their signal electronics locally, that is directly on the sensor itself.

The alternating field generated in and around the Fe core by the coil current I is unable to pass through the short-circuiting ring, since the eddy currents in the ring reduce it practically to zero. In other words, therefore, the eddy currents in the short-circuiting ring limit the extension of the magnetic flux to the space between the coil and the ring. The ring's position thus has a practically linear effect upon the inductance throughout a wide range. Practically the whole length of the sensor can be utilised for measurement.

The mass of the moving short-circuiting ring is very low. Shaping the gap between coil and limb has an effect upon the shape of the curve: Reducing the gap towards the end of the measuring range further improves the already good linearity. Depending upon material and design, operation is mostly in the 5...50 kHz range. This sensor can also be used in very severe conditions, for instance on diesel injection pumps.

This (short-circuiting ring) measuring principle is also highly adaptable to the particular measurement assignment and is available in a very wide variety of different versions (Figs. 4 and 5).

The "half-differential sensor" is very precise. It has two short-circuiting rings, the movable ring being for measurement and the fixed ring serving for reference purposes. It is applied as follows:
- As an inductive voltage divider (evaluation of the inductances L_1/L_2 or $[L_1 - L_2]/[L_1 + L_2]$) or as
- The frequency-determining component of an oscillatory circuit for generaton of a frequency-analog signal (highly resistant to interference, easy to digitize)

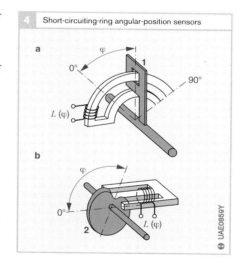

Fig. 4
a Short-circuiting-ring sensor
b Short-circuiting-disc sensor
1 Short-circuiting ring
2 Short-circuiting disc
$L(\alpha)$ Inductance
φ Measured angle

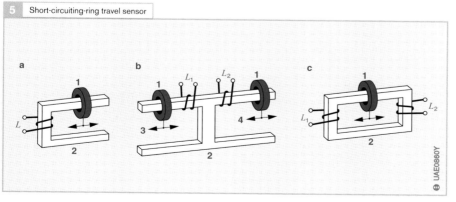

Fig. 5
a Single type
b Half-differential type
c Full-differential type
1 Short-circuiting ring
2 Core
3 Measuring system
4 Reference system (calibration)
L Inductance

The short-circuiting-ring sensors feature a relatively pronounced measuring effect which is typically $L_{max}/L_{min} = 4$. There are a number of simulation programs on the market for calculating the electromagnetic behaviour of a short-circuiting-ring sensor. Results are very realistic and three-dimensional (Fig. 6).

Short-circuiting ring sensors: Examples
- Rack-travel sensors for in-line injection pumps (attached-type load sensor, EDC sensor)
- Angular-position sensor for distributor injection pumps

Solenoid-plunger sensors
Solenoid-plunger sensors (Fig. 7) utilise the fact that a coil's inductance can be varied by means of a movable core. This core can be manufactured from solid iron (wire), rolled Fe sheet, or ferrite, and must be precisely guided (sliding contact). The inherent non-linearity of these sensors can be reduced by using special signal-conditioning circuitry. It is often the case that sensor length considerably exceeds the measured travel.

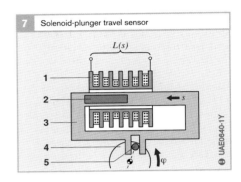

7 Solenoid-plunger travel sensor

Dividing the winding into uneven cavities (Fig. 7) avoids these disadvantages.

The addition of a second plunger coil extends the measuring concept to provide a "differential throttling sensor" which, connected as an AC voltage divider, features better linearity and zero-point stability. If both coils, whose values change in opposite directions, are then not supplied directly but rather from a magnetically coupled, symmetrical-configuration primary coil, it is possible to avoid the negative effects of the copper losses in the coils. This measuring concept is not suitable for angular measurement since the angle of rotation must first of all be mechanically converted to travel, and this is a source of further errors.

Solenoid-plunger sensors: Examples
- Accelerator-pedal sensor (electric vehicles (EV))
- Position proportioning valves

Magnetostatic sensors
Magnetostatic sensors measure a DC magnetic field. In contrast to the magnetically inductive (coil) sensors they are far more suitable for miniaturisation and can be manufactured at reasonable cost using microsystem techniques. Above all, galvanomagnetic effects (Hall and Gaussian effects, Fig. 8) are used, as well as anisotropic **mag**netoresistive (AMR) metallic thin-film elements.

Fig. 7
1 Multiple-cavity coil
2 Ferrite core
3 Plastic extrusion coating with sliding guide
4 Rotating shaft with guide pin
$L(s)$ Inductance at measured travel s
φ Measured angle

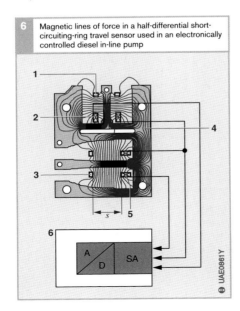

6 Magnetic lines of force in a half-differential short-circuiting-ring travel sensor used in an electronically controlled diesel in-line pump

Fig. 6
1 Short-circuiting reference ring, fixed
2 Reference coil
3 Short-circuiting measuring ring, movable
4 Coupling flux
5 Measuring coil
6 ECU
s Control-rack travel
SA Signal conditioning
A/D A/D converter

Galvanomagnetic sensors

Above all, thin semiconductor wafers are used in such sensors for the evaluation of the *Hall effect.* If current flows through such a wafer which is permeated vertically by a magnetic induction B, a voltage U_H, which is proportional to the field strength, can be picked-off at right angles to the direction of current (Hall effect). At the same time the wafer resistance increases along a roughly parabolic curve (Gaussian effect, magnetoresistor). When silicon is used as the basic material, a signal-conditioning circuit can be integrated on the wafer. Sensors using these principles are very cheap to produce, whereby silicon is by far not the most favorable semiconductor material for Hall-effect sensors. For instance, such "III-V semiconductors" as gallium arsenide (GaAs) or indium antimonide have far better characteristics.

Hall-effect switch

In the most simple case, the Hall voltage is applied to an electronic threshold circuit (Schmitt trigger) which then outputs a digital signal. If the magnetic induction B applied at the sensor is below a given minimum threshold level, the Schmitt trigger's output value corresponds to a logical "0" (release status); if it exceeds a given upper threshold the output value corresponds to a logical "1" (operate status). Since this behaviour is guaranteed across the complete operating-temperature range and for all sensors of a given type, the two threshold values are relatively far apart (approx. 50 mT). In other words, it takes a considerable induction jump (ΔB) to trigger the Hall-effect switch.

Sensors using the "spinning-current" principle

Up to now, the sensor's sensitivity to the unavoidable mechanical strain resulting from the packaging was a disadvantage, and led to unfavorable temperature sensitivity of the offset. It became possible to overcome this by the application of the "spinning-current" principle (Fig. 9). For the first time, Hall ICs were now suitable for analog sensor applications. By means of high-speed, electronically controlled rotation (spinning) of the elec-

8 Galvanomagnetic effect

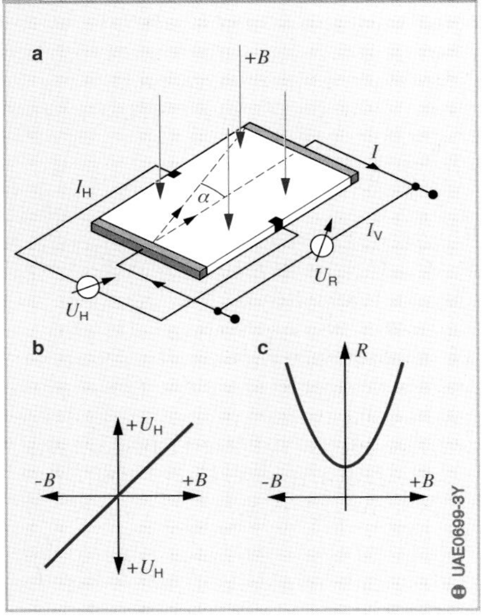

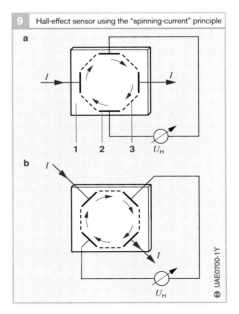

Fig. 8

- a Circuit
- b Curve of Hall voltage U_H
- c Increase of wafer resistance R (Gaussian effect)
- B Magnetic induction
- I Wafer current
- I_H Hall current
- I_V Supply current
- U_R Longitudinal voltage
- α Deviation of the electrons due to the magnetic field

9 Hall-effect sensor using the "spinning-current" principle

Fig. 9

- a Rotary phase φ_1
- b Rotary phase $\varphi_2 = \varphi_1 + 45°$
- 1 Semiconductor wafer
- 2 Active electrode
- 3 Passive electrode
- I Supply current
- U_H Hall voltage

trodes, or their cyclic reversal, and output-signal averaging, it was possible to suppress the mechanical interference effects (piezoresistive effects). These measures though did not result in a reduction of the considerable effects of temperature on the sensor's measurement sensitivity.

Such Hall ICs are suitable above all for the measurement of small travel distances (refer to "Acceleration sensors"), in which they register the fluctuating field strength of a permanent magnet as it approaches.

Differential Hall-effect sensors

For a number of years now, there have been fully integrated duplicate Hall-effect sensors ("Differential Hall sensor configurations", Fig. 10) on the market. Here, two complete Hall systems are located on a single chip at a defined distance from each other. The appropriate electronic circuitry evaluates the difference between the two Hall voltages. The advantage of such sensors lies in the fact that their output signal is for the most part independent of the absolute value of the magnetic-field strength, and as differential sensors they only register the magnetic induction's change in space, in other words the field gradient (thus the common designation "Gradient sensor").

Since the polarity of their output signal is independent of the air gap between rotor and sensor, these sensors are mostly used for rotational-speed measurement. Usually, in order to achieve as high an output signal as possible, the two sensors are each located on the edge of the (elongated) chip, the distance between them corresponding to about half the rotor tooth interval.

The signal maximum is very wide and covers a broad range of variation of the increment spacing. More pronounced variations in the spacing necessitate a highly complex redesign of the sensor.

A gradient sensor must be precisely aligned to the rotor's direction of rotation.

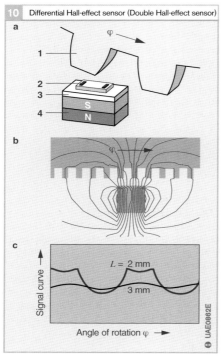

Fig. 10
a Design and construction
b Field-strength distribution (1.5 times increment spacing)
c Signal curve for air-gap widths L
1 Rotor
2 Differential Hall-effect IC
3 Homogenizing wafer (soft iron)
4 Permanent magnet

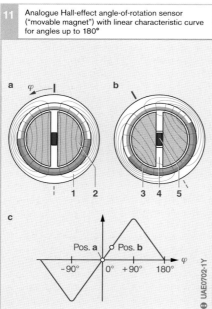

Fig. 11
a Position a
b Position b
c Output signal
1 Magnetic yoke
2 Stator (1.2 soft iron)
3 Rotor
4 Air gap
5 Hall-effect sensor
φ Angle of rotation

Angle-of-rotation sensors in the range up to 180°:
Using a rotatable magnetic ring ("movable magnet"), together with a number of fixed soft-magnetic conductive elements, a linear output signal can be generated for a larger angular range without conversion being necessary (Fig. 11). Here, the movable magnet's bipolar field is directed through a Hall-effect sensor located between semicircular conductive elements. The effective magnetic flux flowing through the Hall-effect sensor is a function of the angle of rotation φ.

The Type ARS1 Hall-effect angle-of-rotation sensor with a measuring range of approx 90° (Fig. 12) is derived from the basic "movable magnet" principle. The magnetic flux from a practically semicircular permanent-magnet disc is returned to the magnet through a pole-shoe, two additional conductive elements each of which contains a Hall-effect sensor in its magnetic path, and the shaft which is also ferromagnetic. Depending upon the angular setting, the flux is led through the two conductive elements to a greater or lesser degree. Using this principle, it is possible to achieve a practically linear characteristic.

The Type ARS2 is a simplified version which does without conductive elements (Fig. 13). In this version, the magnet moves around the Hall-effect sensor in a circular arc. Only a relatively small section of the resulting sinusoidal characteristic curve features good linearity. If the Hall-effect sensor is located slightly outside the center of the circular arc, the characteristic curve increasingly deviates from the sinusoidal, and now features a short measuring range of almost 90°, and a longer measuring range of more than 180° with good linearity.

A great disadvantage though is the low level of shielding against external fields, as well as the remaining dependence on the geometric tolerances of the magnetic circuit, and the fluctuations in magnetic flux density of the permanent magnet as a function of temperature and age. On the other hand, mechanically it is an easy matter to integrate these sensors in an accelerator-pedal module.

12 Hall-effect angle-of-rotation sensor ASR1 ("movable magnet") with linear characteristic for angles up to 90°

13 Hall-effect angle-of-rotation sensor ASR2 ("movable magnet") for angles >180°

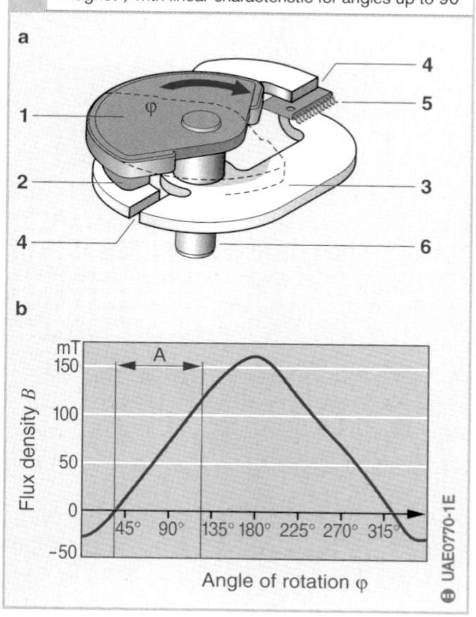

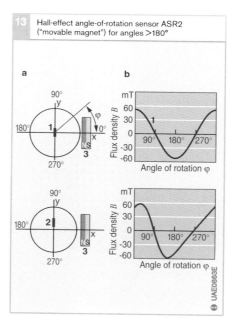

Fig. 12
- a Design and construction
- b Characteristic curve with working range A
- 1 Rotor disc (permanent magnet)
- 2 Pole shoe
- 3 Conductive element
- 4 Air gap
- 5 Hall-effect sensor
- 6 Shaft (soft magnetic)

Fig. 13
- a Principle of operation
- b Characteristic curve
- 1 Hall-IC positioned in the mid-point of the circular path
- 2 Hall-IC located outside the mid-point (linearization)
- 3 Magnet

14 Digital Hall-effect angle-of-rotation sensor for angles up to 360° using a circular, arrangement of simple Hall-effect switches located equidistantly from each other

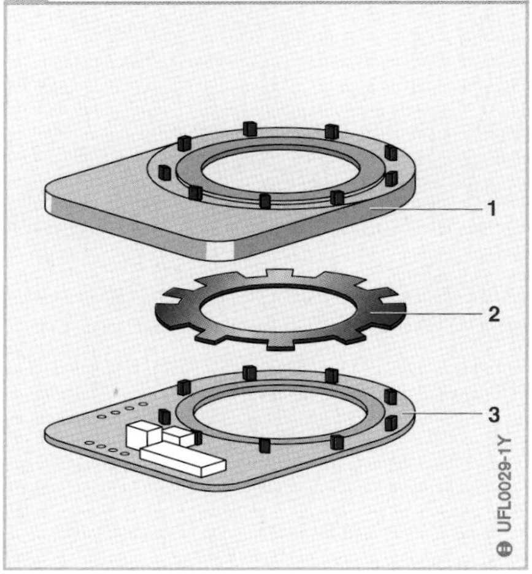

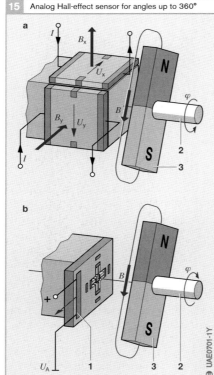

Fig. 14
1 Case cover with permanent magnets
2 Code disc
3 pcb with Hall-effect switches

15 Analog Hall-effect sensor for angles up to 360°

Fig. 15
a Constructed from discrete Hall ICs
b Constructed from planar integrated Hall ICs
1 Signal electronics
2 Camshaft
3 Control magnet
B Induction
I Current
U Voltage
U_A Output voltage
φ Angle of rotation

Angle-of-rotation sensors in the range above 180°

Type LWS3 steering-angle sensors are simple Hall ICs ("Hall-effect switches"), similar to those also used for rotational-speed measurement. In conjunction with small working-point magnets, they can be used as digital angle-of-rotation sensors for angles up to 360°. Here, in order to obtain an n-bit resolution, n Hall-effect switches are arranged in a circle at equal distances from each other (Fig. 14). Depending upon its position, a rotatable soft-magnetic code disc blocks the magnetic field of the individual permanent magnets located above each Hall-effect switch, or opens it when it rotates further so that one after another the Hall-effect switches generate n different code words. To prevent errors in intermediate settings of the code disc, it is expedient to apply the Gray code (cyclic binary code).

For the practical implementation of a steering-angle sensor, the code disc is connected to the steering shaft for instance, and the sensor's non-moving parts to the vehicle body.

In order to measure a number of complete rotations, an additional 3-bit configuration can be used in which the code disc is rotated by a step-down gearing. Such configurations though usually have a resolution of not better than 2.5°.

Fig. 15 shows an analog Hall-effect angle-of rotation sensor with a measuring range of up to 360°. As shown, a permanent magnet rotates in front of 2 Hall-effect sensors arranged at right angles to each other and in parallel to the permanent magnet's rotary axis. When the field-strength vector B rotates past the sensors, therefore, they register its x and y components:

$U_{H1} = U_x = B \sin \varphi$

$U_{H2} = U_y = B \cos \varphi$

Using the trigonometrical relationship $\varphi = \arctan (U_{H1}/U_{H2})$, it is then an easy matter to use these signals for calculating the angle φ in a commercially available evaluation chip belonging to the sensor. Principally speaking, the planar integration of this Hall-effect sensor configuration with "VHD" (Vertical Hall Devices) as shown in Fig. 15 is possible, so that the sensor-chip is vertical to the rotary axis. Monolithic integration also guarantees the high level of precision as required for the 90° arrangement of the two Hall systems.

Differential magnetoresistive sensors
The resistive or Gaussian effect with semiconductor wafers mentioned at the beginning is put to use in magnetoresistors which are manufactured from a "III-V-semiconductor" (crystalline indium antimonide (InSb)). In contrast to the Hall-effect sensors, the optimum wafer shape for a magnetoresistor tends to be shorter and squatter, and represents a very low resistance. In order to arrive at technically applicable resistance values in the kΩ range it is therefore necessary to connect a large number of these wafers in series. This problem is solved elegantly by adding microscopically fine, highly conductive nickel-antimonide needles to the semiconductor crystal. These are located obliquely to the direction of current flow. A further measure is to apply meander techniques to the semiconductor resistor (Figs. 16 and 17a).

Magnetoresistors are usually applied to a ferrite substrate so that the effective air gap can be kept to an absolute minimum when they are installed in a magnetic circuit. Since their temperature sensitivity has a pronounced effect upon their resistance (approx. 50% reduction for 100K), they are usually delivered only in the dual-configuration form in voltage-divider circuits (differential magnetoresistors). For the particular application, each of the two resistor sections must then be magnetically triggered (as far as possible with oppposite polarities). Notwithstanding the high temperature coefficient of the individual resistors, the voltage-divider circuit guarantees good stability of the working-point (that point at which both resistor sections have the same value).

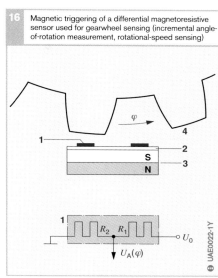

Fig. 16
1 Magnetoresistors R_1, R_2
2 Soft-magnetic substrate
3 Permanent magnet
4 Gearwheel
U_0 Supply voltage
U_A Output voltage for angle of rotation φ

Magnetic triggering of a differential magnetoresistive sensor used for gearwheel sensing (incremental angle-of-rotation measurement, rotational-speed sensing)

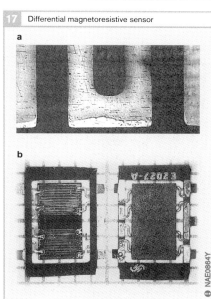

Differential magnetoresistive sensor

Fig. 17
a Microscopic section
b On ferrite substrate in Super-8 film carrier

Magnetoresistors are usually delivered at the production line in "Super-8 film packaging" (Fig. 17b). In this mode, a specially structured copper grid provides an effective connection from the internal semiconductor connection to the external assembly contact.

Taking into account their use in automotive applications, these sensors have operating-temperature limits of 160 °C sustained temperature and 200 °C short-time peak temperature. The dependence of the resistance on the magnetic flux density B follows a square-law function up to inductances of approx. 0.3 T, and above this point it is increasingly linear. There is no upper limit to the control range, and dynamic response can be regarded as practically free from lag.

In order to achieve good measurement sensitivity, it is best to operate the magnetoresistors at a magnetic working point between 0.1...0.3 T. Generally, the required magnetic bias is supplied by a small electromagnet the effects of which can be increased by using a small magnetic return plate. Without such a bias magnet, the sensor's measuring sensitivity would be practically zero. For measurement of displacement or angle, a small conductive element usually moves past the sensor configuration. At its symmetrical mid-point, this element triggers both sensor resistances equally, whereas when it is off-center it unbalances the voltage divider so that the output voltages feature good linearity and lead to high sensitivity. The magnetoresistor nevertheless still features pronounced temperature sensitivity so that it is used almost exclusively in incremental angle-of-rotation and displacement sensors, or in binary limit-value sensors (with switching characteristic).

The magnetoresistor's main advantage is its high signal level which is usually in the volts range. This means that amplification is unnecessary, as well as the local electronic circuitry and the associated protective measures which would otherwise be needed. Furthermore, in their role as passive, resistive components they are highly insensitive to electromagnetic interference and, as a result of their high bias voltage practically immune against external magnetic fields (for examples of application, refer to the Chapter "Speed and rpm sensors").

Magnetoresistive NiFe thin-film sensors
These sensors are otherwise known as AMR sensors (AMR = **A**nisotropic **M**agnetoresis-tive) and are formed from 30...50 nm thick NiFe films (also termed permalloy). They permit the design of highly compact, non-contacting angle-of-rotation sensors. In the AMR, the resistance of the printed conductor track is anisotropic, that is, in the direction of the magnetization vector it is several percent higher than at right angles to it.

Without an external control field being necessary, spontaneous magnetisation is generated in the longitudinal direction of the conductor (form anisotropy). In order to give this magnetisation a clearly defined direction – theoretically, it could be in the other direction – AMR sensors are often provided with weak bias magnets. If external influences are applied to turn the magnetizing vector through the angle φ, the resistance drops gradually until reaching its minimum at $\varphi = 90°$. Here, the resistance depends only on the angle φ which is enclosed by the magnetisation and the current. It has an approximate cosine shape as a function of φ. If the external field is much stronger than the spontaneous generated magnetisation, and this is usually the case when control magnets are used, the effective angle φ is *almost completely a function of the direction of the external field.* The field strength is now irrelevant, and in other words the sensor is now operating in the "saturated state".

Highly-conductive short-circuiting strips (for instance of gold) on the AMR film force the current to flow at an angle of below 45° to the spontaneous magnetisation (longitudinal direction) without the application of an external field. As a result of this "trick", the sensor curve shifts by 45° compared to that of the simple resistor. This results in the so-called "Barber Pole" sensor. This means, therefore, that even with the external field strength at zero, the curve is at the point of maximum sensitivity. The *"striping of two resistors in opposite directions"* (Fig. 18) means that they change their resistances in opposite directions under the effects of the same field. In other words, one of the resistances increases while the other drops. In principle, the oxidized silicon wafers which serve as the substrate material can also incorporate the electronic circuitry for signal conditioning. At present, cost considerations dictate that sensor chip and electronics chip are for the most part manufactured separately, and then mounted for instance on a common "Leadframe" and packaged. The magnetic control field B is usually generated by magnets which rotate above the sensor.

"Barber Pole" AMR angle-of-rotation sensors with extremely limited accuracy and somewhat limited measuring range (max. ± 15°) rely on the unbalance of a magnetoresistive voltage divider comprising an elongated (possibly meander-shaped) permalloy resistor with highly conductive cross stripes of gold (Fig. 18). Although the zero point of such sensors is practically independent of the distance from magnet to sensor, this does not apply to the gradient of their characteristic curve.

Magnetoresistive angle-of-rotation sensors of the "Pseudohall" type utilise the practically 100 % sinusoidal signal that is picked-off at the output terminals of the four-pole, sensor structure, whereby two complete periods of the electrical output signal correspond to a mechanical rotation of the magnets through 360°. Using a second element, which has been turned through 45°, a cosine signal is also generated (Fig. 19). By appling the arc-tangent function for instance, the relationship between the two signal voltages can be applied to determine the measured angle using a microcontroller or an ASIC. This applies throughout a range of 180°, and is for the most part independent of temperature changes and magnetic-field fluctuations (aging, spacing).

The measurement sensitivity of this so-called Pseudohall Element can be considerably increased (without excessively falsifying the sinusoidal shape) by "hollowing out" the element from the inside so that only the "frame" remains. This modification converts

18 Magnetoresistive angle-of-rotation sensor ("barber-pole" version) for measuring ranges of up to ±15°

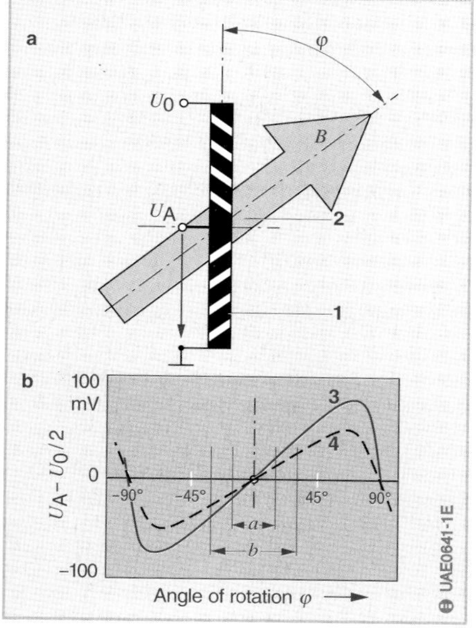

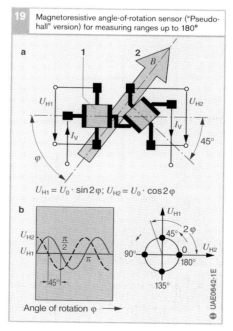

Fig. 18
a Measuring concept
1 Permalloy resistors
2 Rotatable permanent magnet with control induction B
b Characteristic curve
3 Lower operating temperature
4 Higher operating temperature
a Linear measuring range
b Effective measuring range
U_A Output voltage
U_0 Supply voltage (DC 5V)
φ Angle of rotation

19 Magnetoresistive angle-of-rotation sensor ("Pseudo-hall" version) for measuring ranges up to 180°

Fig. 19
a Measuring concept
b Output signals
1 Thin NiFe film (AMR sensor)
2 Rotatable permanent magnet with control induction B
I_V Supply current
U_{H1}, U_{H2} Measurement voltages
φ Angle of rotation

the pseudohall sensor to a full bridge consisting of four AMR resistors (Fig. 20). Even when the bridge resistors are meander-shaped, provided a given minimum conductor width is not dropped below, this still has negligible effect of the signal's sinusoidal shape.

A further prerequisite for the high accuracy of this sensor principle is that the field at both bridges is at least in the same direction (above a given magnitude, field strength is irrelevant). This can only be guaranteed when both bridges are directly above one another. A design was drawn up in which the two bridges, which are at 45° to each other, were interweaved so that they can be regarded as being at the same point and quasi "on top of one another" (Fig. 21). The *major advantage* of these Pseudohall-version sensors is the fact that in contrast to the corresponding "genuine" Hall-sensor versions, they are almost completely independent of the magnitude of the control field. As soon as this has exceeded a given magnitude, the output signal is dependent solely upon the control-field angle. The reason is that these sensors operate in the "saturation region" in which the angle of the spontaneous internal magnetism has switched almost completely to the direction imposed from outside. In other words, it is not necessary to have a constant control-field magnitude. At the measuring point, all that is required is a certain homogeneity of direction. With this sensor principle, neither the aging of the magnets and of the magnetic conductor elements, nor of the air-gap tolerances and fluctuations, plays an important role.

A dual-configuration "pseudohall angle-of-rotation" sensor can be used to measure a number of rotations of a rotating component (for instance, a steering shaft). The shaft's rotating member rotates the two permanent magnets through a step-up unit with a high transmission ratio. Since the two driven, smaller gearwheels differ from each other by one tooth, their respective phase position is a clear measure for the absolute angular position. Furthermore, each sensor's resolution of the angle of rotation is somewhat course. Using such a configuration, it is possible for instance to register the complete steering-wheel

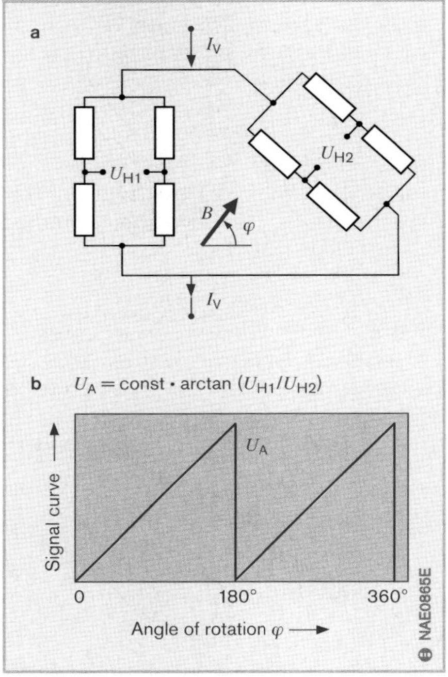

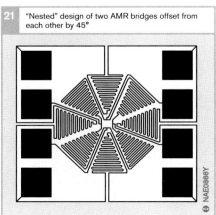

Fig. 20
a Bridge circuit
b Output signal U_A of the evaluation circuit
B Control induction
I_V Supply current
U_{H1} Measurement
U_{H2} voltages
φ Angle of rotation

Fig. 21 "Nested" design of two AMR bridges offset from each other by 45°

$U_A = \text{const} \cdot \arctan (U_{H1}/U_{H2})$

range of four full rotations with a resolution of better than 1°.

GMR sensors

Just lately magnetoresistive (GMR) sensors have appeared on the market which use "nanotechnology" (GMR = **G**iant **M**agneto **R**esistive). These are composed of a number of thin layers applied one on top of the other. These layers have a thickness of only one or two atomic layers (Fig. 23). They are very similar to the AMR sensors, but have a far more pronounced measuring effect (Fig. 22). In contrast to the AMR sensors, the GMR sensor resistance (in the case of a Pseudohall configuration) depends solely upon the angle of rotation and not on the sinus of double the mechanical angle of rotation. This means that a full 360° angle of rotation can be measured.

Compass sensors (earth's-field sensors)

A completely novel type of angle-of-rotation sensor is required by vehicle navigation systems. At least at road junctions and crossings, these must be able to measure the angle steered by the vehicle ("Heading"). This must also be possible even when there is no steering-wheel-angle sensor installed in the vehicle. Until suitable inert sensors (time-integrated yaw-rate sensors) became available, magnetic-field sensors ("saturation core field sensors", Fig. 24) are used which use the geomagnetic field (compass) to determine the direction taken by the vehicle.

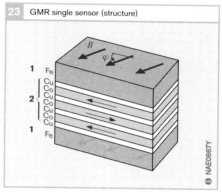

23 GMR single sensor (structure)

Fig. 23
1 Fe layers
2 Thin anti-ferromagnetic CuCo layers
B Control induction
φ Angle of rotation

24 Sensor core of the geomagnetic field sensor

Fig. 24
1 Sensor coil (x-axis)
2 Sensor coil (y-axis)
3 Excitation winding
4 Toriodal core
B Measuring field (horizontal component of the geomagnetic field)

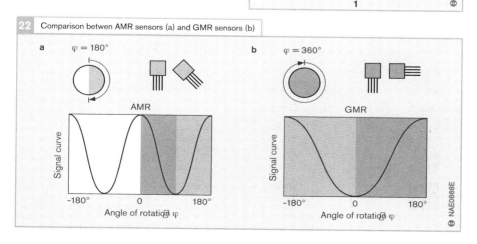

22 Comparison betwen AMR sensors (a) and GMR sensors (b)

Wave-propagation sensors

For vehicle-spacing measurement, ultrasonic running-time methods (near range 0.5...5 m) are suitable, as are running-time or triangulation methods using light in the near-infrared range (Lidar, medium range up to 50 m), and electromagnetic radar (distance zone up to 150 m).

Acoustic sensors (ultrasonic)

Similar to the echo-sounding process, the sensors here transmit ultrasonic pulses with a frequency of around 40 kHz (Fig. 25), and register the time taken for the echo pulse to arrive after having been reflected back from an obstacle. The distance a to the next obstacle is calculated from the propagation time of the first reflected echo pulse t_e to arrive and the speed of sound in air (Fig. 26):

$$a = 0.5 \cdot t_e \cdot c$$

Electromagnetic sensors (radar)

Using a far-ranging radar sensor, ACC systems automatically detect preceding vehicles in the same lane which could eventually necessitate application of the brakes. Here, ACC stands for **A**daptive **C**ruise **C**ontrol. The working frequency is 76 GHz (wavelength approx. 3.8 mm) and permits the relatively low-profile construction as needed for automotive applications. A Gunn-effect oscillator (Gunn diode in a cavity resonator), feeds three Patch antennas arranged adjacent to each other, which also serve to receive the reflected signals again (Fig. 27). Referred to the vehicle axis, a plastic lens (Fresnel lens) concentrates the transmit beam horizontally at an angle of ±5°, and vertically at an angle of ±1.5°. The antenna receive characteristics are aligned in different directions due to the antennas being off-set

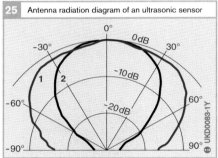

Fig. 25
1. Vertical
2. Horizontal

Antenna radiation diagram of an ultrasonic sensor

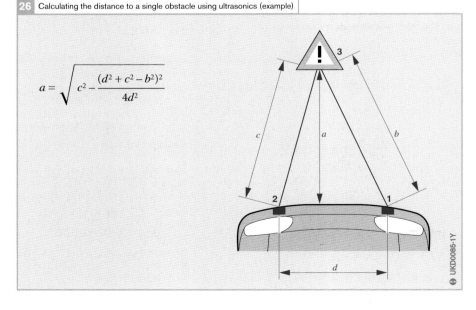

26 Calculating the distance to a single obstacle using ultrasonics (example)

$$a = \sqrt{c^2 - \frac{(d^2 + c^2 - b^2)^2}{4d^2}}$$

Fig. 26
a Distance between the bumper and the obstacle
b Distance sensor 1 to obstacle
c Distance sensor 2 to obstacle
d Distance sensor 1 to sensor 2
1 Transceiver sensor
2 Receiver sensor
3 Obstacle

from the center (6 dB width, 4°). This means that in addition to the calculation of the distance to preceding vehicles and their relative speeds, it is also possible to determine the direction in which they are travelling when detected. Directional couplers are used to separate the transmitted and received reflection signals. By mixing the receive frequency and the transmit frequency, three downstream mixers transpose the receive frequency down to practically zero (0...300 kHz). In order to evaluate them, the low-frequency signals are now digitized and put through a high-speed Fourier (harmonic) analysis to determine the frequencies.

The Gunn-effect oscillator frequency is continually compared with that of a stable DRO reference oscillator (**D**ielectric **R**eso**n**ance **O**scillator), and maintained at a stipulated setpoint frequency. To do so, the Gunn-effect oscillator's supply voltage is adjusted until the frequency is correct again. Via a closed control loop, and following a saw-tooth waveform, the Gunn-effect oscillator frequency is briefly raised and lowered by 300 MHz every 100 ms (FMCW **F**requency-**M**odulated **C**ontinuous **W**ave). The signal reflected from a preceding vehicle is delayed in accordance with the propagation time (in other words, in the positive-going edge by the lower frequency, and in the negative-going edge by the higher frequency). Lower and higher frequencies deviate from the basic frequency by the same amount.

The frequency difference Δf is a direct measure of the distance (e.g. 2kHz/m). If, on the other hand, there is also a given relative speed between the two vehicles, the Doppler principle causes the receive frequency f_e to increase in the positive-going and negative-going edges by a certain proportional amount Δf_d (e.g. 512 Hz per m/s). In other words, this results in two different differential frequencies Δf_1 and Δf_2. Adding these two frequencies provides the distance, subtracting them provides the relative speed between the two vehicles (Fig. 27). This method is used to detect and follow a number of vehicles (as many as 32).

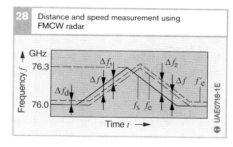

Fig. 28

f_S Transmit frequency
$f_e/f_{e'}$ Receive frequency without/with relative speed
Δf_d Frequency increase due to Doppler effect (relative speed)
Δf_S/ Frequency difference
$\Delta f_{1,2}$ Without/with relative speed

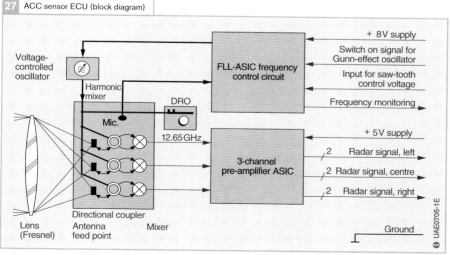

27 ACC sensor ECU (block diagram)

Sensor-plate potentiometer

Application
The sensor-plate potentiometer is used in the air-flow sensor of the KE-Jetronic fuel-injection system to register the position (angle of rotation) of the sensor flap. The rate at which the driver presses the accelerator pedal is derived from the sensor plate's movement, which is only slightly delayed with respect to the throttle-valve movement. This signal corresponds to the change in intake air quantity as a function of time, in other words approximately engine power. The potentiometer inputs it to the ECU which applies it when triggering the electrohydraulic pressure actuator (Fig. 2).

Depending upon the engine's operating state and the corresponding current signal generated by the ECU, the pressure actuator changes the pressure in the vacuum chambers of the differential-pressure valves in the fuel distributor, and with it the amount of fuel metered to the injectors.

Design and operating concept
The potentiometer in the air-flow sensor is produced on a ceramic substrate using film techniques. It is a potentiometer-type angle-of-rotation sensor, which for measurement purposes applies the relationship which exists between the length of a film resistor (printed conductor) and its resistance. The printed-conductor width is varied in order to make the potentiometer characteristic non-linear so that the highest acceleration signal is generated when sensor-plate movement originates from the idle setting. The signal decreases along with increasing engine power output.

The brush wiper slides over the potentiometer tracks (pick-off track and wiper track) and is comprised of a number of very fine wires welded to a lever which is mechanically connected to the sensor-plate shaft (from which it is electrically insulated). The individual wires only apply very light pressure to the potentiometer tracks so that wear remains at a very low level. The large number of wires leads to good electrical contact in case the track surface is very rough and also when the brush is moved very quickly over the track. The wiper voltage is picked-off by a second brush wiper which is connected electrically to the main wiper (Fig. 1).

Damage due to air blowback in the intake manifold is ruled out since the wiper is free to travel far enough beyond the measurement range at both ends of the track. Protection against electrical short circuit is provided by a fixed film resistor connected in series with the wiper.

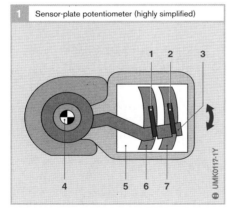

Sensor-plate potentiometer (highly simplified)

Fig. 1
1. Pick-off wiper brush
2. Main wiper brush
3. Wiper lever
4. Air-flow sensor shaft
5. Potentiometer board
6. Pick-off track
7. Measurement track

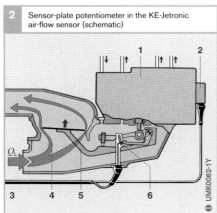

Sensor-plate potentiometer in the KE-Jetronic air-flow sensor (schematic)

Fig. 2
1. Fuel distributor
2. Electrohydraulic pressure actuator
3. To the ECU
4. Air-flow sensor
5. Sensor plate
6. Potentiometer
Q_L Air quantity

Throttle-valve sensor

Application

The throttle-valve sensor registers the angle of rotation of the gasoline-engine throttle valve. On M-Motronic engines, this is used to generate a secondary-load signal which, amongst other things, is used as auxiliary information for dynamic functions, as well as for recognition of operating range (idle, part load, WOL), and as a limp-home or emergency signal in case of failure of the primary-load sensor (air-mass meter). If the throttle-valve sensor is used as the primary-load sensor, the required accuracy is achieved by applying two potentiometers for two angular ranges.

The ME-Motronic adjusts the required engine torque via the throttle valve. In order to check that the throttle valve moves to the required position, the throttle-valve sensor is used to evaluate the valve's position (closed-loop position control). As a safety measure, this sensor is provided with two parallel-operation (redundant) potentiometers with separate reference voltages.

Design and operating concept

The throttle-valve sensor is a potentiometer-type angle-of-rotation sensor with one (or two) linear characteristic curve(s).

The wiper arm is connected mechanically with the throttle-valve shaft, and with its brushes slides across the respective potentiometer tracks. In the process, it converts the rotation of the throttle valve shaft into a voltage ratio U_A/U_V which is proportional to the valve's angle of rotation (Fig. 2). The operating voltage is 5 V. The electrical wiper connection is usually through a second potentiometer track. This has the same surface, but the track itself is formed of a low-resistance printed-conductor material (Figs. 1 and 3).

As a protection against overload, the voltage is applied to the measurement (potentiometer) tracks through small series resistors (also used for zero-point and slope calibration). The shape of the characteristic curve can be adapted by varying the width of the potentiometer track (variation can also apply to sections of the track).

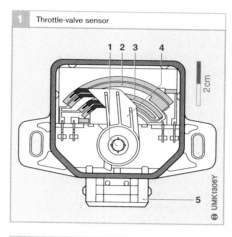

Fig. 1
1. Throttle-valve shaft
2. Resistance track 1
3. Resistance track 2
4. Wiper arm with wipers
5. Electric connection (4-pole)

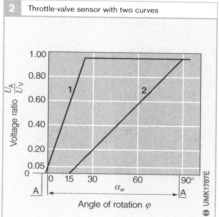

Fig. 2
A Internal stop
1 Curve for high resolution in angular range 0°...23°
2 Curve for angular range 15°...88°
U_0 Supply voltage
U_A Measurement voltage
U_V Operating voltage
$α_W$ Effective measured angle

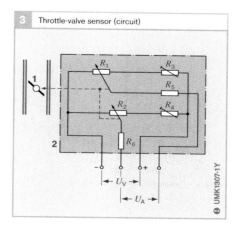

Fig. 3
1. Throttle valve
2. Throttle-valve sensor
U_A Measurement voltage
U_V Operating voltage
R_1, R_2 Resistance tracks 1 and 2
R_3, R_4 Calibration resistors
R_5, R_6 Protective resistors

Half-differential short-circuiting-ring sensors

Application

These sensors are also known as HDK (taken from the German) sensors, and are applied as position sensors for travel or angle, They are wear-free, as well as being very precise, and very robust, and are used as:

- Rack-travel sensors (RWG) for measuring the control-rack setting on in-line diesel injection pumps and as
- Angle-of-rotation sensors in the injected-fuel-quantity actuators of diesel distributor pumps

Design and operating concept

These sensors (Figs. 1 and 2) are comprised of a laminated soft-iron core on each limb of which are wound a measuring coil and a reference coil.

Alternating magnetic fields are generated when the alternating current from the ECU flows through these coils. The copper rings surrounding the limbs of the soft-iron cores screen the cores, though, against the effects of the magnetic fields. Whereas the reference short-circuiting rings are fixed in position, the measuring short-circuiting rings are attached to the control rack or control-collar shaft (in-line pumps and distributor pumps respectively), with which they are free to move (control-rack travel s, or adjustment angle φ).

When the measuring short-circuiting ring moves along with the control rack or control-collar shaft, the magnetic flux changes and, since the ECU maintains the current constant (load-independent current), the voltage across the coil also changes.

The ratio of the output voltage U_A to the reference voltage U_{Ref} (Fig. 3) is calculated by an evaluation circuit. This ratio is proportional to the deflection of the measuring short-circuiting ring, and is processed by the ECU. Bending the reference short-circuiting ring adjusts the gradient of the characteristic curve, and the basic position of the measuring short-circuiting ring defines the zero position.

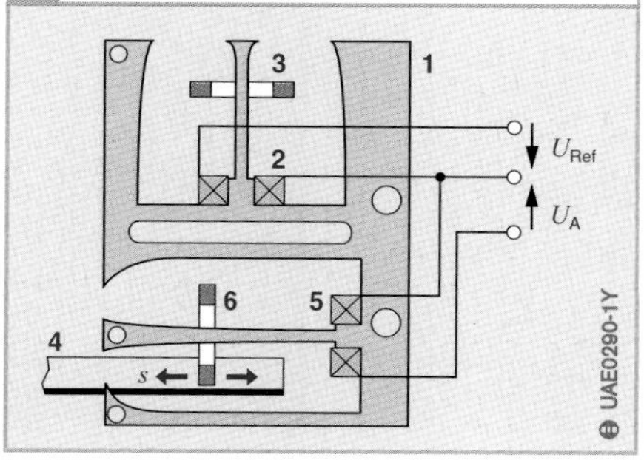

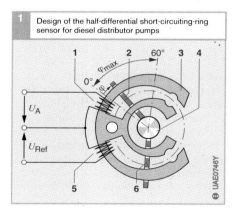

1 Design of the half-differential short-circuiting-ring sensor for diesel distributor pumps

Fig. 1
1 Measuring coil
2 Measuring short-circuiting ring
3 Soft-iron core
4 Control-collar shaft
5 Reference coil
6 Reference short-circuiting ring
φ_{max} Adjustment-angle range for the control-collar shaft
φ Measured angle

2 Design of the rack-travel sensor (RWG) for diesel in-line injection pumps

Fig. 2
1 Soft-iron core
2 Reference coil
3 Reference short-circuiting ring
4 Control rack
5 Measuring coil
6 Measuring short-circuiting ring
s Control-rack travel

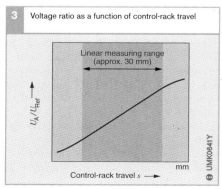

3 Voltage ratio as a function of control-rack travel

Fig. 3
U_A Output voltage
U_{Ref} Reference voltage

Fuel-level sensor

Application

It is the job of the fuel-level sensor to register the level of the fuel in the tank and send the appropriate signal to the ECU or to the display device in the vehicle's instrument panel. Together with the electric fuel pump and the fuel filter, it is part of the in-tank unit. These are installed in the fuel tank (gasoline or diesel fuel) and provide for an efficient supply of clean fuel to the engine (Fig. 1).

Design

The fuel-level sensor (Fig. 2) is comprised of a potentiometer with wiper arm (wiper spring), printed conductors (twin-contact), resistor board (pcb), and electrical connections. The complete sensor unit is encapsulated and sealed against fuel. The float (fuel-resistant Nitrophyl) is attached to one end of the wiper lever, the other end of which is fixed to the rotatable potentiometer shaft (and therefore also to the wiper spring). Depending upon the particular version, the float can be either fixed in position on the lever, or it can be free to rotate). The layout of the resistor board (pcb) and the shape of the float lever and float are matched to the particular fuel-tank design.

Operating concept

The potentiometer's wiper spring is fixed to the float lever by a pin. Special wipers (contact rivets) provide the contact between the wiper spring and the potentiometer resistance tracks, and when the fuel level changes the wipers move along these tracks and generate a voltage ratio which is proportional to the float's angle of rotation. End stops limit the rotation range of 100° for maximum and minimum levels as well as preventing noise. Operating voltage is 5...13 V.

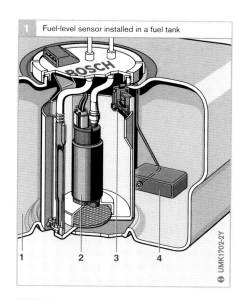

Fig. 1 Fuel-level sensor installed in a fuel tank

Fig. 1
1 Fuel tank
2 Electric fuel pump
3 Fuel-level sensor
4 Float

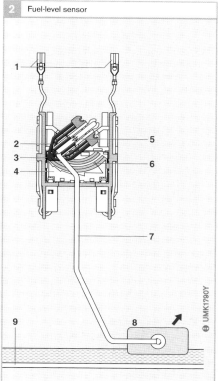

Fig. 2 Fuel-level sensor

Fig. 2
1 Electrical connections
2 Wiper spring
3 Contact rivet
4 Resistor board
5 Bearing pin
6 Twin contact
7 Float lever
8 Float
9 Fuel-tank floor

Accelerator-pedal sensors

Application

In conventional engine-management systems, the driver transmits his/her wishes for acceleration, constant speed, or lower speed, to the engine by using the accelerator pedal to intervene mechanically at the throttle plate (gasoline engine) or at the injection pump (diesel engine). Intervention is transmitted from the accelerator pedal to the throttle plate or injection pump by means of a Bowden cable or linkage.

On today's electronic engine-management systems, the Bowden cable and/or linkage have been superseded, and the driver's accelerator-pedal inputs are transmitted to the ECU by an accelerator-pedal sensor which registers the accelerator-pedal travel, or the pedal's angular setting, and sends this to the engine ECU in the form of an electric signal. This system is also known as "drive-by-wire". The accelerator-pedal module (Figs. 2b, 2c) is available as an alternative to the individual accelerator-pedal sensor (Fig. 2a). These modules are ready-to-install units comprising accelerator pedal and sensor, and make adjustments on the vehicle a thing of the past.

Design and operating concept

Potentiometer-type accelerator-pedal sensor
The heart of this sensor is the potentiometer across which a voltage is developed which is a function of the accelerator-pedal setting. In the ECU, a programmed characteristic curve is applied in order to calculate the accelerator-pedal travel, or its angular setting, from this voltage.

A second (redundant) sensor is incorporated for diagnosis purposes and for use in case of malfunctions. It is a component part of the monitoring system. One version of the accelerator-pedal sensor operates with a second potentiometer. The voltage across this potentiometer is always half of that across the first potentiometer. This provides two independent signals which are used for trouble-shooting (Fig. 1). Instead of the second potentiometer, another version uses a low-idle switch which provides a signal for

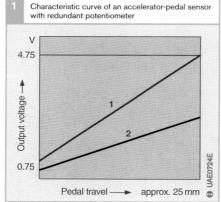

1 Characteristic curve of an accelerator-pedal sensor with redundant potentiometer

1 Potentiometer 1 (master potentiometer)
2 Potentiometer 2 (50% of voltage)

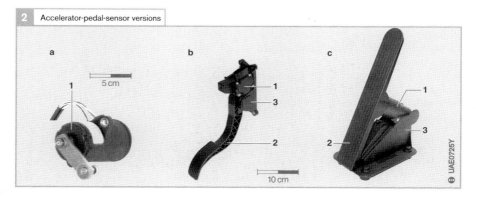

2 Accelerator-pedal-sensor versions

Fig. 2
a Individual accelerator-pedal sensor
b Top-mounted accelerator-pedal module
c Bottom-mounted accelerator-pedal module FMP1
1 Sensor
2 Vehicle-specific pedal
3 Pedal bracket

the ECU when the accelerator pedal is in the idle position. For automatic transmission vehicles, a further switch can be incorporated for a kick-down signal.

Hall-effect angle-of-rotation sensors

The ARS1 (**A**ngle of **R**otation **S**ensor) is based on the movable-magnet principle. It has a measuring range of approx. 90° (Figs. 3 and 4).

A semicircular permanent-magnet disc rotor (Fig. 4, Pos. 1) generates a magnetic flux which is returned back to the rotor via a pole shoe (2), magnetically soft conductive elements (3) and shaft (6). In the process, the amount of flux which is returned through the conductive elements is a function of the rotor's angle of rotation φ. There is a Hall-effect sensor (5) located in the magnetic path of each conductive element, so that it is possible to generate a practically linear characteristic curve throughout the measuring range.

The ARS2 is a simpler design without magnetically soft conductive elements. Here, a magnet rotates around the Hall-effect sensor. The path it takes describes a circular arc. Since only a small section of the resulting sinusoidal characteristic curve features good linearity, the Hall-effect sensor is located slightly outside the center of the arc. This causes the curve to deviate from its sinusoidal form so that the curve's linear section is increased to more than 180°.

Mechanically, this sensor is highly suitable for installation in an accelerator-pedal module (Fig. 5).

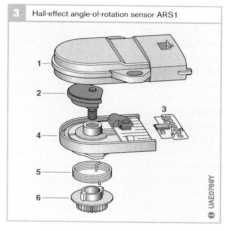

Fig. 3
1 Housing cover
2 Rotor (permanent magnet)
3 Evaluation electronics with Hall-effect sensor
4 Housing base
5 Return spring
6 Coupling element (e.g. gear)

Fig. 4
1 Rotor (permanent magnet)
2 Pole shoe
3 Conductive element
4 Air gap
5 Hall-effect sensor
6 Shaft (magnetically soft)
φ Angle of rotation

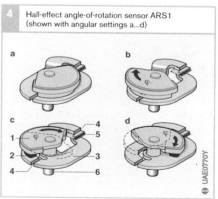

Fig. 5
a Installation in the accelerator-pedal module
b Components
1 Hall-effect sensor
2 Pedal shaft
3 Magnet

Steering-wheel-angle sensors

Application

The Electronic Stability Program (ESP) applies the brakes selectively to the individual wheels in order to keep the vehicle on the desired track selected by the driver. Here, the steering-wheel angle and the applied braking pressure are compared with the vehicle's actual rotary motion (around its vertical axis) and its road speed. If necessary, the brakes are applied at individual wheels. These measures serve to keep the float angle (deviation between the vehicle axis and the actual vehicle movement) down to a minimum and, until the physical limits are reached, prevent the vehicle breaking away.

Basically speaking, practically all types of angle-of-rotation sensors are suitable for registering the steering-wheel angle. Safety considerations, though, dictate that only those types are used which can be easily checked for plausibility, or which in the ideal case automatically check themselves. Potentiometer principles are used, as well as optical code-registration and magnetic principles. Whereas a passenger-car steering wheel turns through ±720° (a total of 4 complete turns), conventional angle-of-rotation sensors can only measure maximum 360°. This means that with the majority of the sensors actually used for this purpose it is necessary to continually register and store the data on the steering wheel's actual setting.

Design and operating concept

There are two absolute-measuring (in contrast to incremental-measuring) magnetic angle-of-rotation sensors available which are matched to the Bosch ECUs. At any instant in time, these sensors can output the steering-wheel angle throughout the complete angular range.

Hall-effect steering-wheel-angle sensor (LWS1)

The LWS1 uses 14 Hall-effect vane switches to register the angle and the rotations of the steering wheel. The Hall-effect vane switch is similar in operation to a light barrier. A Hall-effect element measures the magnetic field of an adjacent magnet. A magnetic code disc rotates with the steering shaft and strongly reduces the magnet's field or screens it off completely. In this manner, with nine Hall ICs it is possible to obtain the steering wheel's angular position in digital form. The remaining five Hall-effect sensors register the particular steering-wheel revolution which is transformed to the final 360° range by 4:1 step-down gearing.

The first item from the top in the exploded view of the LWS 1 steering-wheel-angle sensor (Fig. 1) shows the nine permanent magnets. These are screened individually by the magnetically-soft code disc beneath them when this rotates along with the steering shaft, and depending upon steering-wheel movement. The pcb immediately below the code disc contains Hall-effect switches (IC), and a micro-

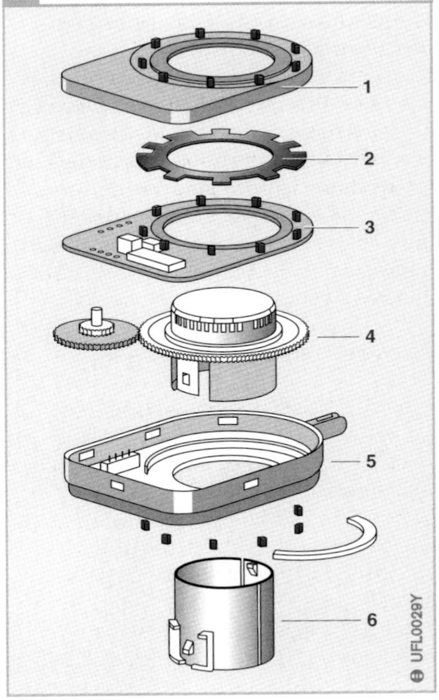

Fig. 1
1. Housing cover with nine equidistantly spaced permanent magnets
2. Code disc (magnetically soft material)
3. pcb with 9 Hall-effect switches and microprocessor
4. Step-down gearing
5. Remaining 5 Hall-effect vane switches
6. Fastening sleeve for steering column

1 Exploded view of the digital LWS1 steering-wheel-angle sensor

One of the basic functions is the conventional Cruise Control which holds the vehicle speed constant, once it has been set. This function remains permanently in operation as long as a preceding vehicle is not detected which is travelling at a speed below that set by the ACC vehicle's driver. If the system picks up such a vehicle inside the radar's detection zone (approx. 100...150 m) which would prevent the set speed being maintained, the speed of the radar-equipped vehicle is adapted to that of the preceding vehicle. In case of only minor differences in speed it suffices to reduce the accelerator-pedal setting. Considerable differences in speed on the other hand necessitate the brakes being applied.

As soon as the difference in speed has been compensated for, the ACC vehicle follows the preceding vehicle with a relatively constant time gap. That is, as speeds increase so does the gap.

Selection of the "right" target vehicle is the most difficult problem with the signal processing in the SCU of the ACC. Here, first of all, those radar reflections must be identified which belong to the already defined vehicles. Then, it must be ascertained whether these vehicles really are in the same lane. Even though the sensors belonging to the ESP (**E**lectronic **S**tability **P**rogram) provide a whole range of important signals which can be used for comparitive purposes, this is particularly difficult before and in bends.

It's up to the reader to judge for himself whether the ACC SCU is to be regarded as an ECU with integrated sensor, or as a sensor with ECU. One thing is quite certain though: More of these systems which monitor the vehicle's surroundings will definitely appear on the market in the years to come and, similar to video "sensors", with only one single device will be able to perform a number of functions.

2 ACC Adaptive Cruise Control (block diagram)

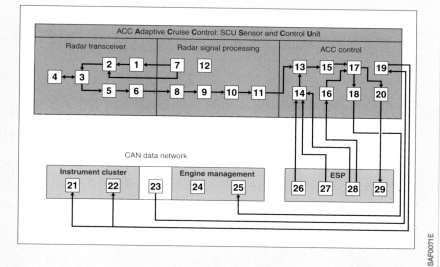

Fig. 2
1 Oscillator
2 Modulator
3 Transmit/receive switch
4 Antenna
5 Demodulator
6 Amplification
7 Radar control
8 Fourier transformation
9 Detection
10 Matching
11 Tracking
12 Radar monitoring
13 Target selection
14 Curve recognition
15 Ranging control
16 Speed control
17 Prioritization
18 Drivetrain-control commands
19 System monitoring
20 Braking-system control commands
21 ACC status display
22 Display of driver's desired speed, time gap
23 Control switch
24 Monitoring logic
25 Torque control
26 Yaw rate
27 Steering-wheel angle
28 Wheel speeds
29 Electronically controlled brake intervention

Speed and rpm sensors

Measured variables

Speed and rpm sensors measure the number of revolutions or the distance travelled per unit of time. When automotive applications are concerned, these are in both cases measured variables which occur between two components or with respect to the road surface or another vehicle. In some cases, it is necessary to measure the absolute rotational speed in space or about the vehicle axes. This is often referred to as yaw rate. For instance, for the Electronic Stability Program (ESP), the yaw rate about the vehicle's vertical (or yaw) axis must be picked-off by "sensing".

In the detection of relative yaw rate, depending upon the number and size of the scanned peripheral rotor markings, one differentiates between the following types of sensor (Fig. 1):
- *Increment sensor* with closely spaced peripheral markings. Up to a certain point, this form of sensor permits instantaneous speed to be measured at points on the circumference, or the registration of very fine angular divisions
- *Segment sensor,* with only a small number of scanned peripheral segments (for instance, equivalent to the number of engine cylinders)
- Simple *rpm sensor,* with only a single scanned marking per revolution, so that only the average rotational speed can be registered

Examples of relative rotational speed are as follows:
- Crankshaft and camshaft speeds
- Wheel speeds (for ABS/TCS/ESP) and
- Speed of the diesel injection pump

Here, measurement usually makes use of an incremental pick-up system comprised of a toothed wheel (rotor) and an rpm sensor.

Newer applications include the following:
- Rotational-speed measurement using an rpm sensor incorporated in the bearing (wheel bearing, or the so-called composite seal with sensor (CSWS) on the crankshaft)
- Speed over ground
- Vehicle yaw rate around the longitudinal (roll) axis and the pitch axis (roll-over protection)

Measuring principles

Conventional sensors used for rotational-speed measurement are based on pronounced measuring effects (e.g. inductive). They are therefore for the most part electrically passive. That is, they are usually not provided with any form of local/on-site electronics. With the newer sensors, however, measurement is based on less-pronounced measuring effects (for instance, the Hall effect), and these sensors thus need local, integrated electronics for signal conditioning. In the broader sense, according to the definition in "Development Trends" at the front of this manual, they belong to the category "intelligent sensors" (which are also often referred to as "active" sensors). In fact, the sensors used for measuring absolute rotating speed (yaw rate) need highly complex electronic circuitry directly at the sensor since the measuring effects used here are not only particularly small, but also require complex signal conditioning.

Incremental rotational-speed measurement takes advantage of a wide variety of different physical effects (some of which can be applied in sensors at a very reasonably price). Optical

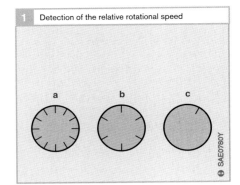

Fig. 1
a Incremental sensor
b Segment sensor
c Rotational-speed sensor

and capacitive sensors, though, are highly unsuitable for the rough operating conditions encountered in the vehicle. Here, magnetic-effect sensors are used almost exclusively.

Presently used sensor shapes

The following shapes of sensor are presently in use (Fig. 2) for speed and rpm sensors:
- Rod sensor
- Fork-shape sensor
- Internal and external toroidal sensor

Due to its ease of mounting and simplicity, the rod sensor is the most widespread sensor version. The rod sensor is located near the rotor (Fig. 2 b), the teeth of which approach it and pass by in close proximity. The fork-shaped sensor (Fig. 2a) is permissible in some cases, and is also in service in the field. This sensor must be roughly aligned to the rotor when installed. The sensor type in which the sensor surrounds the rotor shaft in the form of a ring is practically no longer used. From time to time, an inner-ring sensor is used which is introduced into the end of a hollow shaft featuring an incremental internal structure.

Unfortunately, although it is the most widely used sensor type, the rod-shaped sensor features the lowest measuring sensitivity and is problemetical when air gaps become excessive. The fork-shaped sensor on the other hand is practically impervious to axial and radial play. Regarding toroidal sensors, the most widely used form combines a large measurement-signal output with pronounced insensitivity to geometric tolerances.

New sensor shapes

In many respects, the traditional inductive-type sensors are highly unsatisfactory. Their output signal features an rpm-dependent amplitude, and for this reason they are unsuitable for measuring low rotational speeds. In addition, they only permit comparitively low-level air-gap tolerances and are usually unable to differentiate between air-gap fluctuations and genuine rotational-speed pulses. And at least the sensor tip must be proof against high temperatures (for instance when installed near the brakes). These disadvantages are behind the additional features aimed at with new, innovative sensors:
- Static detection (that is, at zero speed, or at extremely low cranking or wheel speeds)
- Efficient measurement in large air gaps (non-aligned mounting with air gaps >0),
- Small size
- Efficient operation independent of air-gap fluctuations
- Temperature stability ($\leq 200\,°C$)
- Identification of the sense of direction (optional for navigation) and
- Reference-mark identification (ignition)

Magnetostatic sensors (Hall, magnetoresistors, AMR) are highly suitable for complying with the first two demands. And, as a rule, they also permit compliance with the second and third stipulations.

Fig. 3 (next page) shows three basically suitable sensor shapes, which generally are insensitive to air-gap fluctuations. Here, one must differentiate between sensors which sense radially and those which sense tangentially. This means that independent of the air gap, magnetostatic sensors are always able to differentiate between the north and south poles of a magnetically active pole wheel or rotor ring.

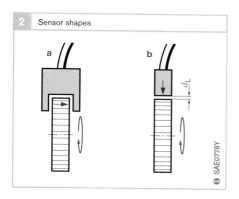

Fig. 2
a Fork shape (vane principle)
b Rod shape (proximity principle)

d_L Air gap

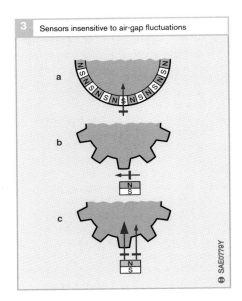

Fig. 3
a Radial-field sensor with pole wheel
b Tangential sensor
c Differential sensor with toothed rotor

In the case of magnetically passive rotors, the sign of the output signal is then no longer independent of the air gap when they register the tangential-field strength (here though, the fact that the air gap is often enlarged due to the rotor is a disadvantage).

Radially measuring differential-field or gradient sensors are often used. These always register only the gradients of the radial-field components, the signs of which do not change with the air gap but only with the angle of rotation (Fig. 3).

Rotors

The rotor is of decisive importance when measuring rotational speed. It is usually provided by the vehicle manufacturer, while the sensor itself comes from a component supplier. Up to now, *magnetically passive rotors* have been used almost exclusively. These are made from magnetically passive materials (usually iron), and are less expensive than magnetically hard rotors (also known as pole wheels). Apart from that, since they are not magnetized they are easier to handle, and there is no danger of mutual demagnetisation (for instance during storage). Unfortunately, this form of rotor is the most difficult to scan, particularly in combination with rod sensors.

As a rule, presuming the same increment width and output signal, the pole wheel's intrinsic magnetism *(a pole wheel is defined as a magnetically active rotor)* permits a considerably larger air gap.

Normally, *passive rotors* are in the form of gearwheels. In many cases, these are already present on the engine (for instance the flywheel ring gear). Otherwise they must be installed in a specific position in order to generate the required signal (as needed for instance for ABS). In the latter case, both planar-toothed and axial-scan versions are in use.

An easily identifiable reference mark is required for each revolution when picking-off the crankshaft speed or position (for instance, at the starter ring gear). This reference mark must ensure interference-free, optimal timing of ignition and fuel injection. This applies in particular when there is no camshaft pick-off available. The reference mark can be in the form of a completely (or partially) removed tooth. Due to the fact that a tooth gap "takes more time", the reference mark is immediately identifiable, particularly since the engine speed can only vary gradually and never abruptly.

In addition to toothed gears, stamped perforated discs or wave-shaped metal rings have been introduced in the meantime as low-priced rotors (ABS).

It was the integration of ABS sensors in the vehicle's wheel bearing which led to the introduction of *pole wheels* some of which also assume the role of a shaft seal (plastic-bound magnetic powder). Small, for the most part encapsulated, tachometer sensors connected through a short flexible shaft with one of the vehicle's wheels, also use pole wheels (with only very few poles) for generating a speed signal. These are usually picked-off by means of integrated Hall-effect sensors.

Relative rpm and speed measurement

Inductive-type sensors
Basics

Inductive-type sensors were available on the market as coil versions long before the first suitable microstructure sensor versions (e.g. using the Hall principle) became available. Such inductive-type sensors use Faraday's law when measuring rotational speed. In other words, they generate a voltage U_A at their two-pole output which is proportional to the change (referred to time) of a magnetixc flux Φ (w = number of turns).

$$U_A = U_{ind} = w \cdot d\Phi/dt$$

The magnetic flux Φ is also a function of the rotational position x and the air gap d_L:

With $\Phi = \Phi(x, d_L)$ and d_L = constant, the following applies:
$$U_A = U_{ind} = w \cdot \partial\Phi/\partial x \cdot dx/dt$$

Whereby, dx/dt represents the (rotating) speed being measured.

The inductive-type sensors' weak point is underlined though by the equation: If it is impossible to keep the air gap d_L constant (due to flutter or other forms of mechanical play), the air-gap fluctuations induce the same change of flux as does a fluctuation of speed. This effect can cause the generation of voltage pulses which are either impossible, or at least difficult to separate from the genuine rpm signals. Since the flux varies exponentially along with the change in air gap, and the air-gap fluctuations are often of the high-frequency type (e.g. brake flutter), these unwanted pulses can easily feature a high voltage amplitude.

Inductive-type sensors are therefore always of the dynamic type, and being as their output signal tends to fall to zero in such cases, they are in principle also unsuitable for the registration of extremely low speeds (quasi-static or static). The only exception here are the coil sensors which are powered by a carrier frequency and which are based on the eddy-current or damping principle. These though are hardly ever used in automotive applications.

Basically, inductive-type sensors are comprised of three important magnetic components (Fig. 1):
- Fixed coil
- Soft-iron component and
- The permanent-magnet component

The change in flux needed for the generation of the output voltage results from the motion or rotation of the hard or soft-iron magnetic component. Sensors which use a DC applied to the induction coil instead of permanent-magnetic flux were formerly referred to as TDC sensors, and were commonly used for the manual adjustment of the ignition.

The inductive-type sensors in use today are preferably composed of a bar magnet (Fig. 1, Pos. 1) with a soft-magnetic pole pin (2) carrying the 2-connection induction coil (3). When a ferromagnetic gear wheel (5), or some form of similar rotor, rotates in front of this sensor (pick-up) the changing magnetic flux (which varies as a function of time) induces a practically sinusoidal voltage.

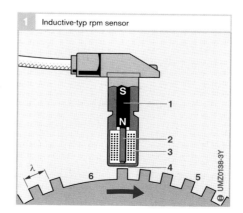

Inductive-typ rpm sensor

Fig. 1
1 Rod magnet
2 Soft-magnetic pole pin
3 Induction coil
4 Air gap d_L
5 Ferromagnetic gearwheel (or rotor or trigger wheel)
6 Rotational or reference mark

λ Tooth interval

For scanning very fine tooth structures, the end of the pole pin is sometimes pointed and acts as a flux-concentrating element. In other words it is shaped like a pole "blade" which usually projects through the metal or plastic housing and is adapted to the increment structure regarding shape and direction.

The rotor can be provided with one or more peripheral markings (6). Fig. 2 shows the flux curve and the voltage induced by a single peripheral or reference mark (slot, cam, or pole pin).

Normally, the steep passage through zero which takes place at the mid-point of maximum flux is utilised for the electronic registration of such a peripheral or reference mark. According to Faraday's Law, in all phases the signal's amplitude is proportional to rotational speed.

In order to ensure adequate, interference-free evaluation in the ECU, the spacing between the peaks of a double pulse (or of a periodic voltage pulse) U_{SS} should be at least 30 mV. The major disadvantage of the inductive-type sensors is the fact that at high rotational speeds their output voltages can reach levels far in excess of 100 V which are difficult to process electronically.

If Zener diodes are used to clip the high voltage peaks, the resulting changes in the sensor's load impedance rapidly lead to considerable phase-angle errors. With camshaft and crankshaft sensors this can have highly undesirable results with regard to the ignition where the correctness of phase relationship must be better than approx. 0.2°.

Normally, the prepulse generated by the magnetic return field can be ignored at low speeds. With some *magnetically passive or active peripheral markings* though, at high speeds the prepulse voltage can increase to such an extent that it exceeds the threshold value of the downstream threshold discriminator and can cause an even greater error (Fig. 2a). For this reason, the threshold values of the ECU input circuit are adapted dynamically to the speed in question.

Provided that the tooth gap is not too narrow, a uniform tooth structure results in the practically sinusoidal voltage curve shown in Fig. 2b. The rotational speed can be taken from the spacing between the passages through zero of this generated voltage. Its amplitude is proportional to the rotational speed.

The signal amplitude is highly dependent (exponentially) upon air gap and tooth size. As is the case with all magnetic increment processes, up to air gaps of d_L, teeth can be efficiently detected as from half or 1/3 of a tooth interval λ.

$$d_L \leq \lambda/(2\ldots3)$$

The conventional toothed rotors for ABS and crankshaft applications cover air gaps of up to 1.5 and 0.8 mm respectively. The reference mark needed for the ignition results from leaving out a tooth or by closing a tooth gap. The reference point is detected when the distance between the passages through zero changes abruptly and causes a far higher signal voltage (corresponding to an apparently larger tooth) which has a negative effect upon the previous and upon the subsequent incremental voltage – this can under certain circumstances be undesirable.

Assessment

Advantages
- Low manufacturing costs
- High-level EMC: Low static internal resistance (dynamic resistance is higher), no local electronic circuitry (electrically passive) which needs protection
- No problems with DC voltage drift (dynamic measuring concept)
- Broad temperature range (limited primarily by the casting-compound characteristics)

Disadvantages
- Conventional coil technology imposes limits on size reduction
- The output signal is rpm-dependent, unsuitable for quasi-static movements
- Sensitive to air-gap fluctuations

Examples of application
- Inductive engine-speed sensor (crankshaft-rpm sensor)
- Inductive-type wheel-speed sensor
- Inductive-type camshaft sensor (transistorized ignition with induction-type pulse generator TC-I)
- Needle-motion sensor (diesel fuel injection)

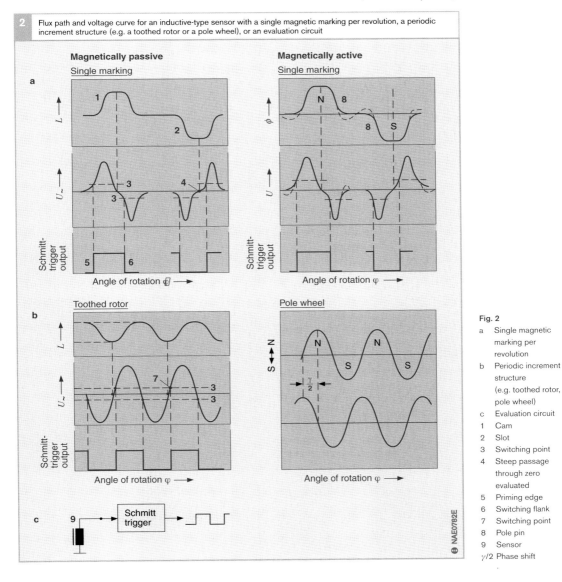

Fig. 2
a Single magnetic marking per revolution
b Periodic increment structure (e.g. toothed rotor, pole wheel)
c Evaluation circuit
1 Cam
2 Slot
3 Switching point
4 Steep passage through zero evaluated
5 Priming edge
6 Switching flank
7 Switching point
8 Pole pin
9 Sensor
$\gamma/2$ Phase shift

Magnetostatic sensors

Overview

Quasi-static speed measurement is best performed using magnetostatic sensors. Their output signal is independent of rotational speed and depends solely on field strength, so that at high rotational speeds their low signal voltages are easier to handle electronically. Furthermore, they not only have the advantage of imposing practically no limits with regard to *size reduction*, but also *signal amplification and/or signal processing* can be integrated locally. The fact that they are very small means that multiple systems such as differential configurations, or arrangements with integrated recognition of direction, present no difficulties.

On the other hand, such active sensors have a serious disadvantage, and that is the fact that their *operating-temperature* range is for the most part defined by the relevant Si (silicon) evaluation electronics which as a rule cannot withstand such high temperatures as the sensor element itself. For some time now, active sensors have been available with the option of a two-pole current output, so that the two-core connection can no longer be regarded as a specific advantage of the inductive-type coil sensor.

Hall-effect vane switches

If Hall-effect Si sensors are to be used for incremental rpm measurement, pronounced manufacturing scatter together with the effects of temperature mean that they must be provided with an adequate induction jump of typically 40...50 mT in order to ensure reliable high-speed switching. With conventional Hall-effect sensors and acceptable air gaps, this was only possible with the sensor in the form of a "Hall-effect vane switch" (for instance when used as ignition-triggering sensors in the ignition distributor). The sensor and its electronic circuitry for supply and signal evaluation are integrated directly on the sensor chip.

This "Hall-IC" (using bipolar techniques, for sustained temperatures up to 150 °C, and for direct connection into the vehicle's on-board electrical supply), is located in a practically closed magnetic circuit comprising permanent magnet and pole pieces (Fig. 3). A soft-magnetic trigger wheel (driven, for instance, by the camshaft) rotates through the air gap. With a trigger-wheel vane in the air gap, the magnetic flux is short-circuited past the sensor element. On the other hand though, when a trigger-wheel opening passes through the air gap the flux is unhindered on its way to the sensor. This principle ensures that the sensor also performs perfectly when the trigger-wheel varies in how far it penetrates into the air gap, or in case the air-gap shifts radially, that is vertically to the direction of rotation.

Since Hall-effect vane switches of this type feature limited peripheral resolution they are mainly used as segment sensors. If the vane slot is too narrow, it is practically impossible for the magnetic flux to pass through, with the result that the required induction jump is not generated.

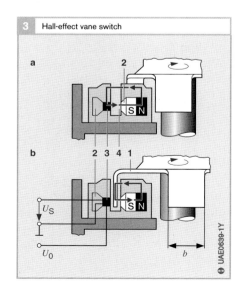

Fig. 3
a Magnetic flux: Unhindered
b Magnetic flux: Short-circuited

1 Vane width b
2 Soft-magnetic conductive element
3 Hall IC
4 Air gap

U_0 Supply voltage
U_S Sensor voltage

Simple Hall-effect rod sensors

In contrast to the Hall-effect vane switch, the working-point magnetization depends too much upon the width of the air gap, and the induction excursion is too small for reliable switching with this configuration. Simple rod-shaped Hall-effect sensors provided with a working-point magnet are therefore unsuitable for static or quasi-static scanning of a *magnetically passive rotor* (Fig. 4a, toothed rotor). The switching point of a downstream threshold comparator (Schmitt trigger) would have to be continually adapted to the varying working point. Such applications are only feasible when DC coupling is dispensed with and therefore also static signal evaluation. The coupling capacitors required for such configurations (also known as $\Delta\Phi$ sensors) involve high costs though and lead to reduced operational reliability.

On the other hand, simple Hall-effect sensors are highly suitable for scanning a *magnetically active rotor* (Fig. 4b, pole wheel). In this case, a working-point magnet is unnecessary. The sensor is triggered by the pole wheel with varying polarity only in the vicinity of the magnetic zero point. Although the magnetic control excursion reduces along with increasing air gap, the position of the working point ($B = 0$) remains unchanged. Being as the working point on new types of Hall-effect sensors is to a great extent thermally stable, the switching points of the downstream threshold comparator can be set relatively close together. This means that relatively wide air gaps become possible. Air-gap fluctuations in this configuration cannot cause missing pulses as they do not lead to polarity change which is the only thing that characterises the continuing measurement motion (rotation).

Gradient sensors

When it comes to scanning passive rotors, gradient sensors (Fig. 5) designed on the basis of differential Hall sensors or differential magnetoresistive sensors, are far better than simple Hall-effect sensors. They are provided with a permanent magnet. This pole face of this magnet facing the rotor is homogenised by a thin ferromagnetic wafer (Pos. 2). There are two galvanometric elements (generic term for Hall-effect sensors and magnetoresistors) located on each wafer about half a tooth gap apart. This means that one of the elements directly faces a tooth gap while the other faces a tooth. The sensor measures the difference in field strength between two neighboring points on

Fig. 4
a Passive rotor
b Active rotor

1 Incremental rotor
2 Simple Hall-effect sensor
3 Permanent magnet
4 Pole wheel
5 Case

φ Angle of rotation

Fig. 5
1 Magnetoresistors R_1, R_2 or Hall-effect elements H_1, H_2
2 Soft-magnetic substrate
3 Permanent magnet
4 Toothed rotor

U_0 Supply voltage
$U_A(\varphi)$ Measurement voltage for a rotational angle of φ

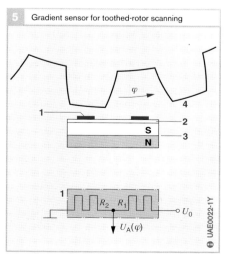

the circumference of the rotor. The output signal corresponds roughly to the field strength derived as a function of the angle at the circumference, and its sign is therefore independent of the air gap. Being as they do not alter the gradient signal's sign, air-gap fluctuations does not cause missing pulses.

For *signal evaluation*, it is a simple matter to connect the two *magnetoresistors* as a voltage divider which is supplied by a constant voltage and whose (usually unloaded) output signal is registered by the ECU. At room temperatures and with the customary air gap, this signal is in the volts range, and even at high temperatures it is suitable for transmission to the ECU without any form of preamplification.

Provided appropriate circuitry is used, measuring the loaded output current of the magnetoresister divider instead of its open-circuit voltage, permits the sensor's pronounced temperature sensitivity to be compensated for to a great extent.

In the case of a gradient sensor based on the Hall effect, the current paths of both Hall-effect elements can be connected in parallel, and their opposite-polarity output voltages in series, so that their differential voltage can be picked off directly and inputted to the downstream amplification and evaluation stages.

Tangential sensors
In contrast to gradient sensors, tangential sensors react to the sign and the intensity of the magnetic-field components which are tangential to the rotor's circumference. Using AMR thin-film techniques, tangential sensors are available as barber-pole or permalloy resistor types in full-bridge or half-bridge circuits (Fig. 6). In contrast to gradient sensors, they need not be matched to the particular tooth pitch of the rotor and can in fact be designed to sense practically at a given point. Local amplification is necessary, even though their measuring effect is 1…2 orders of magnitude larger than that of the silicon-Hall sensors.

In the case of the crankshaft-speed sensor integrated in the bearing (composite seal with sensor), the AMR thin-film sensor is mounted together with an evaluation IC on a common leadframe.

In order to save space and protect against high temperatures, the evaluation IC is turned through 90° and located further away from the sensor tip.

Giant magnetoresistive (GMR) elements
In 1988, Baibich discovered that in multi-layer (CuCo) elements of only a few nanometers (nm) thickness, the resistance changes by 50% when an external magnetic field is applied at low temperatures. This resistance change, which became known as the Giant Magneto Resistance effect (GMR), is considerably more pronounced than on AMR sensors.

The resistance changes because the magnetisation which was originally inverse-parallel aligned, re-orientates itself in parallel when an external magnetic field is applied. The effect reaches saturation at a defined magnetic-field strength.

GMR sensors are already in use as the reading head in high-capacity data disc drives. In the automotive sector, rotational-speed measurement is the priority applica-

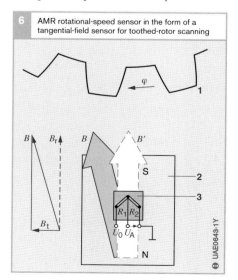

6 AMR rotational-speed sensor in the form of a tangential-field sensor for toothed-rotor scanning

Fig. 6
1 Toothed rotor (Fe)
2 Permanent magnet
3 Sensor

B Control-field strength with tangential components B_t, and radial components B_r (B' off position, $B_t = 0$). R_1, R_2 permalloy thin-film resistors (AMR)
φ Angle of rotation
U_0 Supply voltage
U_A Measurement voltage

tion aimed at at present – it is even more important than the sensor's use for travel/angle measurement.

Application examples:
- Hall-effect sensor (transistorized ignition TI-H)
- Hall phase sensor (camshaft)
- Gearbox Hall-effect sensor (RS50, RS51)
- Active Hall rotational-speed sensor
- Active AMR rotational-speed sensor
- Magnetoresistive sensor (for diesel radial-piston distributor pumps)

Absolute rotating-speed measurement

Oscillation gyroscope
Basics
Mechanical gyroscopes (gyros for short) utilise inertial forces in precisely measuring angular movements in space independent of the reference system. Notwithstanding their pronounced measuring effect, rotating gyroscopes as well as optical sensors based on the interferometric *Sagnac effect* (laser and fiber gyroscopes), are out of the question for automotive applications due to the high costs involved.

On the other hand, with the new automotive systems, the slightly less-severe requirements regarding precision can be complied with by *gyroscopes* manufactured using fine-mechanical and micromechanical processes. Instead of a rotational movement, these units use an equivalent, elastic, oscillatory movement to generate the measuring effect. These sensors are known as tuning-fork sensors and up to now have been used primarily for stabilisation controls. They also comply adequately with other automotive stipulations such as freedom from maintenance, long service life, switch-on time constant etc., not to forget the question of costs.

Oscillation gyroscopes measure the absolute yaw rate Ω_z at the vehicle's vertical axis (yaw axis). This applies for instance in vehicle-dynamics systems (ESP or Electronic Stability Program for the prevention of skidding), and for short-term navigation (for instance at a road junction). Highly advanced systems for triggering roll-over protection systems need data on the yaw rates Ω_x and Ω_y around the vehicle's pitch and roll axes. In principle, these sensors are similar to mechanical gyroscopes. They utilise the so-called *Coriolis accelerations* which are generated when rotation Ω is coupled with an oscillatory component (velocity v). In line with the familiar vector law, these accelerations are vertical with respect to the x and y axes (Fig. 1).

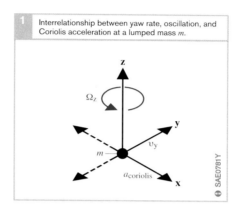

Fig. 1
Ω_z Yaw rate
v_y Velocity of the oscillatory motion
$a_{Coriolis}$ Coriolis acceleration
m Lumped mass

$$\vec{a}_{Coriolis} = a_x = 2 \cdot \vec{v}_y \cdot x \cdot \vec{\Omega}_z \qquad (1)$$

Whereby, the velocity v_y changes sinusoidally in accordance with the oscillatory movement:

$$v_y = \hat{v}_y \cdot \sin \omega \cdot t \qquad (2)$$

Assuming a constant yaw rate Ω_z, therefore, this means that a sinuosoidal Coriolis acceleration $a_{Coriolis}$ of the same freuqncy and phase is also measured. The amplitude is then:

$$\vec{a}_{Coriolis} = 2 \cdot \hat{v}_y \cdot \Omega_z \qquad (3)$$

Hypothetically, this acceleration could be felt and measured by anyone located on the lumped mass m.

To register the yaw rate, the amplitude of velocity of the oscillatory motion is maintained at a constant level by means of appropriate control circuitry. The Coriolis acceleration measured at the oscillating mass m is then subjected to frequency and phase-selective rectification. Here, for instance, a lock-in amplifier can be used. In the process, unwanted acceleration from the outside (e.g. bodywork acceleration) is removed.

An output voltage is generated which is proportional to the yaw rate:

$$U_A = const \cdot \hat{a}_{Coriolis} = const' \cdot \Omega \qquad (4)$$

The acceleration a_y which is also applied to the mass m in the oscillatory direction is usually several orders of magnitude higher than the useful Coriolis acceleration.

$$a_y = \frac{dv_y}{dt} = \omega \cdot \hat{v}_y \cdot \cos \omega \cdot t \qquad (5)$$

The falsifying effect of the acceleration a_y caused by over-response is counteracted by both the directional selectivity of the Coriolis acceleration sensor and its correct mounting position (factor $10^2 ... 10^4$), as well as by the correct-phase rectification of the Coriolis signal. Compared to the useful signal, namely, the interfering oscillatory acceleration is off-phase by 90°. With increasing frequency though, the signal-to-disturbance ratio increases proportionally.

Examples of application
- Piezoelectric yaw-rate sensors
- Micromechanical yaw-rate sensors MM1 and MM2

Radar sensors

On special-purpose vehicles with high levels of drive slip (e.g. agricultural tractors), simple low-cost close-range Doppler-effect radar systems (24...35 GHz) are used to measure the quantity "vehicle speed over ground v_F" (Figs. 1 and 2).

Each side of the vehicle is equipped with a transceiver probe which directs its radar beam onto the ground at an oblique angle of α, and with a frequency f_0. If a receiver were situated at this point on the ground, due to the Doppler effect it would receive this permanent signal at a higher frequency f_1 since the transmitter is moving towards it. This is similar to the acoustic effect experienced when overtaken by an emergency vehicle with its warning siren in operation:

$$f_1 = f_0 \frac{c}{c - v_F \cdot \cos \alpha} \qquad (6)$$

The ground underneath the tractor, though, reflects the signal back to the transceiver probes, and again a higher frequency f_2 is measured at the receiver, since in this case the receiver is moving towards the source.

$$f_2 = f_1 \frac{c + v_F \cdot \cos \alpha}{c} = f_0 \frac{c + v_F \cdot \cos \alpha}{c - v_F \cdot \cos \alpha} \qquad (7)$$

All in all, this results in a frequency shift Δf of:

$$\Delta f_2 = f_2 - f_1 = f_0 \frac{2 \cdot v_F \cdot \cos \alpha}{c - v_F \cdot \cos \alpha} \qquad (8)$$

or, conversely, the speed v_F:

$$v_F = \frac{c}{\cos \alpha} \cdot \frac{f_2 - f_0}{f_2 + f_0} \approx \frac{1}{2} \cdot \frac{c}{\cos \alpha} \cdot \frac{\Delta f_F}{f_0} \qquad (9)$$

This speed v_F is directly proportional to the frequency shift Δf.

The vehicle's rocking or tilting movements slightly change the angle α at which the signal is transmitted to ground. The probe which is rigidly attached to the bodywork utilises a radar system which "looks to the rear". This is known as the *Janus principle* (Fig. 2).

With the tractor horizontal, both systems measure the same value. If the vehicle tilts to the front or rear, since the beam angle increases on the one system by the same amount as it decreases on the other, the systems indicate opposite deviations from this (horizontal) value.

Averaging the two values eliminates the error resulting from tilt. The low measuring effect necessitates a relatively long-term averaging of the signals (approx. 1 s), so that rapid measurements of speed are impossible with this system.

The difference in the signals from the systems on each side of the vehicle provide a good indication of the angle-of-travel actually taken by the vehicle.

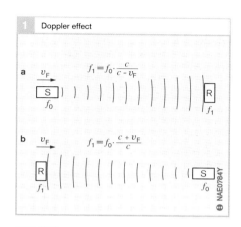

1 Doppler effect

Fig. 1
a Transmitter S moves towards the stationary receiver R
b Receiver R moves towards stationary transmitter S

f_0 Transmit-signal frequency
f_1 Frequency of signal arriving at the receiver
v_F Vehicle speed

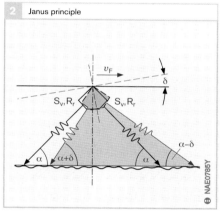

2 Janus principle

Fig. 2
S/R Transmitter/receiver
S_r, R_r Rearward-measuring system
S_v, R_v Forward-measuring system
v_F Vehicle speed
α Alignment angle of the measuring system
φ Vehicle's angle of tilt referred to the ground

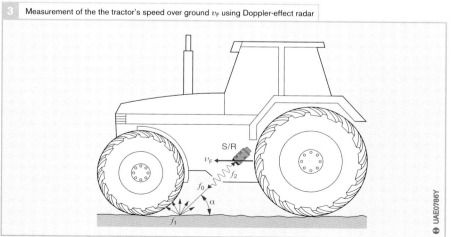

3 Measurement of the the tractor's speed over ground v_F using Doppler-effect radar

Fig. 3
S/R Transmitter/receiver
v_F Vehicle speed
f_0 Transmit-signal frequency
f_1 Frequency of signal arriving at the ground
f_2 Frequency of signal arriving at the receiver
α Alignment angle of the measuring system

Inductive engine-speed sensors

Applications
Such engine-speed sensors are used for measuring:

- Engine rpm
- Crankshaft position (for information on the position of the engine pistons)

The rotational speed is calculated from the sensor's signal frequency. The output signal from the rotational-speed sensor is one of the most important quantities in electronic engine management.

Design and operating concept
The sensor is mounted directly opposite a ferromagnetic trigger wheel (Fig. 1, Pos, 7) from which it is separated by a narrow air gap. It has a soft-iron core (pole pin) (4), which is enclosed by the solenoid winding (5). The pole pin is also connected to a permanent magnet (1), and a magnetic field extends through the pole pin and into the trigger wheel. The level of the magnetic flux through the winding depends upon whether the sensor is opposite a trigger-wheel tooth or gap. Whereas the magnet's stray flux is concentrated by a tooth and leads to an increase in the working flux through the winding, it is weakened by a gap. When the trigger wheel rotates therefore, this causes a fluctuation of the flux which in turn generates a sinusoidal voltage in the solenoid winding which is proportional to the rate of change of the flux (Fig. 2). The amplitude of the AC voltage increases strongly along with increasing trigger-wheel speed (several mV...>100 V). At least about 30 rpm are needed to generate an adequate signal level.

The number of teeth on the trigger wheel depends upon the particular application. On solenoid-valve-controlled engine-management systems for instance, a 60-pitch trigger wheel is normally used, although 2 teeth are omitted (7) so that the trigger wheel has 60 − 2 = 58 teeth. The very large tooth gap is allocated to a defined crankshaft position and serves as a reference mark for synchronizing the ECU.

There is another version of the trigger wheel which has one tooth per engine cylinder. In the case of a 4-cylinder engine, therefore, the trigger wheel has 4 teeth, and 4 pulses are generated per revolution.

The geometries of the trigger-wheel teeth and the pole pin must be matched to each other. The evaluation-electronics circuitry in the ECU converts the sinusoidal voltage, which is characterized by strongly varying amplitudes, into a constant-amplitude square-wave voltage for evaluation in the ECU microcontroller.

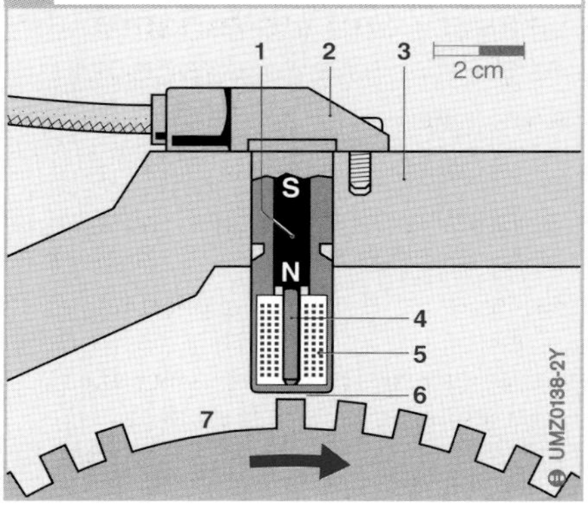

Fig. 1
1 Permanent magnet
2 Sensor housing
3 Engine block
4 Pole pin
5 Solenoid winding
6 Air gap
7 Trigger wheel with reference-mark gap

Fig. 2
1 Tooth
2 Tooth gap
3 Reference mark

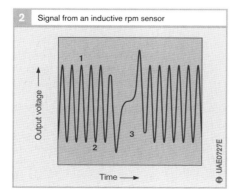

Rotational-speed (rpm) sensors and incremental angle-of-rotation sensors

Application

The above sensors are installed in distributor-type diesel injection pumps with solenoid-valve control. Their signals are used for:

- The measurement of the injection pump's speed
- Determining the instantaneous angular position of pump and camshaft
- Measurement of the instantaneous setting of the timing device

The pump speed at a given instant is one of the input variables to the distributor pump's ECU which uses it to calculate the triggering time for the high-pressure solenoid valve, and, if necessary, for the timing-device solenoid valve.

The triggering time for the high-pressure solenoid valve must be calculated in order to inject the appropriate fuel quantity for the particular operating conditions. The cam plate's instantaneous angular setting defines the triggering point for the high-pressure solenoid valve. Only when triggering takes place at exactly the right cam-plate angle, can it be guaranteed that the opening and closing points for the high-pressure solenoid valve are correct for the particular cam lift. Precise triggering defines the correct start-of-injection point and the correct injected fuel quantity.

The correct timing-device setting as needed for timing-device control is ascertained by comparing the signals from the camshaft rpm sensor with those of the angle-of-rotation sensor.

Design and operating concept

The rpm sensor, or the angle-of-rotation sensor, scans a toothed pulse disc with 120 teeth which is attached to the distributor pump's driveshaft. There are tooth gaps, the number of which correspond to the number of engine cylinders, evenly spaced around the disc's circumference. A double differential magnetoresistive sensor is used.

Magnetoresistors are magnetically controllable semiconductor resistors, and similar in design to Hall-effect sensors. The double differential sensor has four resistors connected to form a full bridge circuit.

The sensor has a permanent magnet, and the magnet's pole face opposite the toothed pulse disc is homegenized by a thin ferromagnetic wafer on which are mounted the four magnetoresistors, separated from each other by half a tooth gap. This means that alternately there are two magnetoresistors opposite tooth gaps and two opposite teeth (Fig. 1). The magnetoresistors for automotive applications are designed for operation in temperatures of ≤ 170 °C (≤ 200 °C briefly).

1 Rotation-speed/angle-of-rotation sensor (principle)

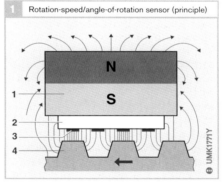

Fig. 1
1 Magnet
2 Homogenization wafer (Fe)
3 Magnetoresistor
4 Toothed pulse disc

2 Rotation-speed/angle-of-rotation sensor (as installed)

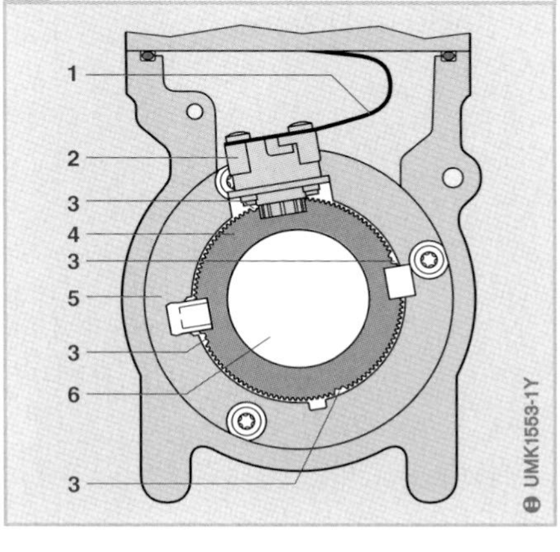

Fig. 2
1 Flexible conductive foil
2 Rotation-speed (rpm)/angle-of-rotation sensor
3 Tooth gap
4 Toothed pulse wheel (trigger wheel),
5 Rotatable mounting
6 Driveshaft

Hall-effect phase sensors

Application

The engine's camshaft rotates at half the crankshaft speed. Taking a given piston on its way to TDC, the camshaft's rotational position is an indication as to whether the piston is in the compression or exhaust stroke. The phase sensor on the camshaft provides the ECU with this information.

Design and operating concept

Hall-effect rod sensors

As the name implies, such sensors (Fig. 2a) make use of the Hall effect. A ferromagnetic trigger wheel (with teeth, segments, or perforated rotor, Pos. 7) rotates with the camshaft. The Hall-effect IC is located between the trigger wheel and a permanent magnet (Pos. 5) which generates a magnetic field strength perpendicular to the Hall element.

If one of the trigger-wheel teeth (Z) now passes the current-carrying rod-sensor element (semiconductor wafer), it changes the magnetic field strength perpendicular to the Hall element. This causes the electrons, which are driven by a longitudinal voltage across the element to be deflected perpendicularly to the direction of current (Fig. 1, angle α).

This results in a voltage signal (Hall voltage) which is in the millivolt range, and which is independent of the relative speed between sensor and trigger wheel. The evaluation electronics integrated in the sensor's Hall IC conditions the signal and outputs it in the form of a rectangular-pulse signal (Fig. 2b "High"/"Low").

Differential Hall-effect rod sensors

Rod sensors operating as per the differential principle are provided with two Hall elements. These elements are offset from each other either radially or axially (Fig. 3, S1 and S2), and generate an output signal which is proportional to the difference in magnetic flux at the element measuring points. A two-track perforated plate (Fig. 3a) or a two-track trigger wheel (Fig. 3b) are needed in order to generate the opposing signals in the Hall elements (Fig. 4) as needed for this measurement.

Such sensors are used when particularly severe demands are made on accuracy. Further advantages are their relatively wide air-gap range and good temperature-compensation characteristics.

Fig. 1
- I Wafer current
- I_H Hall current
- I_V Supply current
- U_H Hall voltage
- U_R Longitudinal voltage
- B Magnetic induction
- α Deflection of the electrons by the magnetic field

Fig. 2
- a Positioning of sensor and single-track trigger wheel
- b Output signal characteristic U_A

1. Electrical connection (plug)
2. Sensor housing
3. Engine block
4. Seal ring
5. Permanent magnet
6. Hall-IC
7. Trigger wheel with tooth/segment (Z) and gap (L)

- a Air gap
- φ Angle of rotation

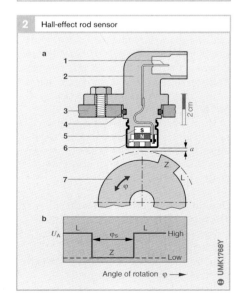

1 Hall element (Hall-effect vane switch)

2 Hall-effect rod sensor

Speed and rpm sensors   Hall-effect phase sensors   401

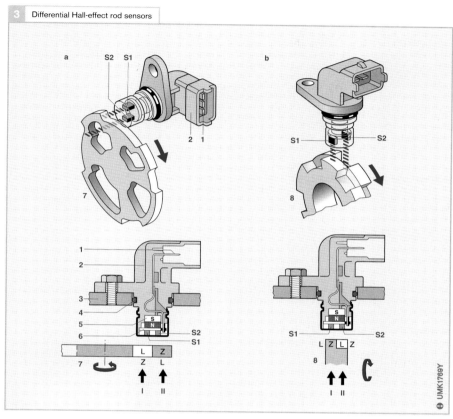

Fig. 3
a Axial tap-off (perforated plate)
b Radial tap-off (two-track trigger wheel)

1 Electrical connection (plug)
2 Sensor housing
3 Engine block
4 Seal ring
5 Permanent magnet
6 Differential Hall-IC with Hall elements S1 and S2
7 Perforated plate
8 Two-track trigger wheel
I Track 1
II Track 2

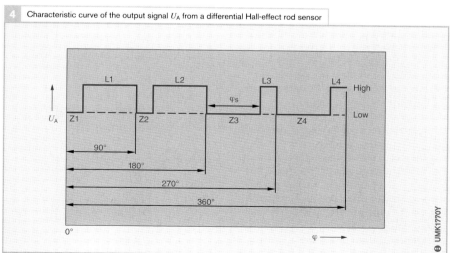

Fig. 4
Output signal "Low":
Material (Z) in front of S1, gap (L) in front of S2

Output signal "High":
Gap (L) in front of S1, material (Z) in front of S2

φ_S signal width

Wheel-speed sensors

Application

It is from the wheel-speed sensor signals that the ABS, TCS, and ESP control units (ECUs) derive the wheel-rotation rates. These wheel speeds are applied in preventing the wheels blocking or spinning so that the vehicle's stability and steerability are maintained. In verhicle navigation systems, the signals are used for calculating the distance travelled.

Design and operating concept

Passive (inductive) wheel-speed sensors
The inductive wheel-speed sensor's pole pin, surrounded by its coil winding, is installed directly above a trigger wheel (rotor) attached to the wheel hub. This soft-magnetic pole pin is connected to a permanent magnet which projects a magnetic field toward and into the trigger wheel. The continuously alternating sequence of teeth and gaps that accompanies the wheel's rotation induces corresponding fluctuations in the magnetic field through the pole pin and its coil winding. These fluctuations induce an alternating current in the coil suitable for monitoring at the ends of its winding.

The frequency and amplitude of this alternating current are proportional to wheel speed, and with the wheel not rotating, the induced voltage is zero. Tooth shape, air gap, rate of voltage rise, and the ECU input sensitivity define the smallest still measurable rotation rate and thus, for ABS applications, the minimum switching speed.

To ensure interference-free signal detection, the gap separating the wheel-speed sensor and the trigger wheel is only approx. 1 mm, and installation tolerances are narrow. The wheel-speed sensor is also installed on a stable mounting to prevent oscillation patterns in the vicinity of the brakes from distorting the sensor's signals. Various pole-pin configurations and installation options are available to adapt the system to the different installation conditions encountered with various wheels. The most common variant is the *chisel-type pole pin* (also called a flat pole pin Fig. 1a) for radial installation at right angles to the pulse rotor. The *rhombus-type (lozenge-shaped) pole pin* (Figure 1b) designed for axial installation is located radially with respect to the trigger wheel. Both pole-pin designs necessitate precise alignment to the trigger wheel. Although precise alignment is not so important with the *round pole pin* (Figure 1c), the trigger wheel must have a large enough diameter, or less teeth.

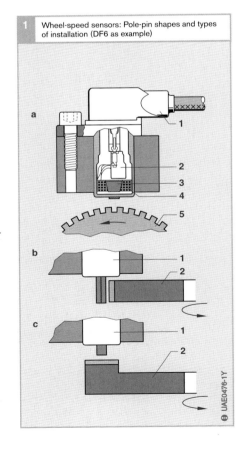

Fig. 1
a Chisel pole pin: Radial installation, radial scan
b Rhombus pole pin: Axial installation, radial scan
c Round pole pin: Radial installation, axial scan

1 Sensor case with electrical connections
2 Permanent magnet
3 Soft-iron core (pole pin)
4 Winding
5 Trigger wheel

Wheel-speed sensors: Pole-pin shapes and types of installation (DF6 as example)

Active wheel-speed sensors

The conventional inductive units are increasingly being replaced by active wheel-speed sensor types in which the function formerly performed by the trigger ring's teeth is taken over by peripheral magnets incorporated around the periphery of a multipole ring so that their polarities alternate (Fig. 2).

The sensor element of such an active wheel-speed sensor is located in the continuously changing fields generated by these magnets. Rotation of the multipole ring is thus accompanied by a continuous alternation in the magnetic flux through the sensor element.

Compact dimensions combine with low weight to make the active wheel-speed sensor suitable for installation on and even within the vehicle's wheel-bearing assemblies (Fig. 3). In the latter case, the bearing seal contains magnetic powder instead of fixed magnets. This means that a second function has been added and the bearing seal now becomes a multipole device.

The most important sensor components are either Hall or magnetoresistive elements, both of which generate a voltage that varies according to the magnetic flux through the measuring element. This voltage is then conditioned by the active wheel-speed sensor. One of the active sensor's advantages is the fact that in contrast to the inductive sensors, its output voltage is independent of the wheel speed. This fact permits monitoring to continue until the wheel is practically stationary.

A typical feature of the active wheel-speed sensor is the local amplifier circuit. Both components – measuring element and amplifier – are integrated in a single sensor casing. The active sensor requires a power supply of between 4.5 and 20 volts, and it is connected to the ECU by a two-conductor wire. The wheel-speed data is impressed on one of the two conductors (supply lines) as load-independent current. As with the inductive wheel-speed sensor, the current's frequency is proportional to wheel speed. This single-wire data-transmission strategy uses pre-conditioned digital signals. These are less sensitive to interference than the signals from the inductive sensor. The concept also features the following options:

- Data transmission identifying the wheel's direction of travel. This option is especially significant for the "hill-holding" feature, which relies on selective braking to prevent the vehicle from rolling backwards when starting off on a hill. Also used in vehicle navigation systems.
- Relay of information on sensor-signal quality, including a display indicating that the driver should have the vehicle serviced in order to check correct sensor functioning.

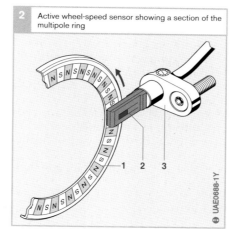

2 Active wheel-speed sensor showing a section of the multipole ring

Fig. 2
1 Multipole ring
2 Sensor element
3 Sensor case

3 Example of sensor installation in the wheel bearing

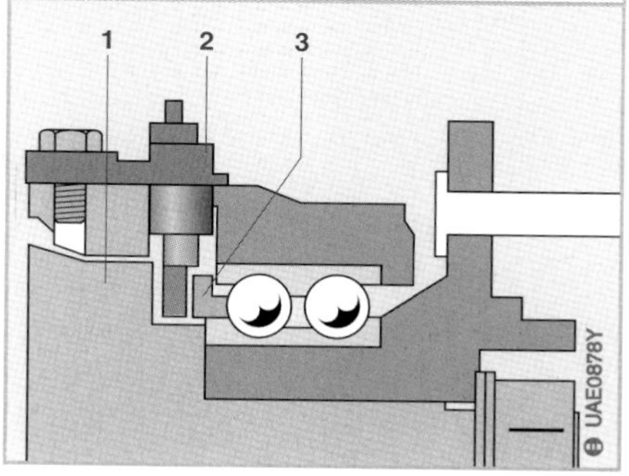

Fig. 3
1 Wheel bearing
2 Sensor
3 Multipole ring/ Bearing seal

Gearbox-rpm sensors

Application
Such gearbox-rpm sensors scan the speeds in automatic-transmissions, automatic shift transmissions and continuously-variable transmissions (AT, AST, and CVT respectively). For such applications, the sensors are designed to be insensitive to ATF gearbox oils. The "packaging concept" provides for sensor integration in the transmission-shift control module or for a "stand-alone" version. The sensor needs a power supply of 4.5...16.5 V and operates in the termperature range $-40...+150$ °C.

Design and operating concept
The active rpm sensor is provided with a differential Hall-effect IC with 2-wire current interface. For operation, the sensor must be connected to a voltage source (supply voltage U_V). It applies the Hall effect when scanning ferromagnetic toothed rotors, punched-sheet rotors, or multipole rings (air-gap range: 0.1...2.5 mm), and generates a constant-amplitude signal which is independent of rotational speed. This means that it is possible to register rotational speeds down to practically $n = 0$. For signal output, the supply current is modulated by the incremental signal. Using a measuring resistor R_M, the current modulation (Low: 7 mA; High: 14 mA) can then be converted in the ECU into a signal voltage U_{RM} (Fig. 1).

There are two different types of gearbox-rpm sensors (Fig. 2):

RS50
Data protocol: rpm information in the form of a rectangular-pulse signal.
Functional scope: A frequency signal triggered by the rotor passing the sensor surface. It is proportional to the rotor speed.

RS51
Data protocol: rpm information in the form of a rectangular-pulse signal with supplementary information which are transmitted using the pulse-width-modulation (pwm) principle.
Functional scope: rpm signal, detection of standstill, direction of rotation, air-gap reserve, and installation position.

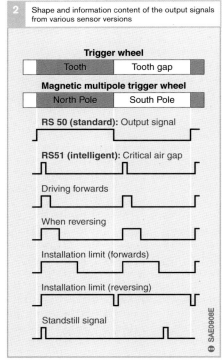

Fig. 2 Shape and information content of the output signals from various sensor versions

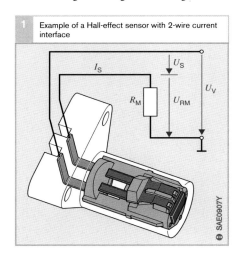

Fig. 1 Example of a Hall-effect sensor with 2-wire current interface

I_S Sensor current (supply and signal)
R_M Measuring resistor (in ECU)
R_{RM} Signal voltage
U_V Supply voltage
U_S Sensor voltage

Nozzle holder with needle-motion sensor

Application
The start-of-injection point is an important parameter for optimum diesel-engine operation. For intance, its measurement permits load and speed-dependent injection timing, and/or control of the EGR rate within a closed control loop. On the inline and distributor injection pumps, a nozzle-holder with needle-motion sensor is used for this purpose (Fig. 2) which outputs a signal as soon as the deedle moves.

Design and operating concept
A current of approx. 30 mA flows through the pick-up coil (Fig. 2, Pos. 11) and generates a magnetic field. The long pressure pin (12) extends into guide pin (9). The so-called immersion dimension "X", defines the magnetic flux in the pick-up coil. Nozzle-needle motion causes a change of flux. This, in turn, generates a velocity-dependent signal voltage which is directly processed in an ECU evaluation circuit. When a given threshold voltage is exceeded, this serves as the signal for the start of injection (Fig. 1).

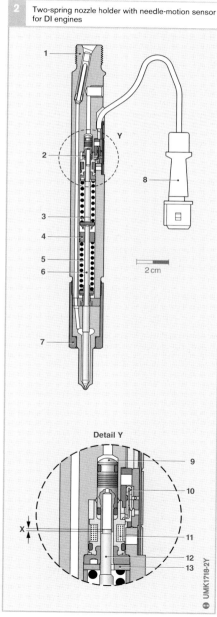

Two-spring nozzle holder with needle-motion sensor for DI engines

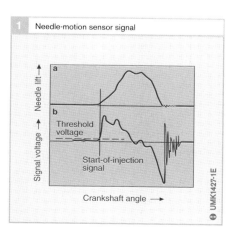

Needle-motion sensor signal

Fig. 1
a Needle-lift curve
b Signal-voltage curve

Fig. 2
1 Mount
2 Needle-motion sensor
3 Spring
4 Guide element
5 Spring
6 Pressure pin
7 Nozzle retaining nut
8 Connection for evaluation circuit
9 Guide pin
10 Contact lug
11 Pick-up coil
12 Pressure pin
13 Spring seat
x Immersion dimension

Induction-type sensors for transistorized ignition

Applications

For ignition-triggering purposes, the TC-I transistorized ignition uses an induction-type sensor which serves as an AC generator. The switch-on point for the dwell angle is defined by comparing its AC signal with that of a voltage signal which corresponds to the current-control time.

Design and construction

The induction-type sensor is incorporated in the ignition-distributor housing in place of the former contact-breaker points (Fig. 1).

The soft-magnetic core of the induction winding is disc-shaped, and together with the permanent magnet and the induction winding, forms a fixed, enclosed subassembly, the stator.

The rotor (trigger wheel) on the distributor shaft rotates past the ends of the stators. Similar to the distributor cam for the former contact breaker assembly, it is firmly attached to the hollow shaft surrounding the distributor shaft.

Core and rotor are produced from soft-magnetic material and have toothed extensions (stator teeth and rotor teeth). The stator teeth are at the ends of the stator "limbs" and bent upwards at right angles. The rotor has similar teeth, but these are bent downwards at right angles.

As a rule, the number of teeth on rotor and stator correspond to the number of cylinders in the engine. The fixed and rotating teeth are separated by a mere 0.5 mm when directly opposite to each other.

Operating concept

The principle of functioning depends upon the air gap between the rotor teeth and the stator teeth, and thus the magnetic flux, changing periodically along with rotation of the rotor. This change in magnetic flux induces an AC voltage in the induction winding whose peak voltage $\pm \hat{U}_S$ is proportional to the rotor's speed of rotation. At low speeds it is approx. 0.5 V and at high speeds approx. 100 V. The frequency f of this AC voltage (Fig. 2) corresponds to the number of ignition sparks per minute (sparking rate). The following applies

$$f = z \cdot n/2$$

where

f Frequency or sparking rate (rpm)
z Number of engine cylinders
n Engine speed (rpm)

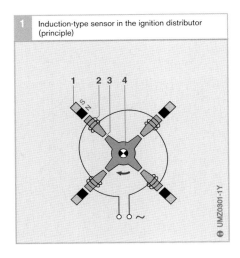

Fig. 1
1 Permanent magnet
2 Induction winding with core
3 Variable air gap
4 Rotor

Induction-type sensor in the ignition distributor (principle)

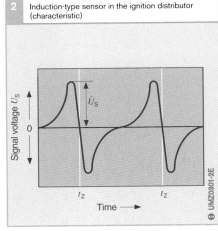

Fig. 2
U_S Signal voltage
$\hat{U}_S$ Peak voltage
t_Z Ignition point

Induction-type sensor in the ignition distributor (characteristic)

Hall-effect sensors for transistorized ignition

Application

The Hall-effect sensor is also used as the ignition-triggering sensor for the TI-H transistorized ignition system. The information contained in the signal from the Hall generator located in the ignition distributor corresponds to that in the signal generated by the breaker points in a conventional breaker-triggered coil-ignition system. Whereas with the conventional ignition system the distributor cam defines the dwell angle via the contact-breaker points, on the transistorized system the Hall-effect sensor in the ignition distributor defines the on/off ratio by means of the rotor (trigger-wheel) vane.

Design and construction

The Hall-effect sensor (Fig. 1) is installed in the ignition distributor, and its vane switch is attached to the movable mounting plate. The Hall IC is mounted on a ceramic substrate and in order to protect it against moisture, dirt, and mechanical damage is encapsulated in plastic at one of the conductive elements. The conductive elements and the rotor are made of a soft-magnetic material. The number of vanes on the rotor corresponds to the number of cylinders in the engine. Depending on the type of ignition trigger box, the width b of the rotor's individual conductive elements can define the ignition system's maximum dwell angle. The dwell angle therefore remains practically constant throughout the Hall sensor's service life and dwell-angle adjustment is unnecessary.

Operating concept

When the ignition-distributor shaft rotates, the rotor vane's pass through the Hall IC air gap without making contact. If the air gap is not occupied by a vane, the magnetic field is free to permeate the Hall IC and the Hall-effect sensor element (Fig. 1). The magnetic flux density is high, the Hall voltage is at its maximum, and the Hall-IC is switched on. As soon as a rotor vane enters the air gap, the majority of the magnetic flux is diverted through the vane and is isolated from the Hall-IC. The magnetic flux density at the Hall sensor element reduces to a negligible level which results from the leakage field, and the Hall voltage drops to a minimum. The dwell angle is defined by the rotor vane's shape as follows: A ramp voltage is generated from the signal voltage U_S (converted Hall voltage, Fig. 2). The switch-on point for the dwell angle is shifted as required along this ramp. The Hall-effect sensor's priniple of operation and its construction permit the ignition to be adjusted with the engine at standstill provided no provision is made for peak-coil-current cut-off.

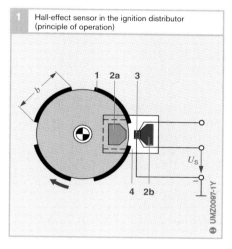

Fig. 1 Hall-effect sensor in the ignition distributor (principle of operation)

Fig. 1
1 Vane with width b
2a Permanent magnet
2b Soft-magnetic conductive element
3 Hall-IC
4 Air gap

U_S Signal voltage (converted Hall voltage)

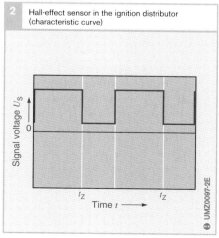

Fig. 2 Hall-effect sensor in the ignition distributor (characteristic curve)

Fig. 2
U_S Signal voltage (converted Hall voltage)
t_z Ignition point

Piezoelectric "tuning-fork" yaw-rate sensor

Application

In order that it can use the digital road map stored on the CD-ROM to calculate the distance driven, the computer in the vehicle's navigation system needs information on the vehicle's movements (composite navigation).

When cornering (for instance at road junctions), the navigation system's yaw-rate sensor registers the vehicles rotation about its vertical axis. With the voltage signal it generates in the process, and taking into account the signals from the tachometer or the radar sensor, the navigation computer calculates the curve radius and from this derives the change in vehicle direction.

Design and construction

The angle-of-rotation sensor is comprised of a steel element shaped like a tuning fork. This incorporates four piezo elements (two above, two below) and the sensor electronics.

This sensor measures very accurately and is insensitive to magnetic interference.

Operating concept

When voltage is applied, the bottom piezo elements start to oscillate and excite the upper section of the "tuning fork", together with its upper piezo elements, which then starts counter-phase oscillation.

Straight-ahead driving
With the vehicle being driven in a straight line there are no Coriolis forces applied at the tuning fork, and since the upper piezo elements always oscillate in counter-phase and are only sensitive vertical to the direction of oscillation (Fig. 1a) they do not generate a voltage.

Cornering
When cornering on the other hand, the Coriolis acceleration which occurs in connection with the oscillation (but vertical to it) is applied for measurement purposes. The rotational movement now causes the upper portion of the tuning fork to leave the oscillatory plane (Fig. 1b) so that an AC voltage is generated in the upper piezo elements which is transferred to the navigation computer by an electronic circuit in the sensor housing. The voltage-signal amplitude is a function of both the yaw rate and the oscillatory speed. Its sign depends on the direction (left or right) taken by the curve.

Fig. 1
a Excursion during straight-ahead driving
b Excursion when cornering

1 Tuning-fork direction of oscillation resulting from cornering
2 Direction of rotation of the vehicle
3 Direction of oscillation resulting from straight-ahead driving
4 Coriolis force
5 Upper piezo elements (sensing)
6 Bottom piezo elements (drive)
7 Excitation oscillation direction

Ω Yaw

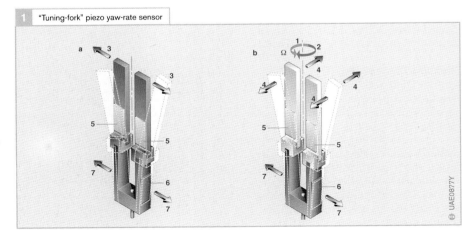

1 "Tuning-fork" piezo yaw-rate sensor

Piezoelectric "oscillating drum" yaw-rate sensors

Applications

In vehicle's with vehicle-dynamics control (ESP), the piezoelectric yaw-rate sensors (otherwise known as gyrometers) register the vehicle's rotation about its vertical axis, for instance when cornering, but also when the vehicle swerves or goes into a skid.

Design and construction

The piezoelectric yaw-rate sensors are high-precision mechanical sensors. Two diametrically opposed piezoceramic elements (Fig. 1, $1 + 1'$) are used to cause sympathetic oscillations in a hollow metal cylinder. Another pair of piezoceramic elements ($2 + 2'$) are used to control and maintain this oscillation at a constant amplitude which has four axially aligned oscillation nodes (offset by 45° to the direction of excitation). Refer to Figs. 1...3.

When rotation takes place at a yaw rate Ω about the cylinder's axis, the nodes are shifted slightly at the circumference due to the effects of Coriolis acceleration. The result is that in the nodes, which otherwise feature zero force, forces are now generated which are proportional to rotational speed and which are detected by a third pair of piezo elements ($3 + 3'$). Using a fourth pair of piezo excitation elements ($4 + 4'$) in a closed control loop, these forces are then controlled back to a reference value $U_{ref} = 0$. The manipulated variable needed here is then carefully filtered and subjected to phase-synchronous rectification before being used as a highly accurate output signal. The selective, temporary change of the desired value to $U_{ref} = 0$ permits an easy check of the overall sensor system ("built-in test"). This sensor's temperature sensitivity necessitates a complex compensation circuit, and the material-based aging of the piezoceramic elements necessitates painstaking preliminary aging.

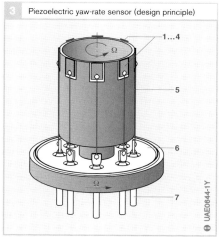

Fig. 1
1....4 Piezo elements
5 Circuit
6 Bandpass filter (phase-locked)
7 Phase reference
8 Rectifier (phase-selective)

U_A Output voltage
Ω Yaw rate
$U_{ref} = 0$ (normal operation)
$U_{ref} \neq 0$ ("built-in" test)

Fig. 3
1....4 Piezo element pairs
5 Oscillatory cylinder
6 Baseplate
7 Connection pins

Ω Yaw rate

Micromechanical yaw-rate sensors

Applications

In vehicles with Electronic Stability Program (ESP), the rotation of the vehicle about its vertical axis is registered by micromechanical yaw-rate (or yaw-speed) sensors (also known as gyrometers) and applied for vehicle-dynamics control. This takes place during normal cornering, but also when the vehicle breaks away or goes into a skid.

These sensors are reasonably priced as well as being very compact. They are in the process of forcing out the conventional high-precision mechanical sensors.

Design and construction

MM1 micromechanical yaw-rate sensor
A mixed form of technology is applied in order to achieve the high accuracies needed for vehicle-dynamics systems. That is, two somewhat thicker oscillating elements (mass plates) which have been machined from a wafer using bulk micromechanics oscillate in counter-phase to their resonant frequency which is defined by their mass and their coupling springs (>2 kHz). On each of these oscillating elements, there is a miniature, surface-type micromechanical capacitive acceleration sensor. When the sensor chip rotates about its vertical axis at yaw rate Ω, these register the Coriolis acceleration in the wafer plane vertical to the direction of oscillation (Figs. 1 and 2). These accelerations are proportional to the product of yaw rate and and the oscillatory velocity which is maintained electronically at a constant value.

To drive the sensor, all that is required is a simple, current-carrying printed conductor on each oscillating element. In the permanent-magnet field B vertical to the chip surface, this oscillating element is subjected to an electrodynamic (Lorentz) force. Using a further, simple printed conductor (which saves on chip surface), the same magnetic field is used to directly measure the oscillation velocity by inductive means. The different physical construction of drive system

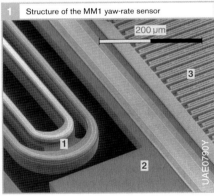

Structure of the MM1 yaw-rate sensor

Fig. 1
1. Retaining/guide spring
2. Part of the oscillating element
3. Coriolis acceleration sensor

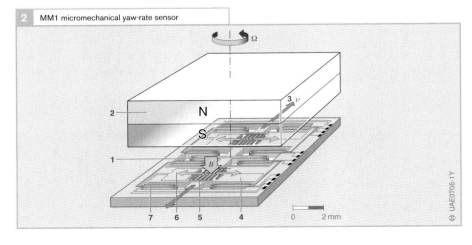

MM1 micromechanical yaw-rate sensor

Fig. 2
1. Frequency-determining coupling spring
2. Permanent magnet
3. Direction of oscillation
4. Oscillating element
5. Coriolis acceleration sensor
6. Direction of Coriolis acceleration
7. Retaining/guide spring

Ω Yaw rate
v Oscillating velocity
B Permanent-magnet field

and sensor system serves to avoid undesirable coupling between the two sections. In order to suppress unwanted external acceleration effects, the opposing sensor signals are subtracted from each other. The external acceleration effects can be measured by applying summation. The high-precision micromechanical construction helps to suppress the effects of high oscillatory acceleration which is several factors of 10 higher than the low-level Coriolis acceleration (cross sensitivity far below 40 dB). Here, the drive and measurement systems are rigorously decoupled from each other.

MM2 micromechanical yaw-rate sensor
Whereas this silicon yaw-rate sensor is produced completely using surface-micromechanic techniques, and the magnetic drive and control system have been superseded by an electrostatic system, absolute decoupling of the power/drive system and measuring system is impossible. Comb-like structures (Figs. 3 and 4) electrostatically force a centrally mounted rotary oscillator to oscillate. The amplitude of these oscillations is held constant by means of a similar capacitive pick-off. Coriolis forces result at the same time in an out-of-plane tilting movement, the amplitude of which is proportional to the yaw rate Ω, and which is detected capacitively by the electrodes underneath the oscillator. To avoid excessive damping of this movement, the sensor must be operated in a vacuum. Although the chip's small size and the somewhat simpler production process result in considerable cost reductions, this miniaturisation is at the expense of reductions in the measuring effect, which in any case is not very pronounced, and therefore of the achievable precision. It also places more severe demands on the electronics. The system's high flexural stability, and mounting in the axis of gravity, serve to mechanically suppress the effects of unwanted acceleration from the side.

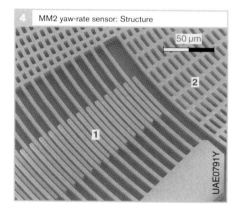

Fig. 4
1 Comb-like structure
2 Rotary oscillator

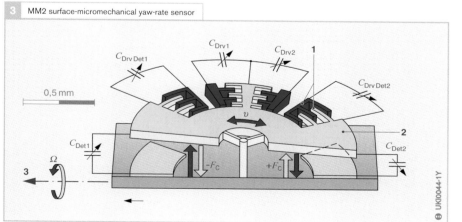

Fig. 3
1 Comb-like structure
2 Rotary oscillator
3 Measuring axis

C_{Drv} Drive electrodes
C_{Det} Capacitive pick-off
F_C Coriolis force
v Oscillatory velocity
$\Omega = \Delta C_{Det}$, measured yaw rate

Acceleration sensors and vibration sensors

Measured variables

Acceleration and vibration sensors are suitable for IC-engine knock control, as well as for triggering passenger-protection and restraint systems (airbag, seat-belt tightener, roll-over bar), and for the registration of the acceleration in a bend and road-speed changes on 4-wheel-drive vehicles equipped with ABS or ESP or chassis control.

Acceleration a is the measured quantity, and is often given as a multiple of the acceleration of free fall g ($1g = 9.81$ m/s²). Typical values encountered in automotive engineering are given in Table 1.

Measuring principles

In principle, all acceleration sensors measure according to the basic law of mechanics: The force F applied to an inert mass m due to the acceleration a, irrespective of whether it is dynamic (vibration sensors) or static:

$$F = m \cdot a \qquad (1)$$

Here, similar to force measurement, systems are available which measure displacement or travel, as well as mechanical strain-measurement systems.

Displacement or travel-measuring systems

Such systems (Fig. 1), are used in particular in applications concerning very low levels of acceleration. They also permit the use of the *compensation method* in which the system excursion caused by acceleration is compensated for by an equivalent restoring force so that ideally the system practically always operates very close to the restoring-force zero point (high linearity, minimum cross sensitivity, high temperature stability).

Due to their closed-loop control, these closed-loop position-controlled systems (Fig. 1b) feature higher stability and have a higher limit frequency than their "excursion"-measuring counterparts (Fig. 1a).

Table 1

1 Acceleration and vibration sensors

Application	Measuring range
Knock control	1 ... 10 g
Passenger-restraint and protection:	
– Airbag, seat-belt tightener	50 g
– Roll-over bar	4 g
– Seat-belt locking	0.4 g
ABS, ESP	0.8...1.2 g
Chassis control:	
– Body/superstructure	1 g
– Axle	10 g

Fig. 1
Schematic:
a Excursion-measuring
b Closed-loop position controlled

a Measured acceleration
x System excursion
F_M Measuring force (inertial force on the mass m)
F_K Compensating force
I_A Output current
U_A Output voltage

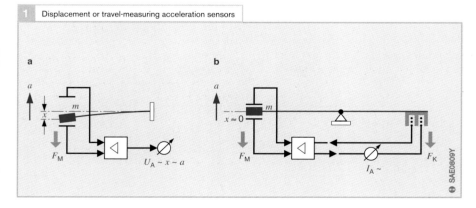

1 Displacement or travel-measuring acceleration sensors

On all acceleration sensors, with the exception of the gravity pendelum, the inert mass is attached flexibly to the body whose acceleration is to be measured. This means that in the static case, the acceleration force is in equilibrium with the restoring force applied to the spring which has been deflected by x:

$$F = m \cdot a = c \cdot x \qquad (2)$$

where
c is the spring constant.

The system's measurement sensitivity S is therefore:

$$S = x/a = m/c \qquad (3)$$

This indicates that a large mass together with low spring stiffness (or constant) result in high measurement sensitivity. If however, Equation 2 is written in full for the static and for the dynamic case, then it becomes apparent that not only the spring's elasticity must be taken into account but also a friction force and an inertial force. These are proportional to the derivations with respect to time of the excursion x (p friction coefficient).

The resulting equation (4) defines a (resonant) system capable of oscillation:

$$F = m \cdot a = c \cdot x + p \cdot \dot{x} + m\ddot{x} \qquad (4)$$

Presuming negligible friction ($p \approx 0$), this system has a resonant frequency of:

$$\omega_0 = \sqrt{\frac{c}{m}} \qquad (5)$$

This means that in accordance with Equation 3, the measuring sensitivity S is directly linked to the resonant frequency ω_0 in the following manner:

$$S \cdot \omega_0^2 = 1 \qquad (6)$$

In other words, it can be expected that sensitivity drops by factor 4 when the resonant frequency is increased by a factor of 2. Of course, it is only below their resonant frequency that such spring-mass systems display adequate proportionality between measured quantity and excursion.

In addition to closed-loop position control, there is another method which can be applied in overcoming the invariable interdependence between measuring sensitivity and bandwidth as defined in Equation 6. This is taken from Equation 4 and can be successfully applied up to at least the system's first harmonic ($2\omega_0$):

Friction and inertia terms can be derived from the excursion term $(c \cdot x)$. If, mathematically, these are added to the excursion term, the resulting sum is a precise measure for the acceleration being measured a – independent of the effects of resonance and damping.

In order to achieve a frequency response which is as constant as possible, and to avoid a disturbing increase of resonant frequency (which can easily lead to system destruction), *damping* is needed which is to be defined as precisely as possible and independent of temperature. If the friction coefficient p is normalized to the other parameters in Equation 4, this results in a standard damping factor D:

$$D = \frac{p}{2 \cdot c} \cdot \omega_0 = \frac{p}{2 \cdot \sqrt{c \cdot m}} \qquad (7)$$

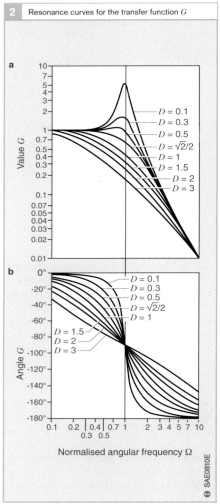

Fig. 2
a Amplitude resonance curve
b Phase resonance curve of the complex transfer function
$G(i \cdot \Omega) = [\underline{x}(i \cdot \Omega)]/[\underline{a}(i \cdot \Omega)]$

$\underline{x}(i \cdot \Omega)$ Deflection amplitude
$\underline{a}(i \cdot \Omega)$ Acceleration excitement amplitude
$\Omega = \omega/\omega_0$ Normalised angular frequency
D Damping

To a great extent, transient response and resonant response are defined by this damping factor. Whereas with periodic excitation, for damping $D > 1/\sqrt{2} = 0.707$, no resonance sharpness results, for values $D > 1$ all oscillating transient response has already disappeared in case of jump excitation. In order to achieve a bandwidth which is as broad as possible, a compromise is usually applied in practice with values of $D = 0.5...0.7$ (Fig. 2).

Compared to the extremely temperature-dependent damping as exhibited by silicone/oil mixtures, air damping using an air gap has proved itself in practice since it has only a very low level of temperature sensitivity. Electrodynamic damping (permanent magnet and conductor plate) has proved to be equally good, but is considerably more expensive and voluminous.

In the case of position-controlled systems, damping can be implemented and adjusted in the electronic (closed-loop) control circuit. Since in operation their deflection is in any case practically zero, in the switched-off, non-damped state, these systems are usually protected with "tight" overload stops to protect them against damage.

Packaging

Similar to the majority of sensors on the market, the so-called "packaging" which is tailor made to suit each individual application, also plays a decisive role for acceleration sensors. Since inertia-type sensors register the measured quantity without any form of movable connection to the outside

Table 2. Acceleration sensors (listed according to electrical pick-off and spring-mass system)

Electrical pick-off	Spring-mass system					
	Ceramic		Metal		Silicon	
	Piezoelectric	Insulating	Steel	CuBe	Bulk	OMM[1]
Voltage measurement						
Piezoresistive (DMS)				▮		▮
Piezoelectric	X					
Travel measurement						
Hall			X			
Capacitive						X

(X) Actually in production. [1] OMM = Surface micromechanics. ▮ Under consideration.

world, it is no problem to encapsulate them hermetically. They need a suitable rigid mechanical coupling to attach them to the structure at which measurement takes place, otherwise flexible or loose elements could lead to falsification of the measurement. On the other hand, this rigid, fixed coupling must not lead to thermal expansion which may occur at the structure being transferred to the sensor so that measurements are falsified.

Table 2 is arranged according to different spring-mass systems and electrical pick-offs. It presents a systematic overview of the possibility of implementing various sensors. The combinations have been marked to indicate either those which are already actually in production and which will be dealt with in more detail in the following (X), or those which are already being closely considered for production (marked in blue):

Often, the spring's own mass is adequate as the seismic mass for achieving adequate measurement sensitivity. If this is not the case, mass must be added (usually of the same material, or in metallic form).

The present-day trend is definitely towards minimum-dimension sensors using *Si-OMM technologies and capacitive signal take-off*. Not least thanks to hermetic encapsulation, this form of pick-off is practically only influenced by the sensor's geometrical parameters and is therefore unaffected by such other material constants and influencing variables as temperature etc. The local electronic circuitry, without which operation is impossible, also provides effective protection against electromagnetic interference (EMC). The threat of damage due to electromagnetic interference, which is common to this type of pick-off, is effectively counteracted by the local electronic circuitry.

Although the measurement capacities which can be generated here are extremely small, the variations that can be achieved due to the effects of acceleration are typically as much as about ±25 %. Further advantages of this technology are its comparatively low current consumption, and the possibility of designing the system as a closed-loop position-controlled system by feeding in electrostatic forces (at the measuring electrodes or at an additional electrode pair).

Systems for measuring mechanical strain and stress

Whereas all other forms of pick-off listed in Table 2, have already been dealt with, *piezoelectric pick-ups* in the vehicle (apart from the piezoelectric yaw-rate sensors) are only used in acceleration sensors. This measuring principle will therefore be discussed here in more detail.

When subjected to mechanical strain/stress which has been caused by outside forces F (Fig. 3), charges Q are generated on the surface of piezoelectric materials provided with electrodes. Crystalline material samples (e.g. quartz crystal) display this feature naturally. Artificially produced materials, on the other hand, such as piezoceramic must first of all be polarized with a strong electric field.

Similar to the magnetic materials, this piezoelectric effect is also subjected to a "Curie temperature" above which the phenomenon disappears completely. For crystals this is reversible, but not for piezoceramic materials. With piezoceramics this so-called "depolarisation" can be caused by intense mechanical

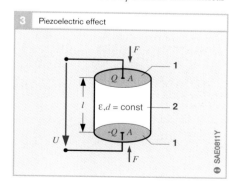

Fig. 3
1 Electrodes
2 Piezoelectric material sample

l Length
A Cross-setional area of the sample
F Force
Q Charge
U Voltage
ε Dielectric coefficient
d Piezoelectric charge coefficient

shock which causes the crystallites in the material to return to their original random positioning.

Whereas the Curie temperature for conventional ceramics is approx. 340 °C, on special quartz sections it can extend to as high as 440 °C. With ceramics, in order to avoid depolaristion during operation the operating temperature must remain a considerable distance from the Curie temperature. On conventional ceramics, this temperature limit is approx. 160 °C.

Not only cemamics, but also special very thin plastic foils also demonstrate piezoelectric characteristics. In contrast to crystalline materials, the man-made piezo materials, which are only used in automotive applications, can be produced very cheaply. On the other hand though, their measurement characteristics (temperature sensitivity, hysteresis, resistance to aging, sensitivity scatter, internal resistance, etc.) are considerably inferior to those of the crystalline materials. Practically all man-made materials demonstrate a very marked, and usually undesirable, pyroelectric effect. Due to this effect, temperature changes generate charges on these materials which are superimposed on the charges generated due to mechanical force.

The generated charges though do not remain the whole time force is applied, but are discharged through the external resistance of the measuring circuit or through the piezo sensor's internal resistance. The time constant of this discharge is the product of the sensor capacity and the effective total resistance. Such sensors cannot measure statically, and are only used where *dynamic* measurement is needed. Whereas high-performance pick-ups can achieve quasi-static measuring times of approx. 15...60 min, maximum measuring times for ceramics are often in the range of only about 1 s ... 1 ms.

The "piezoelectric charge coefficient d" (sometimes referred to as "piezomodule K"), is mainly responsible for the electrical behaviour of these sensors. In the simplest case, taking σ to be the mechanical tension applied during the test, and D to be the dielectric displacement density, the following relationship applies:

$$\sigma = F/A \quad (1) \quad \text{und} \quad D = d \cdot \sigma \quad (2)$$

Using the dielectric coefficients
$$\varepsilon = \varepsilon_r \cdot \varepsilon_o$$
the charge Q and the voltage U at the sensor electrodes can be calculated as follows:

$$Q = A \cdot D = A \cdot d \cdot \sigma = d \cdot F \quad (3)$$

$$U = \frac{Q}{C} = \frac{d \cdot F}{\varepsilon \cdot A} \cdot L = \frac{d}{\varepsilon} \cdot \frac{L}{A} \cdot F \quad (4)$$

$$= g \cdot \frac{L}{A} \cdot F = g \cdot L \cdot \sigma$$

with the piezoelectric voltage coefficient
$$g = d/\varepsilon \quad (5)$$
and an electric field strength in the test sample
$$E = U/L = g \cdot \sigma \quad (6)$$

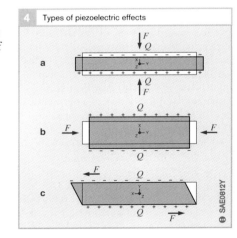

4 Types of piezoelectric effects

Fig. 4
a Longitudinal effect
b Transverse effect
c Tangential force

F Force
Q Charge

In addition to the often-used *longitudinal* piezoelectric effect, *transverse* and *tangential* effects must also be considered (Fig. 4).

Depending upon the material used, these effects can occur singly or, as is often the case in practice all together. If Equation 2 is formulated as a tensor equation this fully defines all the above-named piezoelectric effects.

The *transverse effect* is used for example in "bimorphous plates". These are composed of two oppositely polarised piezoceramic plates joined together and used for the measurement of bending stresses. When the two-layer ceramic composite bends, one half is stretched ($\varepsilon > 0$) and the other compressed ($\varepsilon < 0$). The opposed-polarity of the ceramic plates means that the resulting part voltages U_1 and U_2 now add to form a total voltage U which can be picked-off across the two outside metal layers (Fig. 5). In principle, metallisation is not needed between the two ceramic plates. Bimorphous strips measure their own bending movement, but if they are glued or soldered to a metal diaphragm they also register the diaphragm's deformation (for instance the microphone).

Electrical signal evaluation
Voltage pick-off: Since piezoelectric sensors feature a high internal resistance, when registering their output voltage U it is recommended that a decoupling amplifier is installed as near as possible to the sensor (if practical, inside a hermetically sealed housing together with the sensor). With long feed lines, the parasitic capacity (voltage divider) and the parasitic equivalent resistance falsify the signal (Fig. 6a).

Charge pick-off: With piezoelectric sensors, it is advisable to use a charge amplifier which stores the charge generated by the sensor in a high-precision measuring capacitor C_M and in doing so keeps the sensor itself free of charge and voltage. With this type of signal evaluation, the harmful parasitic influences of a feed line are for the most part suppressed, so that it is not absolutely necessary to integrate the sensor and the amplifier (Fig. 6b).

Examples of application
- Hall-effect acceleration sensors
- Piezoelectric acceleration sensors (bimorphous bending elements, longitudinal elements such as knock sensors)
- Micromechanical acceleration sensors

Fig. 5
a   Non-active state
b   Bent state, upper plate expanded ($\varepsilon > 0$), lower plate compressed ($\varepsilon < 0$)

1   Direction of polarisation

F Measuring force
U Total voltage
U_1, U_2 Part voltages

Fig. 6
a   Voltage pick-off
b   Charge pick-off

1   Feed line
2   Piezoelectric sample with capacity C_P

C_M Measuring capacity
F Measuring force
Q Charge
U Voltage

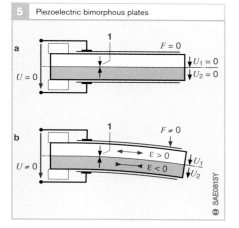

Fig. 5 Piezoelectric bimorphous plates

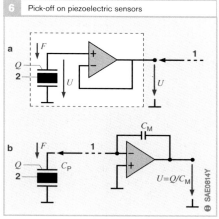

Fig. 6 Pick-off on piezoelectric sensors

Hall-effect acceleration sensors

Applications
Vehicles equipped with the Antilock Braking System ABS, the Traction Control System TCS, all-wheel drive, and/or Electronic Stability Program ESP, also have a Hall-effect acceleration sensor in addition to the wheel-speed sensors. This measures the vehicle's longitudinal and transverse accelerations (depending upon installation position referred to the direction of travel).

Design and construction
A resiliently mounted spring-mass system is used in the Hall-effect acceleration sensors (Figs. 1 and 2).

It comprises an edgewise-mounted strip spring (3) tightly clamped at one end. Attached to its other end is a permanent magnet (2) which acts as the seismic mass. The actual Hall-effect sensor (1) is located above the permanent magnet together with the evaluation electronics. There is a small copper damping plate (4) underneath the magnet.

Operating concept
When the sensor is subjected to acceleration which is lateral to the spring, the spring-mass system changes its neutral position accordingly. Its deflection is a measure for the acceleration. The magnetic flux F from the moving magnet generates a Hall voltage U_H in the Hall-effect sensor. The output voltage U_A from the evaluation circuit is derived from this Hall voltage and climbs linearly along with acceleration (Fig. 3, measuring range approx. 1 g).

This sensor is designed for a narrow bandwidth of several Hz and is electrodynamically damped.

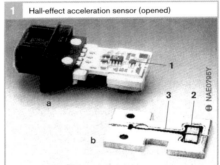

Fig. 1
a Electronic circuitry
b Spring-mass system

1 Hall-effect sensor
2 Permanent magnet
3 Spring

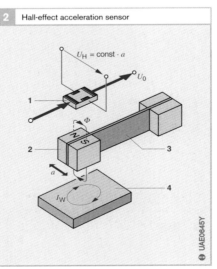

Fig. 2
1 Hall-effect sensor
2 Permanent magnet
3 Spring
4 Damping plate

I_W Eddy currents (damping)
U_H Hall voltage
U_0 Supply voltage
Φ Magnetic flux
a Applied (transverse) acceleration

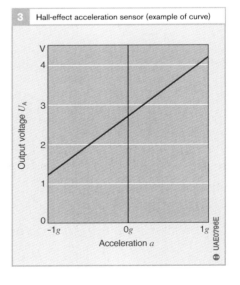

Micromechanical bulk silicon acceleration sensors

Application
Micromechanical bulk silicon acceleration sensors are used in passsenger-restraint systems to register the acceleration values of a frontal or side collision. They serve to trigger the seatbelt tightener, the airbag, and the roll-over bar.

Design and operating concept
Anisotropic and selective etching techniques are used to form the required spring-mass system from the solid wafer (bulk silicon micromechanics), and to shape the spring shoulders. Capacitive pick-offs have proved themselves for the correct measurement of the mass deflection. They require a wafer-thick silicon or glass plate with counterelectrodes (1, 4), on each side of the spring-held mass (Fig. 1, pos. 2), thus forming a triplex construction. Here, the plates with the counter-electrodes also act as an overload protection.

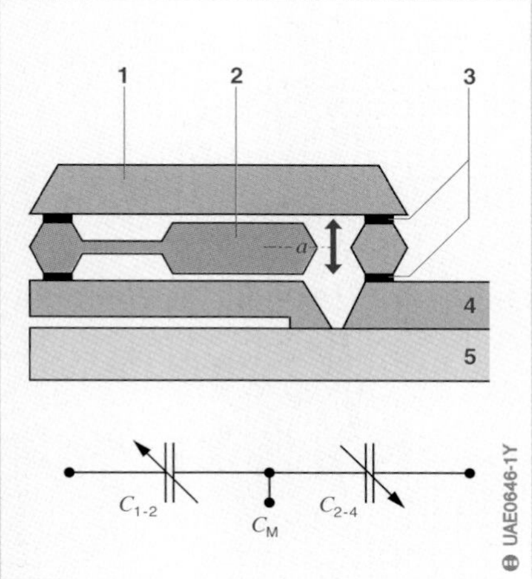

This plate configuration corresponds to a series circuit with two differential capacitors C_{1-2} and C_{2-4} (structure capacities 10...20 pF). Opposed-phase AC voltages are applied across their terminals, and their superimpositions picked-off between the capacitors at C_M (measurement capacity), in other words at the Si center plate (seismic mass).

When acceleration a is applied in the sensing direction, the Si center plate (the seismic mass) is caused to deflect. This causes a change in the spacing to the upper and/or lower plate, and with it a capacitance change in the capacitors C_{1-2} and C_{2-4} which leads to a change in the electrical signal. In the evaluation electronics circuit (CMOS), this change is amplified, and then filtered and digitalised ready for further signal processing in the airbag ECU.

Filling the sensor's hermetically sealed oscillatory system with a precisely metered charge of air leads to a very space-saving, inexpensive form of damping which also exhibits low temperature sensitivity. Today, almost without exception, the three silicon plates are connected together using the "fusion-bonding process". Due to the differences in the expansion due to temperature, the oscillatory system's attachment to the housing base also has a decisive effect on measuring accuracy. Connection to the housing base is therefore practically in a straight-line, and the oscillatory system is unsupported in the sensitive areas.

This type of sensor is above all used in the low acceleration ranges ($< 2\ g$) and necessitates a 2-chip concept:
Sensor chip + CMOS evaluation chip with integral protective function.

The change-over to expanded signal evaluation leads to the seismic mass automatically returning to the zero position, whereby the actuating signal appears as an output quantity.

1 Bulk silicon acceleration sensor with capacitive pick-off

Fig. 1
1 Upper Si plate
2 Center Si plate (seismic mass)
3 Si-oxide
4 Bottom Si plate
5 Glass substrate

a Accelartion in the sensing direction
C_M Measurement capacity

Surface micromechanical acceleration sensors

Application
Surface micromechanical acceleration sensors are used in passenger-restraint systems to register the acceleration values of a frontal or side collision. They serve to trigger the seatbelt tightener, the airbag, and the rollover bar.

Design and operating concept
Although these sensors were initially intended for use with higher accelerations (50...100 g), they also operate with lower acceleration figures when used in passenger-restraint systems. They are much smaller than the bulk silicon sensors (typical edge length: approx. 100...500 µm), and are mounted together with their evaluation electronics (ASIC) in a waterproof casing (Fig. 1). An additive process is used to build up their spring-mass system on the surface of the silicon wafer.

The seismic mass with its comb-like electrodes (Figs. 2 and 3, pos. 1) is spring-mounted in the measuring cell. There are fixed comb-like electrodes (3, 6) on the chip on each side of these movable electrodes. This configuration comprising fixed amd movable electrodes corresponds to a series circuit comprising two differential capacitors (capacity of the comb-like structure: approx. 1 pF). Opposed-phase AC voltages are applied across the terminals C_1 and C_2, and their superimpositions picked-off between the capacitors at C_M (measurement capacity), in other words at the seismic mass.

Since the seismic mass is spring-mounted (2), linear acceleration in the sensing direction results in a change of the spacing between the fixed and movable electrodes, and therefore also to a change in the capacity of C_1 and C_2 which in turn causes the electrical signal to change. In the evaluation electronics circuit, this change is amplified, and then filtered and digitalised ready for further signal processing in the airbag ECU. Due to the low capacity of approx. 1 pF, the evaluation electronics is situated at the sensor and is

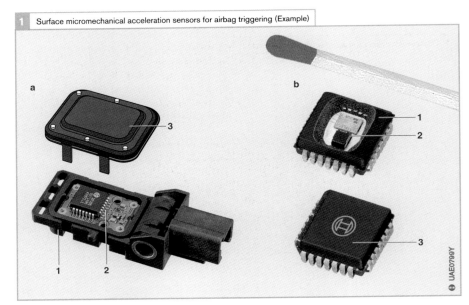

1 Surface micromechanical acceleration sensors for airbag triggering (Example)

Fig. 1
a Side-airbag sensor
b Front-airbag sensor

1 Casing
2 Sensor and evaluation chip
3 Cover

either integrated with the sensor on the same chip, or is located very close to it. Closed-loop position controls with electrostatic return are also available.

The evaluation circuit incorporates functions for sensor-deviation compensation and for self-diagnosis during the sensor start-up phase. During self-diagnosis, electrostatic forces are applied to deflect the comb-like structure and simulate the processes which take place during acceleration in the vehicle.

Dual micromechanical sensors (4) are used for instance in the ESP Electronic Stability Program for vehicle dynamics control: Basically, these consist of two individual sensors, whereby a micromechanical yaw-rate sensor and a micromechanical acceleration sensor are combined to form a single unit. This reduces the number of individual components and signal lines, as well as requiring less room and less attachment hardware in the vehicle.

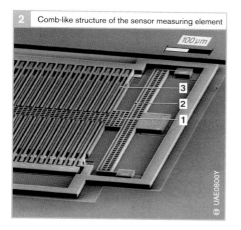

2 Comb-like structure of the sensor measuring element

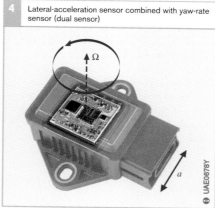

4 Lateral-acceleration sensor combined with yaw-rate sensor (dual sensor)

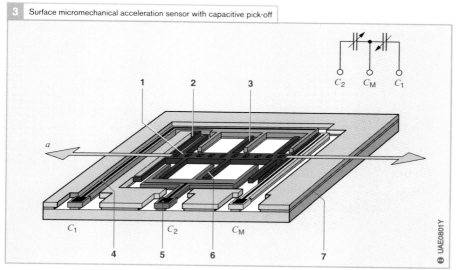

3 Surface micromechanical acceleration sensor with capacitive pick-off

Fig. 2
1 Spring-mounted seismic mass with electrode
2 Spring
3 Fixed electrodes

Fig. 4
a Acceleration in sensing direction
Ω Yaw rate

Fig. 3
1 Spring-mounted seismic mass with electrodes
2 Spring
3 Fixed electrodes with capacity C_1
4 Printed Al conductor
5 Bond pad
6 Fixed electrodes with capacity C_2
7 Silicon oxide

a Acceleration in sensing direction
C_M Measuring capacity

Piezoelectric acceleration sensors

Application

Piezoelectric bimorphous bending elements and two-layer piezoceramic elements are used as acceleration sensors in passenger-restraint systems for triggering the seat-belt tighteners, the airbags, and the roll-over bar.

Design and operating concept

A piezo bending element is at the heart of this acceleration sensor. It is a bonded structure comprising two piezoelectric layers of opposite polarities ("bimorphous bending element"). When subjected to acceleration, one half of this structure bends and the other compresses, so that a mechanical bending stress results (Fig. 1).

The voltage resulting from the element bend is picked off at the electrodes attached to the sensor element's outside metallised surfaces.

The sensor element shares a hermetically-sealed housing with the initial signal-amplification stage, and is sometimes encased in gel for mechanical protection.

1 Bending element from a piezoelectric acceleration sensor

Fig. 1
a Not subject to acceleration
b Subject to acceleration a

1 Piezoceramic bimorphous bending element

U_A Measurement voltage

2 Piezoelectric acceleration sensor (dual sensor for vertical mounting)

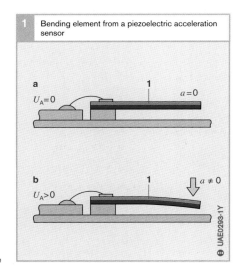

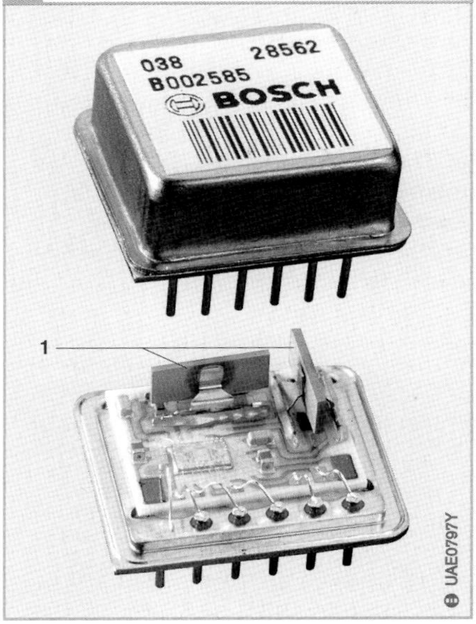

Fig. 2
1 Bending element

For signal conditioning, the acceleration sensor is provided with a hybrid circuit comprised of an impedance converter, a filter, and an amplifier. This serves to define the sensitivity and useful frequency range. The filter suppresses the high-frequency signal components. When subjected to acceleration, the piezo bending elements deflect to such an extent due to their own mass that they generate a dynamic, easy-to-evaluate non-DC signal with a maximum frequency which is typically 10 Hz.

By "reversing" the actuator principle and applying voltage, the sensor's correct operation can be checked within the framework of OBD "on-board diagnosis". All that is required is an additional actuator electrode.

Depending upon installation position and direction of acceleration, there are single or dual sensors available (Fig. 2). Sensors are also on the market which are designed specifically for vertical or horizontal mounting (Fig. 2).

Piezoelectric knock sensors

Application
Regarding their principle of functioning, knock sensors are basically vibration sensors and are suitable for detecting structure-borne acoustic oscillations. These occur as "knock" for instance in a vehicle engine when uncontrolled ignition takes place, and are converted into electrical signals by the sensor and inputted to the ECU. As a rule, 4-cylinder in-line engines are equipped with *one* knock sensor; 5 and 6-cylinder engines, with *two;* and 8 and 12-cylinder engines have *two or more*. They are switched in accordance with the ignition sequence.

Design and operating concept
Due to its inertia, a mass excited by a given oscillation or vibration exerts a compressive force on a toroidal piezoceramic element at the same frequency as the excitation oscillation. Inside the ceramic element, these compressive forces cause a charge transfer so that a voltage appears across the ceramic element's two outer faces which is picked-off by contact discs and inputted to the ECU for processing. Sensitivity is defined as the output voltage per unit of acceleration [mV/g].

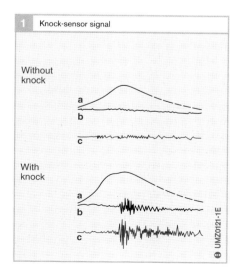

Knock-sensor signal

Without knock

With knock

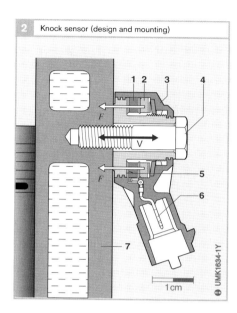

Knock sensor (design and mounting)

1 cm

Fig. 2
1 Piezoceramic element
2 Seismic mass with compressive forces F
3 Housing
4 Fastening screw
5 Contact surface
6 Electrical connection
7 Cylinder block

V Vibration

The sensor's voltage output is evaluated by a high-resistance AC amplifier in the ECU of the ignition or Motronic engine-management system (Figs. 1 and 2).

Mounting
Depending on the particular engine, the knock-sensor installation point is selected so that knock can be reliably detected from each cylinder. The sensor is usually screwed to the side of the engine-cylinder block. In order that the resulting signals (structure-borne oscillations) can be transferred from the measuring point on the engine block and into the sensor without resonant-frequency effects and in agreement with the stipulated characteristic curve, the following points must be observed:
- The fastening bolt must have been tightened with a defined torque
- The sensor's contact surface and bore in the engine block must comply with certain quality requirements and
- No washers of any type may be used

Fig. 1
a Cylinder-pressure curve
b Filtered pressure signal
c Knock-sensor signal

Pressure sensors

Measured variables

Pressure measurement takes place either directly, by way of diaphragm deformation, or using a force sensor. Examples of pressure measurement are given below:
- Intake-manifold pressure (1...5 bar) for gasoline injection
- Braking pressure (10 bar) on electropneumatic brakes
- Air-spring pressure (16 bar) on pneumatic-suspension vehicles
- Tire pressure (5 bar absolute) for tire-pressure monitoring and tire-pressure closed-loop control
- Hydraulic supply pressure (approx. 200 bar) for ABS and power-assisted steering
- Shock-absorber pressure (+200 bar) for chassis-control systems
- Coolant pressure (35 bar) for air-conditioning systems
- Modulation pressure (35 bar) on automatic gearboxes
- Braking pressure in master cylinder and wheel-brake cylinders (200 bar), and automatic yaw-moment compensation on the electronically-controlled brake
- Overpressure/low pressure (0.5 bar) and OBD "On-Board Diagnosis"
- Combustion-chamber pressure (100 bar, dynamic) for detection of misfire and combustion knock
- Element pressure on the diesel fuel-injection pump (1,000 bar, dynamic) for EDC (Electronic Diesel Control)
- Fuel pressure on the diesel Common Rail System (1,500 or 1,800 bar) and
- Fuel pressure on the gasoline Common Rail System (100 bar)

Measuring principles

The measured variable "pressure" is a dynamic effect which occurs in gases and fluids and which is effective in all directions. It propagates well in fluids, and in gel-like substances and soft casting compounds, a fact which is sometimes taken advantage of for a number of reasons. There are static and dynamic pick-ups or sensors for the measurement of pressure.

The *dynamic* pressure sensors include for instance all microphones which, since they are insensitive to static pressures, are used to measure pressure oscillations in gaseous and/or liquid mediums.

Since up to now, practically only static sensors have been used in automotive engineering, these will be dealt with in more detail here.

Direct pressure measurement

Being as all resistors are more or less pressure-dependent (volumetric effect), when very high pressures (>10^4 bar) are to be measured it would suffice theoretically to simply subject an electrical resistor to the pressure medium. On the other hand, they are at the same time more or less temperature-dependent, a characteristic which it is usually very difficult to suppress. Furthermore, the sealed lead-out of their connections from the pressure medium presents difficulties. Encapsulated capacitive measuring modules have more favorable characteristics and, depending upon the particular application, are easier to manufacture.

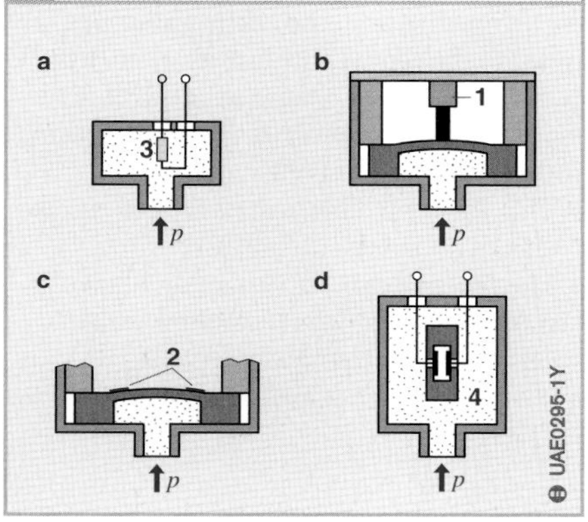

Fig. 1
a Direct measurement, pressure-dependent resistor (3)
b Measurement using a force sensor (1)
c Measuring the diaphragm deformation/DMS (2)
d Capacitive measurement using the deformation of a diaphragm cell

Pressure measurement

Diaphragm-type sensors

The most common method used for pressure measurement (also in automotive applications) uses a thin diaphragm as the intermediate stage. The pressure to be measured is first of all applied to one side of this diaphragm so that this bends to a greater or lesser degreee as a function of the pressure. Within a very wide range, its diameter and thickness can be adapted to the particular pressure range. Low-pressure measuring ranges lead to large diaphragms which can easily deform by as much as 1...0.1 mm. Higher pressures though demand thicker, low-diameter diaphrams which only deform very slightly by a few μm. In case (capacitive) pick-offs for spacing or distance measurements are also required, voltage-measuring methods dominate in the medium-pressure to high-pressure ranges. Here, practically only DMS techniques are used.

Capacitive pick-off

In contrast to their application in inertia sensors (see acceleration/yaw-rate sensors), capacitive pressure sensors are still only rarely encountered even though they could possibly provide similar advantages (particularly with respect to their accuracy). This is more than likely the result of one important difference compared to the other sensors dealt with above:

Pressure sensors need direct contact with the pressure medium, whose dieelectric characteristics practically always affect the calibration of such capacitive pressure sensors. This means that the calibration would then not only be dependent upon the medium in question, but would also be impossible without it (that is, in the "dry" state). Clear separation of the sensor from the pressure medium has up to now only been achieved at the cost of considerable technical outlay.

DMS[1]) pick-off

Table 1 presents a systematic overview of the proven pressure-measurement techniques which to a great extent have already been used in automotive applications. The list is arranged according to the type of diaphragm material and the applied DMS technology. Those combinations are marked which will be dealt with in the following as examples (x) or whose manufacture or purchase have been considered more closely (fields marked in blue):

1 DMS pick-off and diaphragm material

DMS pick-off	Diaphragm material		
	Ceramic	Metal (steel)	Silicon
Foils [1]) (glued)			
Thick-film			
Metal thin-film		x	
Silicon thin-film		x	
Diffusion resistances			x

Table 1

[1]) Unsuitable for large-batch production
x) Present-day examples   ■ Under consideration

With regard to the particular measuring effect's magnitude and type, the DMS techniques listed above have widely varying characteristics. The gauge factor (K) defines the magnitude of the measuring effect of deformation resistors. It gives the relative change in such a resistor's resistance R referred to the relative change in its length l an (Equation 1):

$$K = \frac{\Delta R/R}{\Delta l/l} = 1 + 2 \cdot v + \frac{d\rho/\rho}{\varepsilon}$$

Here, the symbol ε (expansion) is often inserted for the relationship $\Delta l/l$, and in multiples of 10^{-6} (ppm) as "micron" or "micro strain".

v is the material's "transversal-contraction factor", and ρ is its electrical conductivity. v characterises the reduction of cross-section area of the material upon elongation.

---
[1]) DMS = Strain gauge or strain-gauge resistor

In the ideal case of constant volume, $v = 0.5$ (in reality, $v = 0.3...0.4$).

Whereas the conductivity term in Equation 1 is of hardly any importance in the case of metallic resistors, with regard to Si resistors it plays a dominant role.

One refers, incidentally, to a *longitudinal gauge factor* when the resistor is expanded in the direction of current and to a *transverse gauge factor* when it is expanded crosswise to the current direction (Fig. 2). Table 2 provides an overview of typical values for the most important gauge factors.

"Creep" (slight mechanical give under the effects of long-term unidirectional loading) is a highly-feared phenomenon which, when it occurs at all, is only encountered on glued foil-DMS. The other DMS techniques all apply non-glued techniques and are not affected by this phenomena.

To be precise, a diaphragm's deformation depends upon the difference in the pressure applied to its top and bottom sides. This means that there are four different basic pressure-sensor types (Table 3):
- Absolute pressure
- Reference pressure
- Barometric pressure and
- Differential pressure

Transfer to a force sensor

Instead of directly using the force taken up by their diaphram, a number of sensors transfer it to a force sensor whose measuring range can remain constant due to the fact that the purely mechanical diaphragm has already performed the adaptation to the pressure-measuring range. Perfect linkage from measuring diaphragm to force sensor (for instance by a tappet) must be ensured though.

Examples of application
- Thick-film pressure sensors
- Micromechanical pressure sensors
- Si combustion-chamber pressure sensors and
- Metal-diaphragm high-pressure sensors

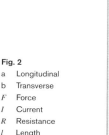

Fig. 2
a Longitudinal
b Transverse
F Force
I Current
R Resistance
l Length
w Width
ε Elongation
K Gauge factor

$K_l = \dfrac{\Delta R/R}{\varepsilon_l}$

$\varepsilon_l = \dfrac{\Delta l}{l}$

$K_t = \dfrac{\Delta R/R}{\varepsilon_t}$

$\varepsilon_t = \dfrac{\Delta w}{w}$

Table 2 Gauge factors for different materials

Material	Gauge factors	
	Longitudinal	Transverse
Foil DMS	1.6...2.0	~0
Thick film	12...15	12...15
Metal thin film	1.4...2.0	−0.5...0
Si thin film	25...40	−25...−40
Si monocrystalline	100...150	−100...−150

Table 3 Basic sensor types for pressure measurement

Pressure on diaphragm bottom side p_u	Pressure on diaphragm top side p_o		
	Measuring pressure	Ambient pressure	Vacuum
Measuring pressure	Difference pressure	Reference pressure	Absolute pressure
Ambient pressure	Reference pressure	–	Barometric pressure
Vacuum	Absolute pressure	Barometric pressure	–

Thick-film pressure sensors

Application
As an alternative to micromechanical pressure sensors, thick-film pressure sensors can sometimes be used (for instance in engine-management systems, M and ME Motronic). These are in the form of a module for installation in the ECU or a stand-alone component. They are used as:

- Manifold-pressure or boost-pressure sensor (pressure range 20...400 kPa or 0.2...4.0 bar) and
- Atmospheric-pressure sensor (pressure range 60...115 kPa or 0.6...1.15 bar)

Design and operating concept
The sensor is subdivided into a pressure-measuring cell and a chamber for the evaluation circuit. Both are arranged on a common ceramic substrate (Fig. 1).

The pressure-measuring cell (Fig. 2) comprises a "bubble-shaped" thick-film diaphragm which encloses a reference pressure of 0.1 bar. The diaphragm deforms as a function of the pressure being measured. There are four deformation resistors on the diaphragm which are connected to form a bridge circuit. Two of these active deformation resistors are located in the center of the diaphragm and change their conductivity when mechanical stress is applied (measured pressure). Two passive deformation resistors are situated on the diaphragm's periphery and function primarily as bridge resistors for temperature compensation. They have little effect upon the output signal.

When pressure is applied, the diaphragm deforms and changes the bridge-circuit balance. The bridge's measurement voltage U_M is therefore a measure of the measured pressure p (Fig. 3). The evaluation circuit amplifies the bridge voltage, compensates for the influence of temperature, and linearises the pressure curve. The evaluation circuit's output voltage U_A is inputted to the ECU.

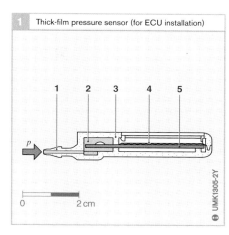

Fig. 1
Measuring range:
1 Pressure connection for the measured pressure p
2 Pressure-measuring cell
3 Sealing web
Signal conditioning:
4 Evaluation circuit
5 Thick-film hybrid on ceramic substrate

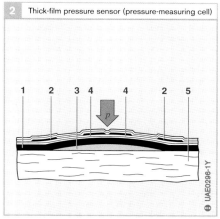

Fig. 2
1 Thick-film diaphragm
2 Passive reference deformation resistor
3 Reference-pressure chamber ("bubble")
4 Active deformation resistor
5 Ceramic substrate
p Measured pressure.

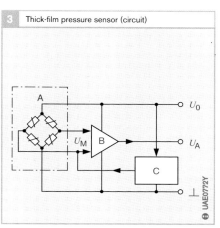

Fig. 3
A DMS pressure-measuring cell
B Amplifier
C Temperature-compensation circuit
U_0 Supply voltage
U_M Measured voltage
U_A Output voltage

Micromechanical pressure sensors

Application

Manifold-pressure or boost-pressure sensor
This sensor measures the absolute pressure in the intake manifold between the supercharger and the engine (typically 250 kPa or 2.5 bar) and compares it with a reference vacuum, not with the ambient pressure. This enables the air mass to be precisely defined, and the boost pressure exactly controlled in accordance with engine requirements.

Atmospheric-pressure sensor
This sensor is also known as an ambient-pressure sensor and is incorporated in the ECU or fitted in the engine compartment. Its signal is used for the altitude-dependent correction of the setpoint values for the control loops. For instance, for the exhaust-gas recirculation (EGR) and for the boost-pressure control. This enables the differing densities of the surrounding air to be taken into account. The atmospheric-pressure sensor measures absolute pressure (60...115 kPa or 0.6...1.15 bar).

Oil and fuel-pressure sensor
Oil-pressure sensors are installed in the oil filter and measure the oil's absolute pressure. This information is needed so that engine loading can be determined as needed for the Service Display. The pressure range here is 50...1,000 kPa or 0.5...10.0 bar. Due to its high resistance to media, the measuring element can also be used for pressure measurement in the fuel supply's low-pressure stage. It is installed on or in the fuel filter. Its signal serves for the monitoring of the fuel-filter contamination (measuring range: 20... 400 kPa or 0.2...4 bar).

Version with the reference vacuum on the component side

Design and construction
The measuring element is at the heart of the micromechanical pressure sensor. It is com-

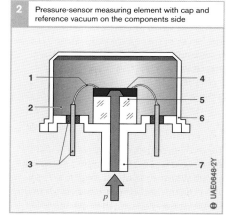

1 Pressure-sensor measuring element with reference vacuum on the components side

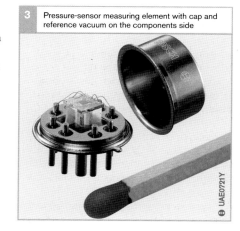

2 Pressure-sensor measuring element with cap and reference vacuum on the components side

3 Pressure-sensor measuring element with cap and reference vacuum on the components side

---

Fig. 1
1 Diaphragm
2 Silicon chip
3 Reference vacuum
4 Glass (Pyrex)
5 Bridge circuit
p Measured pressure
U_0 Supply voltage
U_M Measured voltage
R_1 Deformation resistor (compressed)
R_2 Deformation resistor (extended)

Fig. 2
1, 3 Electrical connections with glass-enclosed lead-in
2 Reference vacuum
4 Measuring element (chip) with evaluation electronics
5 Glass base
6 Cap
7 Input for measured pressure p

prised of a silicon chip (Fig. 1, Pos. 2) in which a thin diaphragm has been etched micromechanically (1). Four deformation resistors (R_1, R_2) are diffused on the diaphram. Their electrical resistance changes when mechanical force is applied. The measuring element is surrounded on the component side by a cap which at the same time encloses the reference vacuum (Figs. 2 and 3). The pressure-sensor case can also incorporate an integral *temperature sensor* (Fig. 4, Pos. 1) whose signals can be evaluated independently. This means that at any point a single sensor case suffices to measure temperature and pressure.

Method of operation
The sensor's diaphragm deforms more or less (10 ... 1,000 µm) according to the pressure being measured. The four deformation resistors on the diaphragm change their electrical resistances as a function of the mechanical stress resulting from the applied pressure (piezoresistive effect).

The four measuring resistors are arranged on the silicon chip so that when diaphragm deformation takes place, the resistance of two of them increases and that of the other two decreases. These deformation resistors form a Wheatstone bridge (Fig. 1, Pos. 5), and a change in their resistances leads to a change in the ratio of the voltages across them. This leads to a change in the measurement voltage U_M. This unamplified voltage is therefore a measure of the pressure applied to the diaphragm.

The measurement voltage is higher with a bridge circuit than would be the case when using an individual resistor. The Wheatstone bridge circuit thus permits a higher sensor sensitivity.

The component side of the sensor to which pressure is not supplied is subjected to a reference vacuum (Fig. 2, Pos. 2) so that it measures the absolute pressure.

The signal-conditioning electronics circuitry is integrated on the chip. Its assignment is to

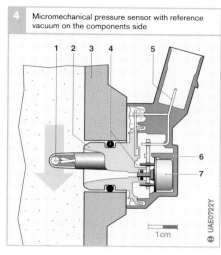

Micromechanical pressure sensor with reference vacuum on the components side

Fig. 4
1 Temperature sensor (NTC)
2 Lower section of case
3 Manifold wall
4 Seal rings
5 Electrical terminal (plug)
6 Case cover
7 Measuring element

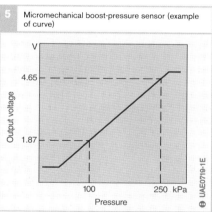

Micromechanical boost-pressure sensor (example of curve)

amplify the bridge voltage, compensate for temperature influences, and linearise the pressure curve. The output voltage is between 0...5 V and is connected through electrical terminals (Fig. 4, Pos. 5) to the engine-management ECU which uses this output voltage in calculating the pressure (Fig. 5).

Version with reference vacuum in special chamber
Design and construction
The *manifold or boost-pressure sensor* version with the reference vacuum in a special chamber (Figs. 6 and 7) is easier to install

than the version with the reference vacuum on the components side of the sensor element. Similar to the pressure sensor with cap and reference vacuum on the components side of the sensor element, the sensor element here is formed from a silicon chip with four etched deformation resistors in a bridge circuit. It is attached to a glass base. In contrast to the sensor with the reference vacuum on the components side, there is no passage in the glass base through which the measured pressure can be applied to the sensor element. Instead, pressure is applied to the silicon chip from the side on which the evaluation electronics is situated. This means that a special gel must be used at this side of the sensor to protect it against environmental influences (Fig. 8, Pos. 1). The reference vacuum is enclosed in the chamber between the silicon chip (6) and the glass base (3). The complete measuring element is mounted on a ceramic hybrid (4) which incorporates the soldering surfaces for electrical contacting inside the sensor.

A *temperature sensor* can also be incorporated in the pressure-sensor case. It protrudes into the air flow, and can therefore respond to temperature changes with a minimum of delay (Fig. 6, Pos. 4).

Operating concept

The operating concept, and with it the signal conditioning and signal amplification together with the characteristic curve, corresponds to that used in the pressure sensor with cap and reference vacuum on the sensor's structure side. The only difference is that the measuring element's diaphragm is deformed in the opposite direction and therefore the deformation resistors are "bent" in the other direction.

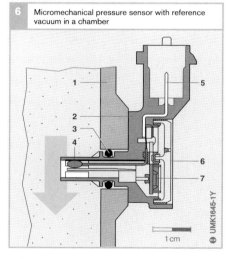

Micromechanical pressure sensor with reference vacuum in a chamber

Fig. 6
1. Manifold wall
2. Case
3. Seal ring
4. Temperature sensor (NTC)
5. Electrical connection (socket)
6. Case cover
7. Measuring element

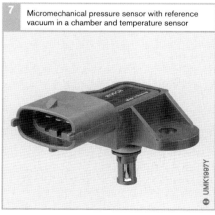

Micromechanical pressure sensor with reference vacuum in a chamber and temperature sensor

Fig. 8
1. Protective gel
2. Gel frame
3. Glass base
4. Ceramic hybrid
5. Chamber with reference volume
6. Measuring element (chip) with evaluation electronics
7. Bonded connection
p Measured pressure

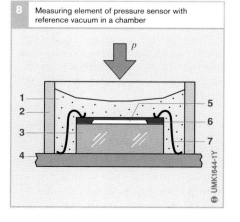

Measuring element of pressure sensor with reference vacuum in a chamber

High-pressure sensors

Application
In automotive applications, high-pressure sensors are used for measuring the pressures of fuels and brake fluids.

Diesel rail-pressure sensor
In the diesel engine, the rail-pressure sensor measures the pressure in the fuel rail of the Common Rail accumulator-type injection system. Maximum operating (nominal) pressure p_{max} is 160 MPa (1,600 bar). Fuel pressure is controlled by a closed control loop, and remains practically constant independent of load and engine speed. Any deviations from the setpont pressure are compensated for by a pressure control valve.

Gasoline rail-pressure sensor
As its name implies, this sensor measures the pressure in the fuel rail of the DI Motronic with gasoline direct injection. Pressure is a function of load and engine speed and is 5...12 MPa (50...120 bar), and is used as an actual (measured) value in the closed-loop rail-pressure control. The rpm and load-dependent setpoint value is stored in a map and is adjusted at the rail by a pressure control valve.

Brake-fluid pressure sensor
Installed in the hydraulic modulator of such driving-safety systems as ESP, this high-pressure sensor is used to measure the brake-fluid pressure which is usually 25 MPa (250 bar). Maximum pressure p_{max} can climb to as much as 35 MPa (350 bar). Pressure measurement and monitoring is triggered by the ECU which also evaluates the return signals.

Design and operating concept
The heart of the sensor is a steel diaphragm onto which deformation resistors have been vapor-deposited in the form of a bridge circuit (Fig. 1, Pos. 3). The sensor's pressure-measuring range depends upon the diaphragm's thickness (thicker diaphragms for higher pressures and thinner ones for lower pressures). When the pressure is applied via the pressure connection (4) to one of the diaphragm faces, the resistances of the bridge resistors change due to diaphragm deformation (approx. 20 µm at 1,500 bar).

The 0...80 mV output voltage generated by the bridge is conducted to an evaluation circuit which amplifies it to 0...5 V. This is used as the input to the ECU which refers to a stored characteristic curve in calculating the pressure (Fig. 2).

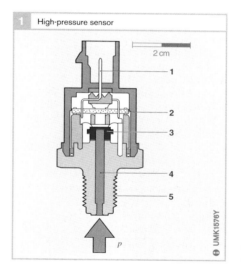

1 High-pressure sensor

Fig. 1
1 Electrical connection (socket)
2 Evaluation circuit
3 Steel diaphragm with deformation resistors
4 Pressure connection
5 Mounting thread

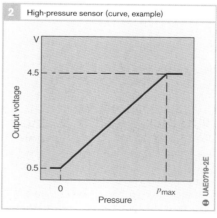

2 High-pressure sensor (curve, example)

Force sensors and torque sensors

Measured quantities

The following list underlines the wide variety of applications for force and torque sensors in automotive engineering:

- In the commercial-vehicle sector, measurement of the coupling force between the towing vehicle and its trailer or semi-trailer for the closed-loop controlled application of the brakes, whereby neither push nor pull forces are active at the drawbar
- Measurement of the shock-absorber force for use in electronic chassis-control systems
- For electronically controlled braking-force distribution, measurement of the axle load on commercial vehicles
- Measurement of the pedal force on electronically-controlled braking systems
- Measurement of the braking force on electrically actuated, electronically-controlled braking systems
- Proximity or non-contact measurement of drive and brake torques
- Proximity or non-contact measurement of steering torque or power-steering torque
- Finger-clamp protection on power windows and electrically operated sunroofs
- Force sensors integrated in the wheel bearing
- Weight measurement of vehicle occupants for occupant-restraint systems

In many cases, initial developments failed to lead to the expected results since generally the costs involved in achieving the stipulated accuracy were excessive for the systems in which the sensors were to be installed.

Contrary to expectations, it proved impossible to force down the costs for the production of good torque sensors below those for pressure and acceleration sensors. In fact the torque sensors cost more.

Matters are aggravated, and this applies particularly to torque sensors, when the measured quantity has to be transferred using non-contact methods from a *rotating shaft* (e.g. steering spindle or drive shaft) to a sensor mounted on the chassis.

Since any form of measurement of only part of the force or torque is very problematical and can easily lead to false results, force and torque sensors must be directly connected into the *power flux* (in other words, the complete measured variable must pass through them). In other words, there is a direct relationship between the extent of the measuring range and the sensor's size, that is, the wider the measuring range the larger the sensor.

Although, in line with automotive-industry demands, there are compact force and torque sensors on the market, these only measure accurately enough when the forces are introduced to the sensor in a precisely defined manner, a stipulation that normally can only be complied with under laboratory conditions. The tolerances and misalignment normally encountered in practice generally dictate the connection of *homogenisation elements* which in turn then lead to the sensors becoming too large

If force and torque-transfer components must be cut in order for sensors to be fitted, this generally results in an *interface problem*. This can only be solved by close cooperation between the sensor supplier and the suppliers of the parts which must be cut in order to install the sensors. This usually involves a large number of different companies and the automaker as well. Up to now, this problem has not been encountered with other sensor types, at least not with this severity and with such wide-ranging implications.

Even if the force and torque-transfer components do not have to be cut, and mechanical elements are used as measuring springs which only need modifying for installation of the sensor elements, very precise alignment is still necessary.

Measuring principles

Basically speaking, when considering force measurement, a difference must be made between static and dynamic measuring principles, and between measuring principles based on displacement and mechanical strain.

For the most part, static sensors have been in demand up to now, whereby for force sensors, non-resilient strain-measuring principles were preferred. In the present-day example of steering-torque sensing though, "soft" resilient sensor systems are acceptable which can also incorporate angle-measurement pick-offs. This is possible, particularly because this characteristic proved to be tolerable in earlier hydraulic systems which were not equipped with sensors. At present, being as the use of microstructural elements for mass production has not yet been clarified, magnetic coil systems still dominate in both sectors.

Force sensors
Magneto-elastic principle

The magneto-elastic effect is based on the anisotropic (directional) behaviour of relative magnetic permeability μ_r, (relationship between magnetic induction and magnetic field strength). This has the same value in all directions when no force is applied from the outside, but under the influence of an applied force its value changes. In the (longitudinal) direction of the applied force, this change (μ_{rl}) is different to that in the transverse direction (μ_{rq}).

In fact, the permeability change in the direction of force is a true reflection of the sign of the force. Even though practically all ferromagnetic materials demonstrate this effect, it can be optimized by using a specific alloy composition. Unfortunately though, the materials which have good linearity, low hysteresis, and low temperature sensitivity, are not identical with the ones which have a high measuring effect. Whereas the maximum measuring effects observed up to the present are about 30% (referred to the fundamental isotropic values), and can be utilised without electronics being necessary, the effects displayed by materials which have been optimised from the point of view of measurement techniques is only in an area of a few percent and needs electronic backup.

The advantages of the magneto-elastic effect are to be found in its broad temperature range and the fact that technical applications are possible up to temperatures as high as approx. 300 °C. On the other hand, it represents a marked volume effect. This means that the coils used for detection not only register a local variation of permeability caused by the application of force, but more or less the effects throughout the whole of the coil's cross section. This makes the sensor somewhat less sensitive to the possibility of force being applied asymmetrically.

Since the changes in permeability as a function of applied force are practically always registered with the help of alternating fields, the penetration depth of these fields, which is highly frequency-dependent, must be observed:

Only those mechanical forces can contribute to the measuring effect which are present in the measuring field's effective penetration depth. In order to be able to put the measuring effect to maximum use, the magnetically active air gap should be kept as small as possible. Often, this means that the magnetically active measuring circuit is

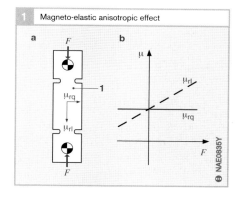

1 Magneto-elastic anisotropic effect

Fig. 1
a Magneto-elastic measurement structure
b Measuring effect
F Force
μ_r Relative magnetic permeability
μ_{rq} Transverse to the direction of force
μ_{rl} In the direction of force

closed with ferromagnetic material, even when this is not included in the power flux.

Figure 2 shows the two most important possibilities of evaluating the magneto-elastic effect: If a coil is arranged on the measurement structure so that its direction of field coincides with the direction of applied force, the change in inductance L can be picked-off and applied directly. Independent of the magnitude of the applied force, the excitation field strength H and the induction B always have the same direction (Figs. 2a and 3a).

If the field strength H of the supply coil is not axially parallel to the applied force, the effect of the latter not only changes the magnetic induction B, but also its direction (due to the anisotropy of the permeability, Fig. 2b).

Assuming that with no force applied the directions of H and B are superposed one upon the other in the normal manner, these assume increasingly different directions when force is applied and increased. In particular, this can be of advantage in varying the magnetic coupling of two measuring coils (Figs. 3b, 3c, and 4).

2 Influence of the magneto-elastic effect

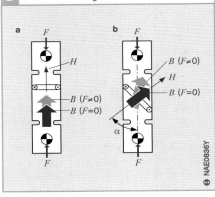

Fig. 2
a With direction of force parallel to the direction of field
b For different directions of field strength H and force F
B Induction
α Enclosed angle

4 Magneto-elastic tensile-force/compressive-force sensor for measuring braking torque

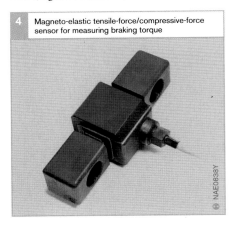

3 Different forms of magneto-elastic force sensors

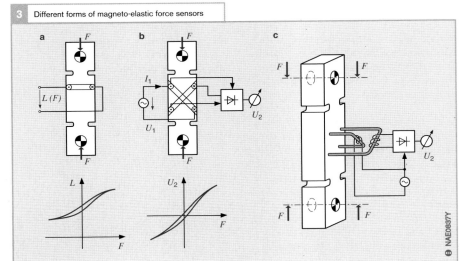

Fig. 3
a Variable inductance
b Variable coupling
c Variable coupling

Strain-gauge (DMS) principle (piezoresistive)

Strain-gauge measuring resistors represent the most widespread, and probably the most reliable and precise method for measuring force and torque (Fig. 5). Their principle is based on the fact that in the zone of the elastic-member material to which Hooke's Law applies there is a proportional relationship between the mechanical strain σ in the member, caused by the introduction of force, and the resulting deformation ε.

$$\varepsilon = \Delta l/l = \sigma/E$$

whereby E is the modulus of elasticity. Since it does not directly measure the strain resulting from the applied force, but rather the resulting deformation, the strain-gauge method can be regarded as an indirect measuring method. For instance, if the modulus of elasticity decreases by 3% above 100 K, a figure which is normal for metals, then the force indicated by the strain-gauge method is 3% too high.

Strain-gauge resistors in the form of film resistors are so closely bonded to the surface of the selected elastic member that they follow its surface deformation perfectly. The change in resistance resulting from the resistor's deformation is defined by the particular gauge factor K of the resistor in question (refer to "pressure sensors"):

$$\Delta R/R = K \cdot \varepsilon$$

On metal film resistors, the K factor rarely exceeds 2. Strain gauges are designed that as far as possible (in combination with a given elastic material and its thermal expansion) they are temperature insensitive ($TK_R \approx 0$). Any residual temperature sensitivity is usually eliminated by depositing the resistors on the elastic member in the form of a half or full bridge. Since temperature effects result in same-direction changes on the strain gauge this results in no output signal.

The auxiliary bridge resistors can be (but need not be) located within the elastic member's deformation zone. They can also be fitted as purely compensation resistors (Fig. 5c). It must be noted that often the K-factor itself also has a temperature coefficient (TK_K). Usually, this decreases along with increasing temperature, which means that in favorable cases it can compensate for the signal increase caused by the E-module.

5 Strain-gauge force sensors

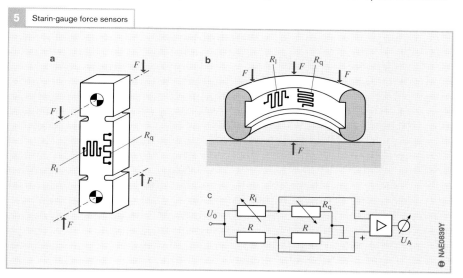

Fig. 5
a   Rod-shaped
b   Toroidal-shaped
c   Electronic evaluation
F   Force
$R_{l, q}$ Metal film resistors, lengthways, crosswise
R   Auxiliary bridge resistors
U_0 Supply voltage
U_A Output voltage

Apart from this, signal reductions caused by the K-factor are usually compensated for by a bridge supply voltage which increases accordingly.

Disadvantages/Limitations
Notwithstanding their high levels of accuracy and reliability, since the deformation and therefore also the changes in resistance (at least in the case of metal film resistors) are only in the percentage range of the original state, the strain-gauge sensors only generate output voltages in the mV range so that in general local amplification is required. A further disadvantage of small strain-gauge resistors is the fact that they only measure the strain at precisely that point at which it is applied, and no averaging takes place to arrive at a figure which applies to the (larger) elastic member as a whole. This, of course, no longer holds true in cases in which the strain-gauge structure is distributed across the elastic member's complete surface. This necessitates extremely precise and reproducible application of the strain to be measured if measuring errors due to uneven strain introduction is to be avoided.

Applications
As a rule, in order to carry out force measurements, very small strain-gauge resistors must be attached to larger force-carrying elastic members. The traditional method of applying the strain-gauge resistors to the elastic member by means of a foil backing (as applied in such devices as high-precision scales), is not inexpensive enough for large-batch "low-cost" production. First attempts are therefore being made to apply low-cost, large batch film/layer techniques by depositing the strain-gauge resistors on small metal wafers which are then pressed into, or welded onto the elastic element.

Orthogonal printed resistors
Practically all electrical film resistors change their resistance not only under the influence of laterally applied deformation strain but also when compression is applied vertical to the film plane (orthogonal). Here, the so-called "conductive plastic", commonly used in potentiometers, features very high sensitivity. "Cermet" and "carbon layers" are also highly sensitive (Fig. 6). Up to a certain limit, the resistance of the above materials decreases along with increasing compression. The values that can be achieved with-

6 Piezoresistive behaviour of various resistance materials when orthogonal compression is applied

7 Force sensor with orthogonally compressed strain-gauge resistors

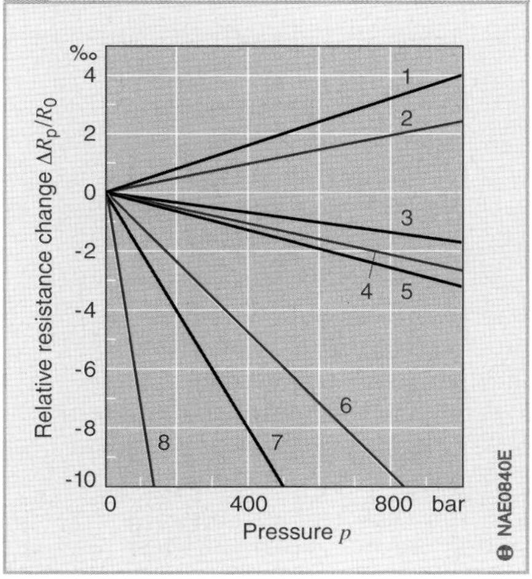

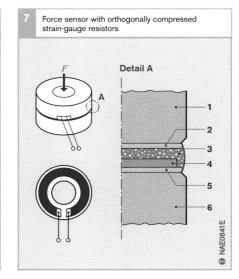

Fig. 6
1 84.5Ag15.5Mn
2 Manganin
3 Cu
4 Au
5 Ag
6 Carbon film/layer
7 Cermet
8 Conductive plastic

Fig. 7
1 Force-application ring
2 Insulation
3 Adhesive layer/ glass layer
4 Sensor layer
5 Insulation
6 Support ring

out permanent resistance change are similar to those which apply for lateral deformation. In both cases, the limit is a function of substrate strength and not of the resistance material. Of course, sensors of this type are almost only suitable for loading by compression but not by lateral strain.

Unfortunately, the majority of materials feature a relatively high temperature sensitivity referred to their deformation dependency. This makes such sensors unsuitable for static measurements. For bridge circuits, pressure-free zones can be provided on the substrate on which pressure-independent auxiliary bridge resistors or temperature-dependent compensation resistors can be located.

When the measuring resistors are designed as thick-film force-sensing discs, they are deposited on a hard substrate (for instance steel), and joined to form a solid body by means of a force-application ring (overglazed or cemented, Figs. 7 and 8).

The pressure-sensitive resistors though can also be applied on a carrier foil (Fig. 9) which can be located in a space-saving manner between the force-carrying components (for instance in a vehicle seat). Even though such foil-type resistors are not highly precise force sensors, they are very suitable for the construction of multiple sensor configurations ("sensor arrays") with which the load distribution over a given surface area can be measured, or even the shape of an object with a certain weight. Some sensor arrays also use materials which demonstrate a far more pronounced measuring effect together with a practically exponential characteristic curve. Here, the elastic foil often assumes the function of a switch which closes when pressure is applied and switches in the measuring resistor. Unfortunately, there is no foil encapsulation which is 100% sealed against moisture so that long-term stability cannot be counted on with such resistors.

Examples of application for force sensors
- Magneto-elastic bearing-pin sensors
- Magneto-elastic braking-force sensors and
- Seat mat (vehicle-passenger weight)

Torque sensors

Basically, the methods used for torque measurement differ from those for angle and strain measurement. In contrast to strain-measurement methods (strain-gauge resistors, magneto-elastic), angle-measurement methods (e.g. eddy-current) require a certain length l of the torsion shaft via which the torsion angle Φ (approx. 0.4...4°) can be picked-off. The mechanical tension σ, which

8 Thick-film force-sensing disc

9 Orthogonally loaded strain-gauge resistors on foil substrate (commercial name "Flexiforce")

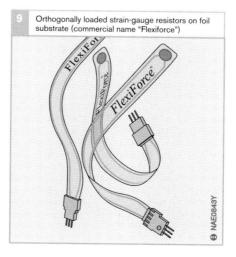

Fig. 8
1 Orthogonally loaded strain-gauge resistors
2 Overglazed force-application ring

is proportional to the torque, is at an angle of less than 45° to the shaft axis (Fig. 10).

The principles described below are all suited for the non-contacting (proximity) transfer of measured values, even from rotating shafts. In the case of steering-torque measurement, it is also desirable for the employed measuring system to measure the steering angle very accurately (through a full rotation of 360°). This is in the sense of modular integration and is to be implemented with only slight modification.

Strain-measuring sensors

Even though *magneto-elastic sensors* which enclose the shaft are available on the market, these involve very high costs. Since it is often impossible to optimise the shaft material with regard to its magneto-elastic properties, a search is being made for methods to coat the measuring shaft with a magneto-elastic layer. Such a coating, which has to exhibit good measuring qualities, has still not been found. The *strain-gauge principle* has therefore also come to the forefront here (Fig. 11):

A strain-gauge bridge which is powered using transformer principles (rectifier and control-electronics circuitry on the shaft, independent of the air gap), is used to register the mechanical stress. Further local electronic components on the shaft permit the measured signal to be amplified and converted to an AC form which is independent of the air gap (for instance, frequency-analog) which can then also be outputted using transformer principles. When large-batch quantities are involved, the electronic circuitry on the shaft can be integrated in a single chip. The deformation sensors can be inexpensively formed on a prefabricated steel blank (for instance, using thin-film techniques) which is then welded to the shaft. Considering the fact that such configurations are inexpensive to manufacture, they nevertheless permit high accuracies.

Angle-measuring (torsion-measuring) sensors

Sensors for measuring angular difference
It is a relatively easy matter to define the torsion angle when two mutually-independent incremental rotational-speed sensors, or an absolute-measuring (analog or digital) non-contacting angular-movement pick-off, are

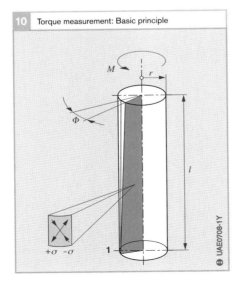

10 Torque measurement: Basic principle

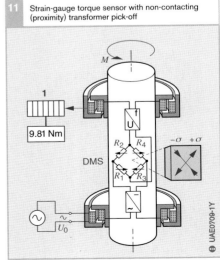

11 Strain-gauge torque sensor with non-contacting (proximity) transformer pick-off

Fig. 10
1 Torque rod
Φ Torsion angle
σ Torsional stress
M Torque
r Radius
l Rod length

Fig. 11
1 Torque-display device
σ Torsional stress
M Torque
U_0 Supply voltage
$R_1...R_4$ Deformation resistors

provided at each end of the 5..10 cm-long section subject to torsion measurement (Fig. 12). The difference in their outputs $\varphi_2 - \varphi_1$ is a measure for the torsion angle. Up to now, since adequate accuracy demands extremely precise bearings, together with the necessity to provide correspondingly accurate angular or incremental subdivision around the complete periphery, this method has been regarded as too complicated. Nevertheless work is forcing ahead with solving this problem (magnetically or optically), because such a system would incorporate two distinct advantages:

- Possibility of simultaneous measurement of the angle of rotation with the same system.
- Possibility of measurement without far-reaching modifications to the torsion shaft being necessary, so that essentially the sensor could be in the form of a plug-in sensor providing an efficient interface for a supplier component.

Eddy-current sensors
Two slotted sleeves are attached to each end of a sufficiently long section of the measuring shaft. These are inserted one inside the other (Fig 13, Pos. 1). Each sleeve is provided with two rows of slots so that when the shaft is twisted, it becomes increasingly visible through one row of slots, and is hidden more and more by the other row.

This leads to increasing, or decreasing, damping of the two high-frequency coils (approx. 1 MHz) situated above each row of slots so that coil inductance also varies accordingly. The slotted sleeves must be precisely manufactured and assembled in order to achieve the stipulated accuracy. The electronic circuitry is located as close as possible to the coils.

Application examples for torque sensors

- Strain-measuring strain-gauge torque sensors and
- Angle-measuring eddy-current torque sensors (electric power tools)

12 Defining torque by measuring angular difference

13 Eddy-current torque sensor

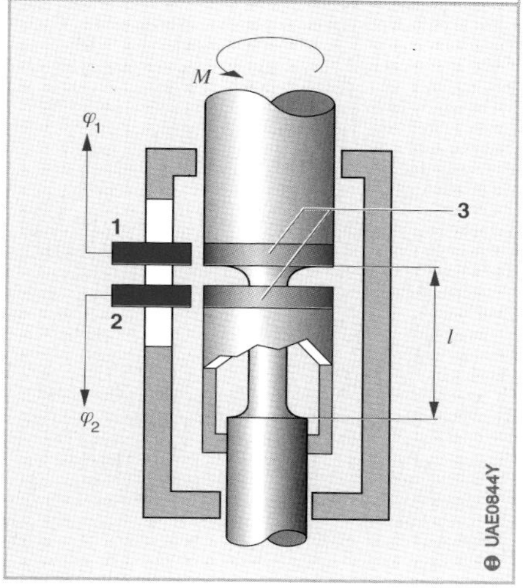

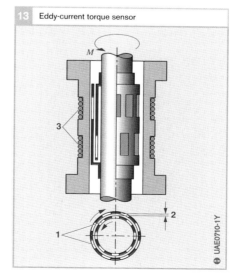

Fig. 12
1, 2 Angle/rotational-speed sensors
3 Angle markings
l Torsion-measurement section
M Torque to be measured
$\varphi_{1,2}$ Angle signals

Fig. 13
1 Slotted sleeves
2 Air gap
3 HF coils
M Torque

Occupant classification (OC) and detection of child's safety seat

Assignment
Following introduction of the airbag for the front-seat passenger, safety and actuarial considerations made it necessary to detect whether the front-seat passenger's seat is occupied or not. Otherwise, when an accident occurs and both front airbags are deployed, unnecessary repair costs result if the passenger seat is unoccupied.

The development of the so-called "Smart Bags" marked an increase in the demand for the ability to detect occupation of the driver-seat and front-passenger seat. The smart bag should feature variable deployment adapted to the actual situation and occupation of the seats. In certain situations, airbag triggering must be prevented when deployment would be injurious to one of the vehicle's occupants (for instance, if a child is sitting in the seat next to the driver, or a child's safety seat is fitted). This led to further development of the "simple" seat-occupation detection to form the "intelligent" **O**ccupation **C**lassification (OC). In addition, the automatic detection of a child's safety seat is integrated as a further sensory function. This can detect whether the seat is occupied or not, provided the seat is equipped with transponders.

Design and construction
A so-called sensor mat and ECU incorporated in the vehicle's front seats (Figs. 1 and 2) registers the information on the person in the seat and sends this to the airbag ECU. These data are then applied when adapting the restraint-system triggering to the current situation.

Operating concept
Measuring concept
This relies upon the classification of passengers (OC) according to their physical characteristics (weight, height, etc.), and applying this data for optimal airbag deployment. Instead of directly "weighing" the person concerned, the OC system primarily applies the correlation between the person's weight and his/her anthropometric[1] characteristics (such as distance between hip-bones). To do so, the OC sensor mat measures the pressure profile on the seat surface. Evaluation indicates first of all whether the seat is occupied or not, and further analysis permits the person concerned to be allocated to a certain classsification (Fig. 3).

[1]) The study of human body measurements, especially on a comparative basis.

Fig. 1 Sensor mat with OC-ECU

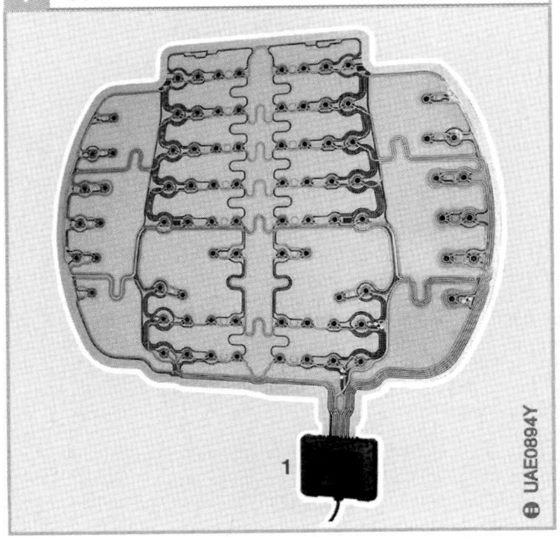

Fig. 1
1 ECU

Fig. 2 Installation of the OC sensor mats in the front seats

Fig. 2
1 OC-ECU
2 Airbag ECU

Sensor technology

Basically, the OC sensor mat comprises pressure-dependent FSR resistance elements (FSR: Force-Sensitive Resistance), the information from which can be selectively evaluated. A sensor element's electrical resistance drops when it is subjected to increasing mechanical load. This effect can be registered by inputting a measuring current. The analysis of all sensor points permits definition of the size of the occupied seat area, and of the local points of concentration of the profile.

A sensing antenna and two receive antennas in the OC sensor mat serve to implement the child's safety-seat detection function. During the generation of a sending field, transponders in the specially equipped child's seats are excited so that they impose a code on the sending field by means of modulation. The data received by the receive antenna and evaluated by the electronic circuitry is applied in determining the type of child's seat and its orientation.

ECU

The ECU feeds measuring currents into the sensor mat and evaluates the sensor signals with the help of an algorithm program which runs in the microcontroller. The resulting classification data and the information on the child's safety seat are sent to the airbag ECU in a cyclical protocol where, via a decision table, they help to define the triggering behaviour.

Algorithm

Among other things, the following decision criteria serve to analyse the impression of the seating profile:

Distance between hip-bones:
A typical seating profile has two main impression points which correspond to the distance between the passsenger's hip-bones.

Occupied surface:
Similarly, there is a correlation between the occupied surface and the person's weight.

Profile coherence:
Consideration of the profile structure.

Dynamic response:
Change of the profile as a function of time.

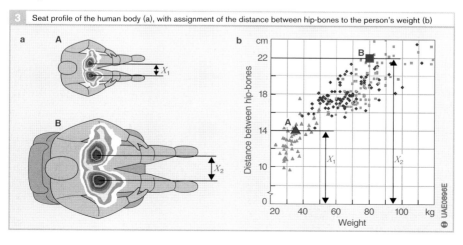

Fig. 3
a Seating profile
b Diagram
A Child with distance between hip-bones X_1
B Adult with distance between hip-bones X_2

3 Seat profile of the human body (a), with assignment of the distance between hip-bones to the person's weight (b)

Flow meters

Measured quantities

Flow measurement is only required at a few points on the vehicle:
To register the delivered fuel quantity, and in particular to measure the amount of air drawn in for combustion.

Fuel-flow measurement

On electronically controlled fuel-injection systems, the fuel quantity is metered to the IC engine precisely (without specific flow measurement), either intermittently or continuously. The required fuel quantity is injected precisely thanks to the evaluation of such variable/adjustable parameters as injection duration, setting of the metering unit, injection pressure, fuel temperature etc.

The fuel-flow meters were developed principally during periods of intense fuel scarcity and were used on the IC engines of the time, which were not yet electronically controlled. They indicated the fuel consumption in liters for a given distance of 100 km (60 miles). The difference between the fuel delivered from the tank and the amount of fuel which flowed back to it (whereby, particularly at idle, the amount of fuel returned to the tank was considerable) was applied as the basis for calculating the amount of fuel that had actually been used (Problem: This was the difference between two large quantities).

Since there is presently no actual necessity for such fuel flow meters, and since they are practically no longer in use, no further space will be devoted to them here.

Airflow measurement

As such, the often-used term *"air quantity"* is incorrect because it does not stipulate whether volume or mass is concerned. Since the chemical processes involved in fuel combustion are clearly based on mass relationships, the measurement must apply to the mass of air drawn in or the mass of supercharged air. In other words, the *"air mass"*. At least on IC engines, the air-mass flow rate is the most important load parameter. The sensors which are used for measuring air quantity or gas flows in general are also referred to as "anemometers".

Depending upon engine power, the average maximum air-mass flow rate to be measured is between 400 and 1,200 kg/h. Due to the low air requirements at engine idle, the ratio of minimum to maximum flow is 1:90...1:100. The severe emissions and fuel-consumption requirements dictate accuracies of 1...2% of the measured value. Referred to the measuring range, this can easily correspond to a measuring accuracy of 10^{-4}, a figure which is unusually high for the automobile.

The air though, is not drawn in continuously by the engine, but rather in time with the opening of the intake valves. Particularly with the throttle wide open (WOT), this leads to considerable pulsation of the air-mass flow, also at the measuring point which is always in the intake tract between air filter and throttle valve (Fig. 1). Intake-manifold resonance leads to the pulsation in the manifold sometimes being so pronounced that brief return flows can occur. This applies in particular to 4-cylinder engines in which there is no overlap of the air-intake phase and the charge phase. An accurate flow meter must be capable of registering these return flows with the correct direction.

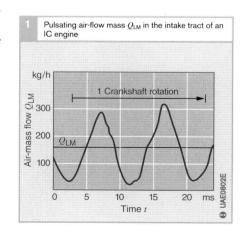

1 Pulsating air-flow mass Q_{LM} in the intake tract of an IC engine

Fig. 1
At WOT and
$n = 3{,}000$ rpm;
intake-manifold pressure
$p_s = 0.96$ bar;
mean air-flow rate
$Q_{LMm} = 157.3$ kg/h

On a 4-cylinder engine, the pulsations are generated at twice the crankshaft speed. This means that they can easily be in the range 50...100 Hz. With an air-mass meter featuring a linear characteristic curve and a narrower frequency bandwidth than above, it would suffice for it to follow the mean value of these rapidly fluctuating air flows. The mean value is in any case positive, so that the meter need not necessarily detect the correct sign.

Practically all of the air-mass meters actually in use feature a characteristic curve which is far from linear, so that the measurement signal must be linearized electronically before it can be evaluated. If averaging takes place before linearisation, this can lead to considerable errors ("mean-value errors"). Being as the pulsations mostly have a pronounced non-sinusoidal characteristic, they therefore also have a considerable harmonic content. This fact means that such air-mass meters must be able to follow the pulsations rapidly enough. This necessitates a *bandwidth* of about 1,000 Hz. Apart from this considerable bandwidth, the air-mass meters must also have a high switch-on time constant in order for them to be able to measure correctly during the engine start phase.

Similar to all flow meters, the versions used in the automobile are calibrated for "tubular" flow with a symmetrical flow profile, in other words for a flow whose velocity vector v at practically every point in the flow cross-section of area A is only a function of the radius to the center line. The flow profile (laminar or turbulent, Fig. 2) is directly related to the Reynolds number R_e.

$$R_e = v \cdot D/\eta$$

Where
D = Typical cross-section, and
η = kinematic viscosity of the medium.

Flow is laminar or turbulent when the Reynolds number R is below or above approx. 1,200. If the transition is in the center of the measuring range, marked irregularity of the characteristic curve can be expected at this point. As far as automotive applications are concerned, a purely turbulent flow (rectangular profile: v = const$_r$) can be reckoned with. This turbulence is sometimes provoked on purpose by means of a special grid element which also serves to protect the measuring system against damage. Assuming a homogeneous density ρ, the flow is simple to calculate as follows:

$Q_V = v \cdot A$ Volume flow rate
$Q_M = \rho \cdot v \cdot A$ Mass flow rate

Whereas in measurement techniques, long, straight, advance and overshoot sections of constant cross-section are stipulated in order to guarantee a symmetrical profile, such

2 Flow profiles

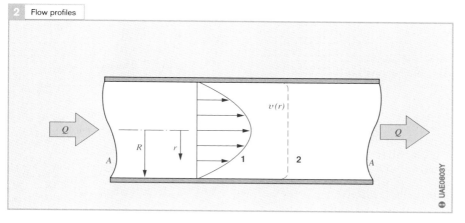

Fig. 2
1 Laminar flow profile
2 Turbulent flow profile

A Cross-section area of the tube
Q Flow
R Tube radius
r Distance from the tube center
$v(r)$ Flow profile

conditions cannot be complied with in a vehicle's cramped under-hood installation space. If pronounced asymmetries occur, the flow meter must be calibrated as a function of the actual installation conditions.

Impact pressure gauges, whose function will be dealt in more detail below, react to the pressure drop (Δp) at a special restriction (metering orifice) in the flow cross-section and measure a flow which corresponds neither to the volume flow rate nor the mass flow rate. Instead this flow value is the geometrical mean of the two:

$$Q_{St} = const \cdot \sqrt{\rho \cdot v} = const. \cdot \sqrt{Q_V \cdot Q_M}$$

Whereby, the pressure loss at the flow meter (above all at WOT) is not to exceed 20…30 mbar.

Measuring principles

Up to now, of the practically unlimited variety of flow meters on the market, only those which operate according to the impact-pressure principle have come to the forefront for air-quantity measurement in the vehicle. This principle still depends upon mechanically moving parts, and in principle correction measures are still needed to compensate for density fluctuations.

Today, air-mass meters are increasingly being used which use a thermal method without moving parts. Their "hot-wire" or "hot-film" principle enables them to follow sudden flow changes with a minimum of delay.

Variable orifice plates (sensor plates)

The calculation of the pressure drop across fixed orifice plates is based on two physical laws:

Continuity equation:

$$\rho_1 \cdot v_1 \cdot A_1 = \rho_2 \cdot v_2 \cdot A_2 = const$$

Bernoulli's equation:

$$p_1 + \frac{1}{2} \cdot \rho_1 \cdot v_1^2 = p_2 + \frac{1}{2} \cdot \rho_2 \cdot v_2^2 = const$$

These laws are to be applied for two different measuring cross-sections A_1 and A_2 (Fig. 3).

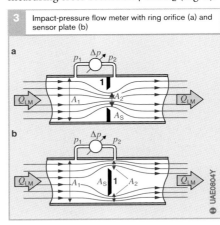

Fig. 3
a Ring orifice
b Sensor plate

1 Orifice plate

A_S Plate diameter
$A_{1,2}$ Measuring cross-section
$p_{1,2}$ Measurement pressure
Δp Pressure drop
Q_{LM} Air-mass flow

3 Impact-pressure flow meter with ring orifice (a) and sensor plate (b)

Assuming constant density $\rho = \rho_1 = \rho_2$, this results in the pressure drop:

$$\Delta p = Q_V^2 \cdot \rho \cdot \left(\frac{1}{A_2^2} - \frac{1}{A_1^2}\right)$$

This pressure drop can be measured either directly with a differential-pressure flow meter, or by means of the force acting against a so-called sensor plate (Fig. 3).

Due to their r.m.s. relationship to the flow, fixed orifice plates permit only a 1:10 variation of the measured-variable. When larger ranges are to be covered, several orifice plates must be used, or such versions which automatically adapt themselves to the measuring range by opening up a larger flow cross-section A_2 in line with the increasing impact pressure.

With such variable, moving sensor plates it is an easy matter to increase the variation to 1:100. Here, the increasing air flow causes the sensor plate to be deflected (usually against a constant counterforce) into an area whose cross section is specifically shaped so that the resulting deflection/angle relationship complies with the desired characteristic. In other words, linear for K-Jetronic and non-linear for L-Jetronic. The sensor plate's (Fig. 4) setting is then a measure for the air flow which is in relationship to the impact pressure defined above. The limit frequency for such air-mass meters is approx. 10 Hz.

Such sensor plates though are unable to follow the high pulsation frequencies which often occur. From the point of view of the pulsation, they can be regarded as fixed orifice plates with a square-law curve. Under certain load conditions this leads to considerable mean-value errors which can only be compensated for roughly by the use of suitable software.

Here, when the density ρ of the drawn-in air changes due to temperature fluctuations or changes in altitude, the measured signal changes by merely $\sqrt{\rho}$. An air-temperature sensor and a barometric pressure sensor are needed in order to register the density fluctuation in full.

Hot-wire/Hot-film anemometers

When current I_H flows through a thin wire with electrical resistance R, its temperature increases. If at the same time a medium with density ρ, flows across it at velocity v, a balance is set up between the electrical power input P_{el} and the power P_V drawn off by the air flow, whereby

$$P_{el} = I_H^2 \cdot R = P_V = c_1 \cdot \lambda \cdot \Delta\vartheta$$

Here, the power drawn off by the air flow is proportional to the temperature difference $\Delta\vartheta$ and the coefficient of thermal conductivity λ. The following applies in close approximation:

$$\lambda = \sqrt{\rho \cdot v} + c_2 = \sqrt{Q_{LM}} + c_2$$

Although λ is primarily a function of the mass flow Q_{LM}, with the medium at standstill ($v = 0$) a certain heat loss takes place (convection) represented by the additive constant c_2. This results in the familiar interrelationship

$$I_H = c_1 \cdot \sqrt{(\sqrt{Q_{LM}} + c_2)\sqrt{\frac{\Delta\vartheta}{R}}}$$

between the heating current I_H and the mass flow Q_{LM}.

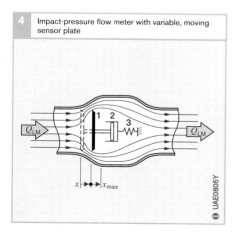

Fig. 4
1 Sensor plate
2 Damping device
3 Soft return spring

Q_{LM} Air-mass flow
x = $x(Q_{LM})$ sensor-plate setting dependent on flow

4 Impact-pressure flow meter with variable, moving sensor plate

With the application of constant heating power ($I_H^2 R$), which presents no problems, a reciprocal temperature increase $\Delta\vartheta$ would occur which decreases at a rate corresponding to the square root of the air-mass flow Q_{LM}. If on the other hand, the heating current I_H is controlled such that a constant temperature increase (for instance, $\Delta\vartheta = 100$ K) is maintained even when the flow rate increases, this will lead to a heating current which increases at the fourth root of the mass flow, and at the same time serves as a measure for the mass flow.

The essential advantage of such a control circuit lies in the fact that the electrical heater resistor always remains at the same temperature so that its calorific content need not be changed by means of time-wasting heat transfer. In fact, with a 70 μm platinum wire for instance, it is possible to achieve *time constants* in the 1 ms range for changes in air-flow rate. In cases where closed-loop control is not used the time constants would be 40...100 times higher (Fig. 5).

If the heater temperature were to be maintained constant simply by keeping its (temperature-dependent) resistance constant, with constant mass flow and higher medium temperature, this would result in a current drop and therefore a false measurement. In practice, this error is avoided by using a bridge circuit containing a second high-ohm "compensation resistor" R_K of the same type (e.g. platinum). Here, the heater resistor is kept at a *constant overtemperature* $\Delta\vartheta$ compared to the medium (Fig. 6). In case of a sudden jump in the medium temperature, the sensor reacts with a long time constant since in this case the calorific content of the heater wire must be changed.

The heater resistors in the first air-mass meters (anemometers) used for automotive applications were of very fine *platinum wire*. This wire was mounted in trapezoidal form across the flow cross-section so that it was able take the mean of irregularities in the flow profile. Service lives which were acceptable from the technical viewpoint only became possible when the platinum wire was stabilised by alloy additives so that its resistance no longer changed due to deposits and cracks on its surface. This meant though that the deposits on the heater wire had to be burned-off following every operating phase (approx. 1,000 °C).

Notwithstanding a number of functional advantages, this sensor concept was far too costly. Although a *thick-film version* (HFM2) was able to combine all the resistors concerned with the measurement on a single ceramic substrate, this failed to bring the

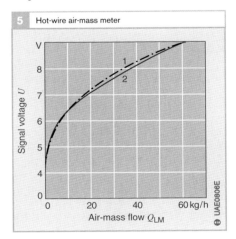

Fig. 5
1 Theoretical characteristic
2 Experimental characteristic

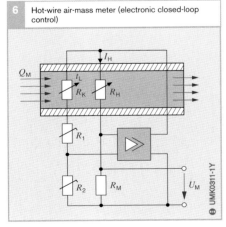

Fig. 6
Q_{LM} Air-mass flow
U_M Measurement voltage
R_H Hot-wire resistor
R_K Compensation resistor
R_M Measuring resistor
$R_{1,2}$ Trimming resistor

hoped-for advantages with regard to the costs. Due to the substrate's considerable thermal capacity, it was difficult not to exceed the maximum permissible switching constants. Furthermore, a complicated saw cut had to be made to reduce the undesirable heat coupling between heating and compensation resistors. On the other hand though, this version permitted the burn-off process to be dispensed with since the special flow conditions no longer led to unwanted deposits.

In contrast to both its predecessor types, a further *silicon-based micromechanical* version (HFM5) fulfilled practically all expectations. In particular, this version is able to measure in both directions with the correct sign (Fig. 7). This means that the brief return flows that occur as a result of pulsation no longer lead to measuring errors (Fig. 8).

To this end, in addition to the heater control circuit used in the previous versions, a temperature sensor is located on each side of the heater resistor, in other words upstream and downstream. This principle is similar to the "Thomas process" often encountered in literature. When there is no flow ($Q_{ML} = 0$), each of these sensors indicates the same temperature. When flow starts though, since the upstream sensor is cooled by the medium and the other is heated by it, the higher the flow the higher the temperature difference between the two sensors. The output signal derived from the temperature difference has a similar characteristic to the anemometers used up to now, whereby its sign is a clear indication of the flow direction.

Due to its small size, the micromechanical flow meter is only a partial-flow meter. In other words, it is no longer in any way able to average-out any non-homogeneity in the flow velocity as a function of the flow cross-sectional area. Rather, this flow meter must ensure that the partial flow it measures represents the same fraction of the total flow throughout the whole measuring range. This is not always an easy matter.

Examples of application
- Sensor-plate air-mass meter LMM
- Hot-wire air-mass meter HLM
- Hot-film air-mass meter HFM2 and
- Hot-film air-mass meter HFM5

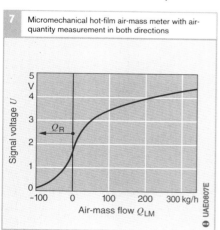

7 Micromechanical hot-film air-mass meter with air-quantity measurement in both directions

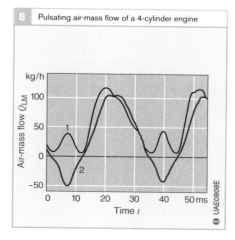

8 Pulsating air-mass flow of a 4-cylinder engine

Fig. 7
Q_R Return flow

Fig. 8
At WOT and speed
$n = 900$ rpm

1 Hot-wire air-mass meter
2 Hot-film air-mass meter

Sensor-flap (impact-pressure) air-flow sensor LMM

Application
The sensor-flap air-flow sensor is still in operation in a number of engines equipped with certain versions of the L-Jetronic or M-Motronic. It is installed between the air filter and the throttle valve and applies the sensor-flap principle in registering the air flow Q_L drawn in by the engine (Figs. 1 through 3).

Design and operating concept
The air-flow sensor's pivoting sensor flap (Fig. 1, Pos. 1) forms a variable orifice plate. The incoming air Q_L deflects the flap against the constant return force of a spring, whereby the free cross-section area increases along with increasing air flow the more the plate is deflected.

The change of the free air-flow-sensor cross section as a function of the sensor-flap setting has been selected so that there is a logarithmic relationship between the sensor-flap angle and the air quantity drawn in by the engine. This leads to high air-flow sensor sensitivity, a valuable asset in the case of small air quantities which necessitate high measuring accuracy. The stipulated measuring accuracy is 1...3% of the measured value throughout a range defined by
$Q_{max} : Q_{min} = 100 : 1$.

The sensor-plate angle is picked-off by a potentiometer (4) which converts it into an output voltage U_A (Fig. 4) which is used as an input to the ECU. In order to eliminate the effects of potentiometer aging and temperature coefficient on accuracy, the ECU only evaluates resistance ratios.

A further phenomenon which must be taken into account are the intake or induction strokes from the individual cylinders. These generate oscillations in the intake manifold, which the air-flow sensor can only follow up to about 10 Hz. To keep these effects down to a minimum, the measuring flap has a compensation flap attached to it which, in combination with a damping chamber (5), serves to damp the oscillations of the pulsating intake air.

Instead of the desired mass flow which is proportional to the product from $\rho \cdot v$, measurement according to the impact-pressure principle only measures a flow which is proportional to the product $\sqrt{\rho} \cdot v$. This means that density compensation (air temperature, air pressure) is required in order to achieve precise fuel metering.

The intake air's density changes along with its temperature. This fact is taken into account by the ECU calculating a correcting quantity from the temperature-dependent resistance of a temperature sensor integrated in the air-flow sensor (2). M-Motronic versions always feature barometric-pressure compensation. Here, a manifold-pressure sensor is connected pneumatically to the intake manifold so that it can pick-off the absolute manifold pressure. It is either integrated directly in the ECU (connected by hose to the intake manifold), or located in the vicinity of the intake manifold, or attached directly to it.

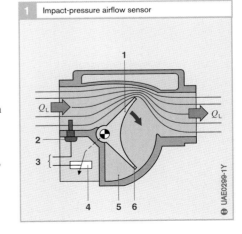

1 Impact-pressure airflow sensor

Fig. 1
1 Sensor plate
2 Air-temperature sensor
3 To ECU
4 Potentiometer
5 Damping chamber
6 Compensation flap

Q_L Intake-air flow

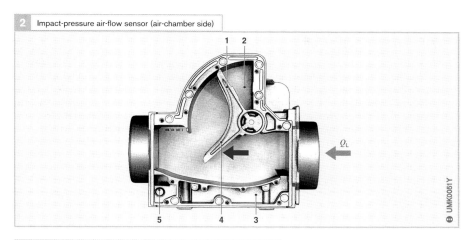

Fig. 2
1 Compensation flap
2 Damping chamber
3 Bypass
4 Sensor plate
5 Idle-mixture adjusting screw

Q_L Intake-air flow

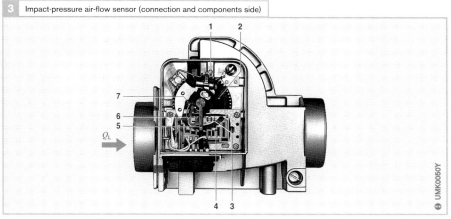

Fig. 3
1 Ring gear for spring preload
2 Return spring
3 Wiper track
4 Ceramic plate with resistors and printed conductors
5 Wiper pick-off
6 Wiper
7 Pump contact

Q_L Intake-air flow

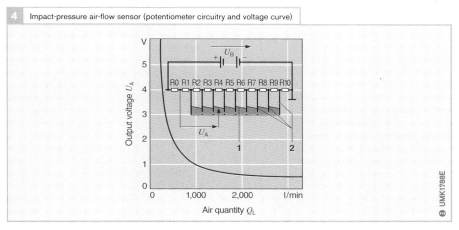

Fig. 4
1 Wiper track
2 Conductor segments (data points)

$U_A \sim 1/Q_L$ applies for the curve

Hot-wire air-mass meter HLM

Application
The HLM hot-wire air-mass meter is installed as a "thermal" load sensor between air filter and throttle plate in a number of LH-Jetronic or M-Motronic gasoline engines. It registers the air-mass flow Q_M drawn in by the engine, and applies this to determine the engine load. Being as it is able to follow average fluctuations of up to 1 Hz, the HLM is the fastest of the air flowmeters at present in use.

Design and construction
The intake air drawn in by the engines flows through the tubular HLM housing which is protected at each end by a wire mesh. A heated, 70 µm thin platinum wire element is suspended across the HLM measuring tube. It is suspended trapezoidally so that in good approximation it is able to cover the whole of the flow cross-section. A temperature-compensation (thin film) resistor projects into the air flow just upstream of the hot wire. Both of these components (hot wire and resistor) are integral parts of a closed-loop control circuit, where they function as temperature-dependent resistors. The control circuit is basically a bridge circuit and an amplifier (Figs. 1 and 2).

Operating concept
Before the air flowing through the HLM cools down the hot wire, its temperature is measured by means of the temperature-compensation resistor. A closed-loop control circuit regulates the heater current so that the hot wire assumes a temperature which is held at a constant level above that of the intake air. Since the air density influences the amount of heat dissipated to the air by the hot wire, this measuring concept must adequately take it into account. The heating current I_H is therefore a measure of the air-mass flow, and across a precision measuring resistor (R_M) it generates a voltage signal U_M, for input to the ECU, which is proportional to air-mass flow. The HLM, on the other hand, cannot detect the direction of air flow.

In order to prevent measurement-result drift due to deposits on the platinum wire, this is heated for about 1 second to a burn-off temperature of approx. 1,000 °C every time the engine is switched off. Here, the dirt deposits evaporate or flake off and leave the hot wire in a clean state.

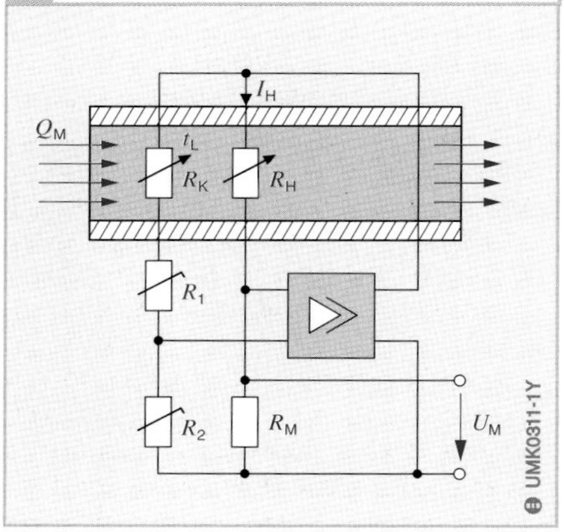

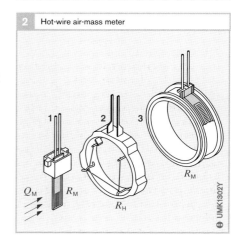

Fig. 1
- R_K Temperature-compensation resistor
- R_H Hot-wire heater resistor
- R_M Measuring resistor
- $R_{1,2}$ Bridge balance resistors
- U_M Measurement voltage
- I_H Heating current
- t_L Air temperature
- Q_M Air-mass flow

Fig. 2
1. Temperature-compensation resistor R_K
2. Sensor ring with hot wire R_H
3. Precision measuring resistor (R_M)

Q_M Air-mass flow

Hot-film air-mass meter HFM2

Application
The HFM2 **h**ot-**f**ilm air-**m**ass **m**eter is a thick-film sensor which is installed as a "thermal" load sensor between air filter and throttle plate in a number of LH-Jetronic or M-Motronic gasoline engines. It very accurately registers the air-mass flow Q_M drawn in by the engine, and applies this to determine the engine load.

The HFM2 electronic circuitry converts this voltage into the voltage UM which it adapts to make it suitable for input into the ECU. The computer than uses this to calculate the air mass drawn in by the engine for every working cycle. The HFM2 cannot determine the direction of air flow.

The long-term measuring accuracy of ±4 % referred to the measured value applies even without the burn-off of dirt deposits.

Design and construction
Together with bridge resistors, the electrically heated HFM2 platinum heater resistor R_H is located on a ceramic chip (substrate, Fig. 1).

The bridge also incorporates a temperature-dependent resistor R_S (flow sensor) which registers the heater temperature. Separation of the heater and the flow sensor is advantageous for the (closed-loop) control circuit. The heater element and the air-temperature compensation sensor (resistor R_K) are decoupled thermally by two saw cuts (Fig. 2).

Since the dirt is deposited mainly on the front edge of the sensor element, the components which are decisive for the heat transition are situated downstream on the ceramic substrate. Furthermore, the sensor is so constructed that air flow around the sensor remains unaffected by dirt deposits.

Operating concept
The electrically heated platinum heater resistor projects into the intake-air flow which cools it down. A closed-loop control circuit regulates the heater current so that the hot wire assumes a temperature which is held at a constant level above that of the intake air. Since the air density, just as much as the flow rate, is decisive regarding the amount of heat dissipated to the air by the hot wire, this measuring concept takes it into account to the appropriate degree. The heating current I_H, and the voltage at the heater, is thus a non-linear measure for the air-mass flow Q_M.

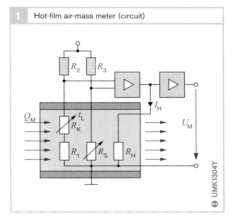

Hot-film air-mass meter (circuit)

Fig. 1
R_K Temperature-compensation sensor (resistor)
R_H Heater resistor
R_S Sensor resistor
R_1, Bridge resistors
R_2,
R_3
U_M Measurement voltage
I_H Heating current
t_L Air temperature
Q_M Air-mass flow

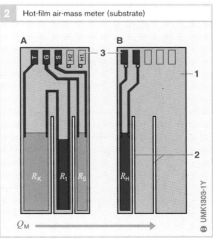

Hot-film air-mass meter (substrate)

Fig. 2
A Front side
B Rear side

1 Ceramic substrate
2 Two saw cuts
3 Contacts

R_K Temperature-compensation sensor (resistor)
R_H Heater resistor
R_S Sensor resistor
R_1 Bridge resistor

Hot-film air-mass meter HFM5

Application
For optimal combustion as needed to comply with the emission regulations imposed by legislation, it is imperative that precisely the necessary air mass is inducted, irrespective of the engine's operating state.

To this end, part of the total air flow which is actually inducted through the air filter or the measuring tube is measured by a hot-film air-mass meter. Measurement is very precise and takes into account the pulsations and reverse flows caused by the opening and closing of the engine's intake and exhaust valves. Intake-air temperature changes have no effect upon measuring accuracy.

Design and construction
The housing of the HFM5 **hot-**film air-mass meter (Fig. 1, Pos. 5) projects into a measuring tube (2) which, depending upon the engine's air-mass requirements, can have a variety of diameters (for 370...970 kg/h). This tube is installed in the intake tract downstream from the air filter. Plug-in versions are also available which are installed inside the air filter.

The most important components in the sensor are the sensor element (4), in the air intake (8), and the integrated evaluation electronics (3). The partial air flow as required for measurement flows across this sensor element.

Vapor-deposition is used to apply the sensor-element components to a semiconductor substrate, and the evaluation-electronics (hybrid circuit) components to a ceramic substrate. This principle permits very compact design. The evaluation electronics are connected to the ECU through the plug-in connection (1). The partial-flow measuring tube (6) is shaped so that the air flows past the sensor element smoothly (without whirl effects) and back into the measuring tube via the air outlet (7). This method ensures efficient sensor operation even in case of extreme pulsation, and in addition to forward flow, reverse flows are also detected (Fig. 2).

Operating concept
The hot-film air-mass meter is a "thermal sensor" and operates according to the following principle:

A micromechanical sensor diaphragm (Fig. 3, Pos. 5) on the sensor element (3) is heated by a centrally mounted heater resistor and held at a constant temperature. The temperature drops sharply on each side of this controlled heating zone (4).

The temperature distribution on the diaphragm is registered by two temperature-dependent resistors which are mounted upstream and downstream of the heater resistor so as to be symmetrical to it (measuring points M_1, M_2). Without the flow of incoming air, the temperature characteristic (1) is the same on each side of the heating zone ($T_1 = T_2$).

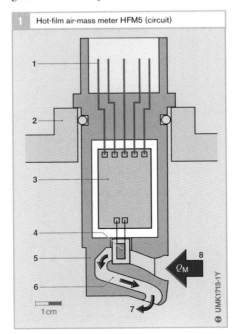

Fig. 1
1. Electrical plug-in connection
2. Measuring tube or air-filter housing wall
3. Evaluation electronics (hybrid circuit)
4. Sensor element
5. Sensor housing
6. Partial-flow measuring tube
7. Air outlet for the partial air flow Q_M
8. Intake for partial air flow Q_M

1 Hot-film air-mass meter HFM5 (circuit)

As soon as air flows over the sensor element, the uniform temperature distribution at the diaphragm changes (2). On the intake side, the temperature characteristic is steeper since the incoming air flowing past this area cools it off. Initially, on the opposite side (the side nearest to the engine), the sensor element cools off. The air heated by the heater element then heats up the sensor element. The change in temperature distribution leads to a temperature differential (ΔT) between the measuring points M_1 und M_2.

The heat dissipated to the air, and therefore the temperature characteristic at the sensor element is a function of the air mass flow. Independent of the absolute temperature of the air flowing past, the temperature differential is a measure of the air mass flow. Apart from this, the temperature differential is directional, which means that the air-mass meter not only registers the mass of the incoming air but also its direction.

Due to its very thin micromechanical diaphragm, the sensor has a highly dynamic response (<15 ms), a point which is of particular importance when the incoming air is pulsating heavily.

The evaluation electronics (hybrid circuit) integrated in the sensor convert the resistance differential at the measuring points M_1 and M_2 into an analog signal of 0...5 V which is suitable for processing by the ECU. Using the sensor characteristic (Fig. 2) programmed into the ECU, the measured voltage is converted into a value representing the air mass flow [kg/h].

The shape of the characteristic curve is such that the diagnosis facility incorporated in the ECU can detect such malfunctions as an open-circuit line. A temperature sensor for auxiliary functions can also be integrated in the HFM5. It is located on the sensor element upstream of the heated zone.

It is not required for measuring the air mass. For applications on specific vehicles, supplementary functions such as improved separation of water and contamination are provided for (inner measuring tube and protective grid).

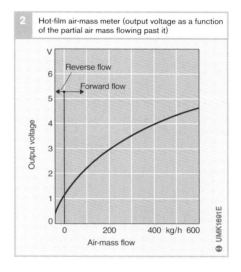

2　Hot-film air-mass meter (output voltage as a function of the partial air mass flowing past it)

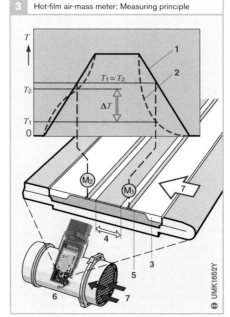

3　Hot-film air-mass meter: Measuring principle

Fig. 3
1　Temperature profile without air flow across sensor element
2　Temperature profile with air flow across sensor element
3　Sensor element
4　Heated zone
5　Sensor diaphragm
6　Measuring tube with air-mass meter
7　Intake-air flow
M_1, M_2　Measuring points
T_1, T_2　Temperature values at the measuring points M_1 and M_2
ΔT　Temperature differential

Gas sensors, concentration sensors

Measured quantities

The concentration of a given material or medium defines the mass or volume percent of a given material in another given material or in a mixture or combination of other materials. With a concentration sensor (also known as a concentration probe), the important thing is that in the ideal case it is sensitive to only one medium, while at the same time practically "ignoring" all other mediums. Of course, in practice, every concentration sensor has its own cross sensitivity to other mediums even though, as is often the case, "temperature" and "pressure" are maintained constant.

In the vehicle, the following parameters must be measured:
- Oxygen content in the exhaust gas (closed-loop combustion control, catalytic-converter monitoring).
- Carbon-monoxide and nitrogen-oxide content, as well as air humidity inside the vehicle (air quality, misting of vehicle windows).
- Humidity in the compressed-air braking system (air-drier monitoring).
- Dampness of the outside air (black-ice warning).
- Concentration of soot in diesel-engine exhaust gas. A still unsolved problem. In contrast to the above-mentioned gas concentrations, this is a particle concentration. The difficulties inherent in the measuring assignment are further aggravated by the possibility of the sensor being blocked by particles so that it no longer functions.

The introduction of the fuel cell as an automotive drive means that further gas sensors will have to be developed, for instance for the detection of hydrogen.

Measuring principles

Measured mediums occur in gaseous, liquid, or solid state, so that in the course of time countless measuring methods have been developed. For automotive applications, up till now only the gas-analysis area, and in particular the measurement of gaseous humidity, has been of any interest. Table 1 presents an overview of the processes applied in general measurement techniques.

Gas measurement in general
Gas sensors are usually in direct unprotected contact with the measured medium (in other words with foreign matter) so that the danger exists of irreversible *damage*. This form of damage is referred to as sensor "contamination". For instance, the lead that may be contained in fuel can make the electrolytic oxygen concentration sensor (Lambda oxygen sensor) unusable.

Moisture measurement
In addition to the outstanding significance of the Lambda oxygen sensor in dealing with exhaust gases, moisture measurement also plays an important role.

1 Gas-analysis processes (without particular attention being paid to the moisture-measurement process). (X) = For automotive applications.

Physical process	Physical-chemical process		Chemical process
Thermal conductivity	Catalytic effect		Selective absorption
Magnetic process	Absorption warmth		Selective absorption with prior chemical conversion
Radiation absorption	Characteristic color reaction		
Gas chromatography	Electrical conductivity	X	
Radioactive process	Electrochemical process	X	

Table 1

In the broader sense, moisture can be said to be present in gaseous, liquid, or solid form. In the narrower sense, we are dealing here with the gaseous-water (water vapor) content in gaseous mediums – above all in the air.

When a damp gas is cooled in an isobaric process, it reaches its saturation point at a specific temperature (known as the dew point τ).

A number of important definitions are given below in connection with humidity measurement (refer also to Fig. 1):

m_w Mass of water
m_s Mass of water in the saturation state
m_{tr} Mass of dry gas

M_w Mole mass of water
M_{tr} Mean mole mass of the dry gas
p Total pressure of the gas mixture
p_w Partial pressure of the water vapor

p_s Saturated vapor pressure (vapor pressure of the water at mixture temperature)

Absolute humidity:

$$\chi = \frac{m_w}{m_{tr}} = \frac{M_w}{M_{tr}} \cdot \frac{p_w}{p-p_w} \quad \text{(in \%)}$$

$$f_a = \frac{m_w}{V_{tr}} \quad \text{(volume-related)}$$

Relative humidity:

$$\Phi = \frac{p_w}{p_s} \quad \text{(in \%)}$$

For low-cost applications (for instance in the vehicle), resistive and capacitive sensors are used almost exclusively. They are provided with hygroscopic layers which can store water as a function of the relative humidity (and release it again), and thus trigger a usually drastic change in a resistor's value or in the value of a planar capacitor.

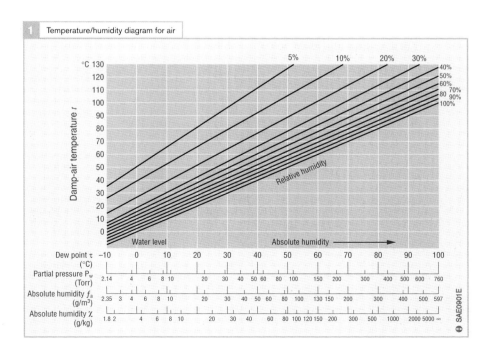

Figure 1: Temperature/humidity diagram for air

Table 2

Method/Process	Running number		Measuring instrument	Measuring method
Saturation method	1	X	Dew-point hygrometer	Direct method
	2	X	LiCi dew-point hygrometer	(Measurement of absolute humidity)
Evaporation method	3	X	Psychrometer	
Absorption method	4		Volume hygrometer	
	5	X	Electrolysis hygrometer	
	6		Condensate-quantity hygrometer	
Energy method	7	X	Infrared hygrometer	
	8		Microwave hygrometer	
	9		Electrical-discharge hygrometer	
	10		Diffusion hygrometer	
Hygroscopic method	11	X	Electrical conductive-foil hygrometer	Indirect method
	12	X	Capacitor hygrometer	(Measurement of relative humidity)
	13	X	Hair hygrometer	
	14		Two-strip hygrometer	
	15		Color hygrometer	
	16		Quartz hygrometer	
	17		Gravimetric hygrometer	

Humidity-measurement procedures. X = technologically important

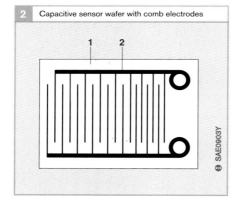

Fig. 2
1 Plastic wafer
2 Comb-shaped, embossed gold-foil electrodes

Capacitive sensor wafer with comb electrodes

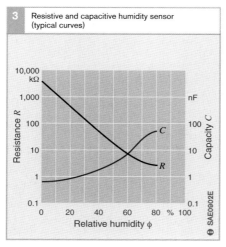

Resistive and capacitive humidity sensor (typical curves)

On capacitive humidity sensors, a hygroscopic insulating layer (e.g. Al_2O_3, or a polymer plastic), which can possibly also be the mounting wafer, serves as the dielectric of a capacitor. Either one of the electrodes is permeable to water, or the electrodes have a comb shape (Fig. 2). Along with increasing humidity, the dieeletric absorbs more water and the sensor's capacitance increases considerably (relative dielectric constant of water $\varepsilon_{rW} \approx 81$, Fig. 3).

In the case of the *resistive* sensor, an insulation substrate is coated with a layer of hygroscopic salt (LiCi) held in a paste binder and located between an electrode pair. The layer's conductivity changes drastically along with the relative humidity (Fig. 3). Unfortunately this resistance change is also highly dependent upon temperature so that normally compensation is needed. If the air temperature (NTC) is then measured, the dew point can be defined and with it the absolute humidity. The typical time constants of these sensors are about 30 s.

Table 2 gives an overview of the numerous measurement processes that have been developed in the course of time for humidity measurement.

Air-quality sensors

Application

Air-quality sensors (Fig. 1) continually monitor air quality at the point where ventilation air enters the vehicle. In particular, these sensors respond to toxic exhaust-gas components such as CO (mainly from gasoline engines) and NO_X (mainly from diesel engines).

A further assignment is to prevent the misting-up of the windows. Here, a humidity sensor registers the air's water-vapor level.

Design and operating concept

The sensors incorporated in the air-quality ECU (Figs. 1 and 2) are comprised of thick-film resistors containing tin oxide. As soon as the measured medium collects there (process is reversible), the resistors in some cases change their electrical resistance drastically (e.g. 1...100 kΩ). The sensor resistors are all grouped on a common ceramic substrate which is heated from the rear to an operating temperature of approx. 330 °C by a heating conductor. Due to the high operating temperature, there is an air gap between components and substrate.

The CO sensor measures concentrations in the 10...100 ppm range (**ppm = parts per million**) and the NO_X sensor in the 0.5...5 ppm range. As soon as the concentration of pollution gases is excessive (sometimes almost 100 times higher than in clean air), the air-quality ECU closes the fresh-air inlet flaps. This serves to prevent the driver breathing in these gases so that he/she does not tire so quickly. The activated carbon filters also last longer when not loaded by these pollution gases.

Rough protection is provided by a metal cover, underneath which there is a Teflon diaphragm for both sensor chambers which permits passage of the measured gases and gaseous vapors while at the same time holding back fluid humidity. Even though the measured gases must first pass through the Teflon diaphragm, such sensors have a response time of milliseconds.

The latest air-quality ECUs also incorporate a humidity sensor (Fig. 3). Together with the interior temperaure as measured by an NTC resistor, its signal is used for calculating the dew point which is an important factor with regard to the misting-up of the vehicle's windows.

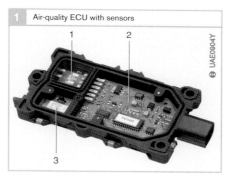

Fig. 1
1 NO_X/CO measuring element
2 Evaluation electronics
3 Humidity sensor

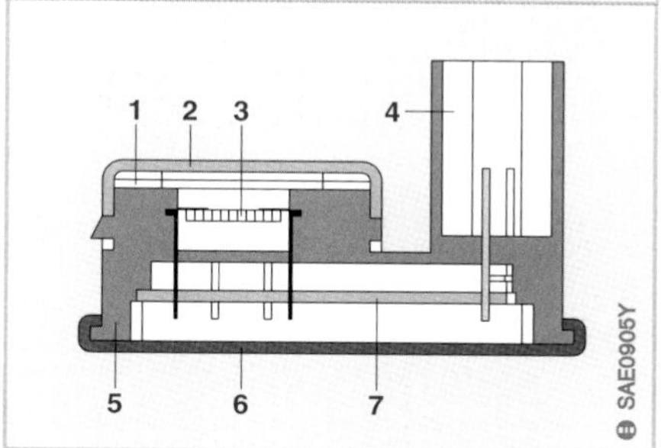

Fig. 2
1 Teflon diaphragm
2 Cover (gas-permeable)
3 NO_X/CO measuring element
4 Plug
5 Housing
6 Cover with gasket
7 pcb

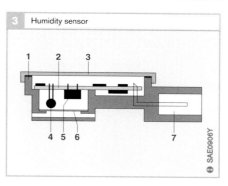

Fig. 3
1 Housing
2 pcb
3 Cover with gasket
4 Temperature sensor
5 Humidity measuring element
6 Teflon diaphragm
7 Plug

Two-step Lambda oxygen sensors

Application

These sensors are used in gasoline engines equipped with two-step Lambda control. They extend into the exhaust pipe between the engine's exhaust manifold and the catalytic converter, and register the exhaust-gas flow leaving each cylinder. Thanks to the Lambda sensor being heated, it can be installed further away from the engine so that long periods of full-load (WOT) engine operation are unproblematical. The LSF4 sensor is also suitable for operation with exhaust-gas control systems featuring a number of sensors (for instance with OBDII).

"Two-step sensors" compare the residual-oxygen content in the exhaust gas with that of the reference atmosphere inside the sensor. They then indicate whether the A/F mixture in the exhaust gas is "rich" ($\lambda < 1$) or "lean" ($\lambda > 1$). The sudden jump in the characteristic curve of these sensors permits A/F control to $\lambda = 1$ (Fig. 1).

Design and construction

LSH25 tube-type (finger) sensor
Sensor ceramic with protective tube
The solid-state electrolyte is a ceramic element and is impermeable to gas. It is a mixed oxide comprising the elements zirconium and yttrium in the form of a tube closed at one end (finger). The inside and outside surfaces have each been provided with a porous platinum coating which serves as an electrode.

The platinum electrode on the outside surface of the ceramic body protrudes into the exhaust pipe, and acts as a catalytic converter in miniature. Exhaust gas which reaches this electrode is processed catalytically and brought to a stoichiometric balance ($\lambda = 1$). In addition, the outside of the sensor which is in contact with the exhaust gas is provided with a porous multiple ceramic (Spinel) layer to protect it against contamination. The ceramic body is also protected against mechanical impact and thermal shocks by a slotted metal tube. A number of slots in the protective tube are specially shaped so that on the one hand they are particularly effective against extensive thermal and chemical stresses, while on

Fig. 1
a Rich A/F mixture (air deficiency)
b Lean A/F mixture (excess air)

Fig. 2
1 Sensor ceramic element
2 Electrodes
3 Contacts
4 Housing contact
5 Exhaust pipe
6 Ceramic protective coating (porous)
7 Exhaust gas
8 Outside air
U_S Sensor voltage

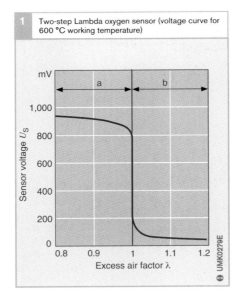

1 Two-step Lambda oxygen sensor (voltage curve for 600 °C working temperature)

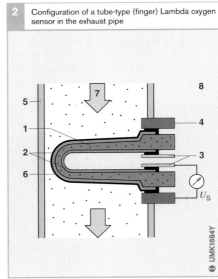

2 Configuration of a tube-type (finger) Lambda oxygen sensor in the exhaust pipe

the other hand preventing a sharp drop in sensor-ceramic temperature when the exhaust is "cool".

The sensor's "open" inner chamber is isolated from the exhaust gas and connected to the surrounding air which acts as a reference gas (Fig. 2).

Sensor body with heater element and electric connection

A ceramic support tube and a disc spring hold the active finger-shaped sensor ceramic in the sensor housing and seal it off. A contact element between the support tube and the active sensor ceramic provides the contact between the inner electrode and the connection cable.

A metal seal ring connects the outer electrode with the sensor housing. The sensor's complete internal structure is located and held in place by a protective metal sleeve which at the same time acts as the support for the disc spring. The protective sleeve also guards against contamination of the sensor's interior. The connection cable is crimped to the contact element which protrudes from the sensor, and is protected against damp and mechanical loading by means of a special temperature-resistant cap.

The tube-type (finger) Lambda sensor (Fig. 3) is also equipped with a electrical heater element. On this sensor, at low engine loads (e.g. low exhaust-gas temperatures) the ceramic element's temperature is defined by the electrical heater, and at high loads by the exhaust-gas temperature. Depending on the heater element's power rating, exhaust-gas temperatures of a low as 150...200 °C suffice to bring the sensor ceramic up to operating temperature. Thanks to its external electrical heating, the sensor heats up so quickly that it has already reached operating temperature 20...30 s after the engine has started so that the Lambda closed-loop control can come into operation. In fact, the sensor heating ensures that above the func-

3 LSH25 heated tube-type (finger) Lambda oxygen sensor (view and section)

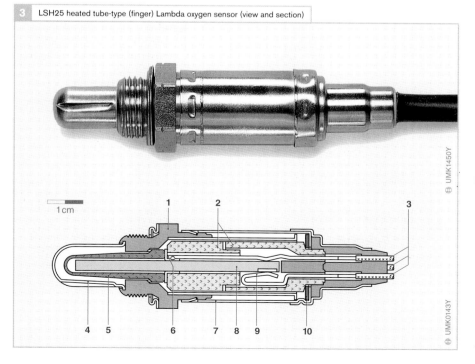

Fig. 3
1 Sensor housing
2 Ceramic support tube
3 Connection cable
4 Protective tube with slots
5 Active sensor ceramic
6 Contact element
7 Protective sleeve
8 Heater element
9 Clamp-type heater-element connections
10 Disc spring

tion limit (350 °C) the sensor is always at optimum operating temperature, and it also contributes to low and stable exhaust-gas emission figures.

Initially the LS21 Lambda sensor was unheated. This version is only rarely found in older vehicles as a replacement part.

LSF4 planar Lambda oxygen sensor
Regarding their function, planar Lambda sensors correspond to the heated finger sensors with their voltage-jump curve at $\lambda = 1$.

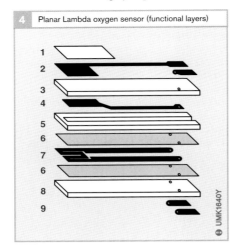

Fig. 4 Planar Lambda oxygen sensor (functional layers)

Fig. 4
1 Porous protective layer
2 Outer electrode
3 Sensor foil
4 Inner electrode
5 Reference-air-passage foil
6 Insulation layer
7 Heater
8 Heater foil
9 Connection contacts

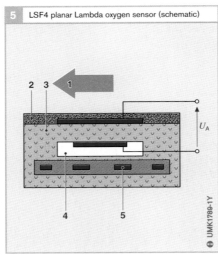

Fig. 5 LSF4 planar Lambda oxygen sensor (schematic)

Fig. 5
1 Exhaust gas
2 Porous protective ceramic layer
3 Measuring element with microporous noble-metal coating
4 Reference-air passage
5 Heater
U_A Output voltage

On the planar sensor though, the solid-state electrolyte is comprised of a number of individual laminated foils stacked one on top of the other (Fig. 4). The sensor is protected against thermal and mechanical influences by a double-walled protective tube.

The sensor's planar ceramic element (measuring element and heater are integrated) is shaped like a long stretched-out wafer with rectangular cross-section.

The measuring element's surfaces have a microporous noble-metal coating. On the exhaust-gas side this has an exrtra microporous ceramic coating to protect it against the erosive effects of the exhaust-gas components. The heater is a wave-shaped element containing noble metal. It is integrated in the ceramic wafer and insulated from it, and even though it has a low power input ensures that the sensor heats up quickly.

The LSF4 sensor (Figs. 5 and 6) operates as a reference-gas sensor. The reference-air passage inside this sensor has a direct connection to the surrounding air. This enables it to compare the residual oxygen content of the exhaust gas with the oxygen content of the reference gas. In other words, with the air inside the sensor. This means that in the case of the planar sensor, the sensor voltage also features a characteristic jump (Fig. 1) in the area of stoichiometric A/F mixture ($\lambda = 1$).

Operating concept

The two-step sensors operate in accordance with the principle of the galvanic oxygen-concentration cell with solid-state electrolyte (Nernst principle). As from about 350 °C, the sensor ceramic becomes conductive for oxygen ions (the most efficient and reliable functioning is at temperatures >> 350 °C). Since there is a sharp change in the oxygen content on the exhaust-gas side in the area of $\lambda = 1$ (e.g. $9 \cdot 10^{-15}$ Vol% for $\lambda = 0.99$, and 0.2 Vol% for $\lambda = 1.01$), this leads to the generation of a voltage between the sensor's boundary layers due to the different oxygen concentrations inside and outside the sensor. This means that the exhaust-gas oxygen content can be applied as a measure for the A/F ratio. The integral heater ensures that the sensor is operational at exhaust-gas temperatures as low as 150 °C.

The sensor's output voltage U_S is a function of the oxygen content in the exhaust gas. In the case of a rich A/F mixture ($\lambda < 1$) it reaches 800...1,000 mV. For a lean mixture ($\lambda > 1$), only approx. 100 mV are generated. The transition from the rich to the lean area at U_{reg} is at about 450...500 mV.

The ceramic structure's temperature influences its ability to conduct the oxygen ions, and therefore also the shape of the output-voltage curve as a function of the excess-air factor λ (the values in Fig. 1 apply for about 600 °C). Apart from this, the response time for a voltage change when the A/F mixture changes is also highly dependent upon temperature.

Whereas response times at ceramic temperatures below 350 °C are in the seconds range, at optimum temperatures of around 600 °C the sensor responds in less than 50 ms. When the engine is started therefore, the Lambda closed-loop control is switched off until the minimum operating temperature of about 350 °C is reached. During this period, the engine is open-loop controlled.

Excessive temperatures reduce the sensor's useful life. This means that the Lambda sensor must be installed so that the operating temperatures 850 °C (LSH25), and 930 °C (LSF4), are not exceeded.

6 LSF4 planar Lambda oxygen sensor (view and section)

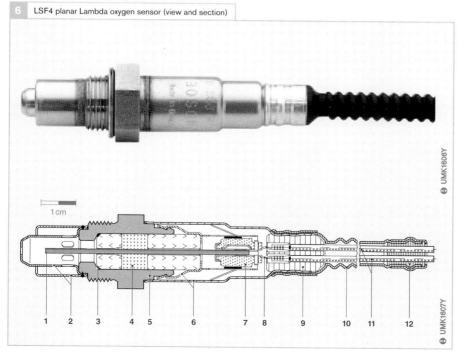

Fig. 6
1. Planar measuring element
2. Double protective tube
3. Seal ring
4. Seal packing
5. Sensor housing
6. Protective sleeve
7. Contact holder
8. Contact clip
9. PTFE sleeve
10. Shaped PTFE sleeve
11. Five connection cables
12. Seal

LSU4 planar broad-band Lambda oxygen sensor

Application

As its name implies, the broad-band Lambda oxygen sensor is used across a very extensive range to determine the oxygen concentration in the exhaust gas. The figures provided by the sensor are an indication of the air-fuel (A/F) ratio in the engine's combustion chamber. The excess-air factor λ is used when defining the A/F ratio.

The sensor protrudes into the exhaust pipe and registers the exhaust-gas mass flow from all cylinders. Broad-band Lambda sensors make precise measurements not only at the stoichiometric point $\lambda = 1$, but also in the lean range ($\lambda > 1$) and in the rich range ($\lambda < 1$). In combination with electronic closed-loop control circuitry, these sensors generate an unmistakable, continuous electrical signal (Fig. 3) in the range from $0.7 < \lambda < \infty$ (= air with 21% O_2). This means that the broad-band Lambda sensor can be used not only in engine-management systems with two-step control ($\lambda = 1$), but also in control concepts with rich and lean air-fuel (A/F) mixtures. This type of Lambda sensor is therefore also suitable for the Lambda closed-loop control used with lean-burn concepts on gasoline engines, as well as for diesel engines, gaseous-fuel engines, and gas-powered central heaters and water heaters (this wide range of applications led to the designation LSU: Lambda Sensor Universal (taken from the German), in other words Universal Lambda Sensor).

In a number of systems, several Lambda sensors are installed for even greater accuracy. Here, for instance, they are fitted upstream and downstream of the catalytic converter as well as in the individual exhaust tracts (cylinder banks).

Design and construction

The LSU4 broad-band Lambda sensor (Fig. 2) is a planar dual-cell limit-current sensor. It features a zirconium-dioxide/ceramic (ZrO_2) measuring cell which is the combination of a Nernst concentration cell (sensor cell which functions the same as a two-step Lambda sensor) and an oxygen-pump cell for transporting the oxygen ions. The oxygen-pump cell (Fig. 1, Pos. 8) is so arranged with respect to the Nernst concentration cell (7) that there is a 10...50 μm diffusion gap (6) between them in which there are two porous platinum electrodes: A pump electrode and a Nernst measuring electrode. The gap is connected to the exhaust gas through a gas-access passage (10). The porous diffusion barrier (11) serves to limit the inflow of oxygen molecules from the exhaust gas.

Fig. 1
1 Exhaust gas
2 Exhaust pipe
3 Heater
4 Control electronics
5 Reference cell with reference-air passage
6 Diffusion gap
7 Nernst concentration cell with Nernst measuring electrode (on the diffusion-gap side), and reference electrode (on the reference-cell side)
8 Oxygen-pump cell with pump electrode
9 Porous protective layer
10 Gas-access passage
11 Porous diffusion barrier

I_P Pump current
U_P Pump voltage
U_H Heater voltage
U_{Ref} Reference voltage (450 mV, corresponds to $\lambda = 1$)
U_S Sensor voltage

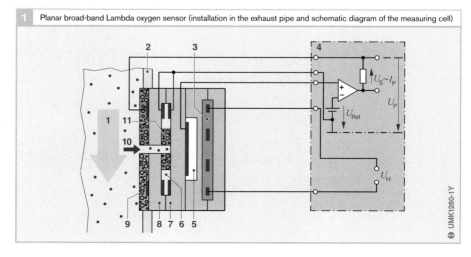

1 Planar broad-band Lambda oxygen sensor (installation in the exhaust pipe and schematic diagram of the measuring cell)

On the one side, the Nernst concentration cell is connected to the atmosphere by a reference-air passage (5), on the other it is connected to the exhaust gas in the diffusion gap.

An electronic closed-loop control circuit is needed in order to generate the sensor signal and for the sensor temperature control.

An integral heater (3) heats up the sensor quickly so that it soon reaches its operating temperature of 650...900 °C needed for generating a usable signal. This functions decisively reduces the effects that the exhaust-gas temperature has on the sensor signal.

Operating concept

The exhaust gas enters the actual measuring chamber (diffusion gap) of the Nernst concentration cell through the pump cell's gas-access passage. In order that the excess-air factor λ can be adjusted in the diffusion gap, the Nernst concentration cell compares the gas in the diffusion gap with that in the reference-air passage.

The complete process proceeds as follows: By applying the pump voltage U_P across the pump cell's platinum electrodes, oxygen from the exhaust gas is pumped through the diffusion barrier and into or out of the diffusion gap. With the help of the Nernst concentration cell, an electronic circuit in the ECU controls the voltage (U_P) across the pump cell in order that the composition of the gas in the diffusion gap remains constant at $\lambda = 1$. If the exhaust gas is lean, the pump cell pumps the oxygen to the outside (positive pump current). On the other hand, if it is rich, due to the decomposition of CO_2 and H_2O at the exhaust-gas electrode the oxygen is pumped from the surrounding exhaust gas and into the diffusion gap (negative pump current). Oxygen transport is unnecessary at $\lambda = 1$ and pump current is zero. The pump current is proportional to the exhaust-gas oxygen concentration and is this a non-linear measure for the excess-air factor λ (Fig. 3).

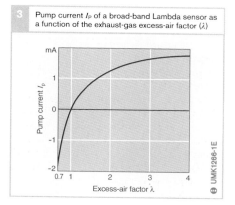

3 Pump current I_P of a broad-band Lambda sensor as a function of the exhaust-gas excess-air factor (λ)

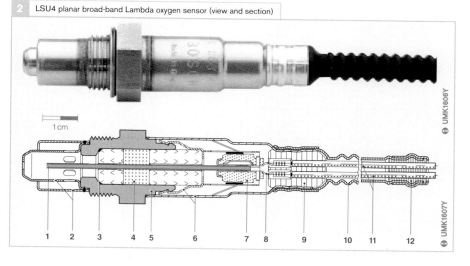

2 LSU4 planar broad-band Lambda oxygen sensor (view and section)

Fig. 3
1 Measuring cell (combination of Nernst concentration cell and oxygen-pump cell)
2 Double protective tube
3 Seal ring
4 Seal packing
5 Sensor housing
6 Protective sleeve
7 Contact holder
8 Contact clip
9 PTFE sleeve (Teflon)
10 PTFE shaped sleeve
11 Five connecting leads
12 Seal ring

Temperature sensors

Measured quantities

Temperature is defined as a nondirectional quantity which characterises the energy state of a given medium, and which can be a function of time and location:

$$T = T(x, y, z, t) \tag{1}$$

Where x, y, z are the space coordinates, t is time, and T is measured according to the Celsius or Kelvin scale.

Generally speaking, with measured mediums which are in gaseous or liquid form, measurements can be taken at any point. In the case of solid bodies, measurement is usually restricted to the body's surface. With the most commonly used temperature sensors, in order for it to assume the medium's temperature as precisely as possible, the sensor must be directly in contact with the measured medium (direct-contact thermometer). In special cases though, proximity or non-contacting temperature sensors are in use which measure the medium's temperature by means of its (infrared) thermal radiation (radiation thermometer = Pyrometer, thermal camera).

Generally speaking, a temperature sensor should reflect this dependency correctly, in other words, it should reflect as accurately as possible the local distribution of the temperature and its change as a function of time.

In special cases, possibly for functional reasons, this stipulation can be relaxed somewhat. The call for high local resolution and for high-speed response both demand that the sensor should be as small as possible, since it should not falsify the temperature readings by itself absorbing heat. In other words it should have a low thermal capacity.

In order to ensure that the temperature assumed by the sensor remains independent of the usually very different temperature of its mounting, it should be thermally well insulated from it. Since this also falsifies the measurement, the heat developed by the majority of sensors in the active state should be kept to a minimum (e.g. <1 mW).

The temperature sensor's dynamic response is given the time constant τ. This defines the time taken by the sensor to reach 63%, 90%, or 99% of its final reading when subjeted to a jump in temperature. This time depends not only upon the sensor's thermal capacity, but also to a great extent upon the heat-transfer coefficient between the sensor and the medium. The higher it is, the faster the sensor reaches its final reading. Naturally, this figure is far higher for liquid mediums than for gaseous mediums. It must also be taken into account that in case of medium flow, the heat-transfer coefficient is highly influenced by the flow rate v (refer to "hot-film air-mass meter"), and increases by about $\sqrt{v}$. In other words, the time constant of a temperature sensor should always be specified with reference to the defined flow rate of a defined medium.

Temperature measurement in the vehicle makes use almost exclusively of the temperature-dependence of electrical resistance material with **p**ositive (PTC) or **n**egative (NTC) **t**emperature **c**oefficient. The direct-contact thermometers apply this phenomena. For the most part, the conversion of the temperature change to an analog voltage takes place by adding a second resistor to form a voltage divider (which also has a linearising effect). The latter can be either neutral with regard to temperature, or temperature-dependent in the other direction. Recently though, non-contacting (pyrometric) temperature sensing has been considered for passenger-protection (passenger-position monitoring for airbag triggering), and for comfort and convenience (climate control in accordance with skin-temperature measurement, prevention of windscreen mist-up). It needed the advent of microsystem technology for this (pyrometric) method to become feasible from the costs viewpoint. Table 1 presents a listing of the temperatures which have to be measured in the vehicle.

Table 1. Temperatures in the vehicle

Measuring point	Temperature range °C
Intake air/Charge air	−40…170
Vehicle surroundings	−40…60
Passenger compartment	−20…80
Fan outlet air/Heater	−20…60
Evaporator (air-conditioner)	−10…50
Coolant	−40…130
Engine oil	−40…170
Battery	−40…100
Fuel	−40…120
Tire air	−40…120
Exhaust gas	100…1,000
Brake caliper	−40…2,000

Not only the highly differing temperature ranges demand a variety of different sensor concepts and technologies, but also the different accuracies and speeds which are required but not listed in the Table. In many cases, the temperature is measured and applied as an auxiliary quantity in order to be able to compensate for it as a cause of defect or as an unwanted influencing variable.

Measuring principles

Direct-contact sensors
The fact that practically all physical processes are temperature-dependent means that there are almost just as many methods for making temperature measurements. The preferable methods though are those in which the temperature effect is very distinctive and dominant and as far as possible features a linear characteristic. Furthermore, the measuring elements should be suitable for inexpensive mass-production, whereby they should be adequately reproducible and non-aging. Taking these considerations into account, the following sensor techniques have come to the forefront, some of which are also applied in automotive technology:

Resistive sensors
In the form of 2-pole elements, temperature-dependent electrical resistors are particularly suitable for temperature measurement, no matter whether in wire-wound, sinter-ceramic, foil, thin-film, thick-film, or monocrystalline form. Normally, in order to generate a voltage-analog signal they are combined with a fixed resistor R_V to form a voltage divider, or load-independent current is applied (Fig. 1). The voltage-divider circuit changes the original sensor characteristic $R(T)$, to another characteristic $U(T)$:

$$U(T) = U_0 \cdot \frac{R(T)}{R(T) + R_V} \quad (2)$$

On the other hand, the application of a load-independent current I_0 permits the resistor curve to be reproduced exactly:

$$U(T) = I_0 \cdot R(T) \quad (3)$$

The measurement sensitivity is reduced more or less depending upon the construction of the voltage-divider circuit. Notwithstanding this fact, in the case of resistance characteristics which display a slightly progressive curve it does have a linearisation influence (an effect which is usually very welcome). Very often, in this connection, the auxiliary resistor is so dimensioned that it equals the measuring resistor at a given reference temperature T_0 (e.g. 20 °C):

$$R_V \approx R(T_0) \quad (4)$$

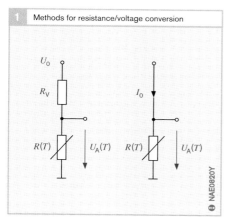

Fig. 1
I_0 Current supply
U_0 Suppy voltage
R_V Temperature-dependent series resistor
$R(T)$ Temperature-dependent measuring resistor
$U_A(T)$ Output voltage

Methods for resistance/voltage conversion

If the precision is inadequate, a resistance sensor can be calibrated to the desired value both with regard to its resistance (referred to a reference temperature), and its temperature coefficient (TK), by means of an adjustable parallel resistor R_P and a series resistor R_S (Fig. 2). Of course, when resistors are added the TK is reduced and the characteristic changes somewhat.

Sintered-ceramic NTC resistors

As a result of their pronounced measuring effects and inexpensive manufacture, the most common semiconductor resistances in use are based on heavy-metal oxides and oxidised crystals. These are sintered in bead form or disc form (Fig. 3) and have a polycrystalline structure. They are often referred to as NTC thermistors. To a good approximation, and by applying the exponential law, their characteristic curve can be defined as follows:

$$R(T) = R_0 \cdot e^{B \cdot (\frac{1}{T} - \frac{1}{T_0})} \quad (5)$$

Where $R_0 = R(T_0)$,
$B = 2{,}000\ldots5{,}000$ K = constant,
T absolute temperature

Here, the characteristic-curve gradient (TK) or the percentage resistance change as a function of temperature, is highly dependent upon the working point. This means that it can only be defined at given points:

$$TK = - B/T^2 \quad (6)$$

It reduces considerably along with increasing temperature, and often the resistance varies by 4...5 powers of ten. For instance, a typical variation would be from several 100 kΩ to 50...60 Ω. This pronounced temperature-dependence means that applications are restricted to a "window" of about 200 K. This though can be selected in the −40...approx. 850 °C range. Tighter toler-

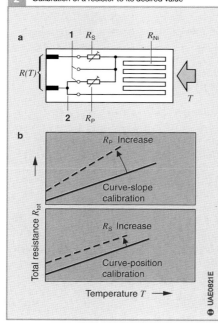

Fig. 2
a Resistance sensor
b Characteristic curves
1 Auxiliary contacts
2 Bridge
R_{Ni} Nickel film resistor
$R_{tot}(T)$ Total resistance referred to temperature
R_P Adjustable parallel resistor
R_S Adjustable series resistor

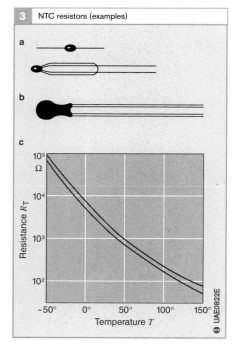

Fig. 3
a Pearl-form
b Disc form
c Characteristc curve with limits of variation

ances of up to ±0.5 K at a given reference point are complied with by using a selection process, or possibly even by grinding in oil, a method which of course has an affect upon costs. The ageing stability of these sensors has been vastly improved compared to earlier versions which means that it is quite possible for the very close tolerances to apply throughout the sensor's useful life.

PTC thin-film/thick-film metallic resistors
The thin-film metallic resistors are integrated with two additional temperature-neutral trimming resistors on a common substrate chip. Since these resistors have a close-tolerance characteristic curve, can be manufactured with long-term stability, and are suitable for fine trimming using laser cuts (Fig. 4), they feature very high accuracy. Thanks to the film technology applied, it is possible to adapt the masking layer as used for protection against the measured medium, and the substrate material, to the particular measuring assignment. The substrate material can be ceramic, glass, or plastic foil, and the masking layer can use plastic moulding, paint, welded foil, glass or ceramic materials. Compared to oxide-ceramic semiconductor sensors, metallic layers feature lower temperature-dependencies, but a more favorable characteristic regarding linearity and reproducibility. The following applies for the computational definition of these sensors:

$$R(T) = R_0 \left(1 + \alpha \cdot \Delta T + \beta \cdot \Delta T^2 + ..\right) \quad (7)$$

Where $\Delta T = T - T_0$ und
$T_0 = 20\,°C$ (reference temperature),
α = Linear temperature coefficient (TC),
β = Quadratic temperature coefficient.

Even though the β coefficient is very small for metals it cannot be ignored completely. This is why the measurement sensitivity of such sensors is usually characterised by means of a mean TC, the "TC 100". The TC 100 corresponds the mean curve gradient between 0 °C and 100 °C (Table 2 and Fig. 5).

Where $TK100 = \dfrac{R(100\,°C) - R(0\,°C)}{R(0\,°C) \cdot 100K}$ \quad (8)

Although platinum (Pt) resistors have the lowest TC, they are not only the most precise resistive temperature sensors, but can also boast the best ageing stability. They are available on the market under the designation "PT 100" or "PT 1,000" (100 Ω or

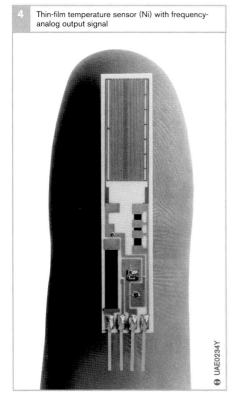

Figure 4: Thin-film temperature sensor (Ni) with frequency-analog output signal

Table 2: Temperature Coefficient TC 100

Sensor material	TK 100 $10^{-3}/K$	Characteristic curve	Measuring range
Nickel (Ni)	5.1	Slightly progressive	−60...320
Copper (Cu)	4.1	Slightly progressive	−50...200
Platinum (Pt)	3.5	Slightly degressive	−220...850

1,000 Ω nominal resistance at a reference temperature of 20 °C) in a variety of different tolerance classes (up to 0.1 °C, Fig. 6). At temperatures up to about 1,000 °C, thick-film Pt sensors are suitable whose Pt layer has been stabilized by special additives.

Thick-film resistors (PTC/NTC)

Thick-film pastes with high specific resistance (low surface area), and with positive and negative temperature coefficients are mainly used as temperature sensors for compensation purposes. They have a non-linear characteristic (which though is not as "bent" as the curve of the solid NTC resistors), and are suitable for laser-trimming. The measuring effect is improved by forming voltage-divider circuits from NTC and PTC materials.

Monocrystalline silicon semiconductor resistors (PTC)

Basically speaking, with temperature sensors of monocrystalline semiconductor materials such as silicon (Si), it is possible to incorporate further active and passive circuit elements on the sensor chip (initial signal conditioning at the measurement point). Due to the close tolerances involved, their production uses the "spreading resistance" principle (Fig. 7a). Current flows through the measuring resistor via a surface point contact and into the Si bulk material from where it spreads in a fan shape to a counter-electrode covering the base of the sensor chip. In addition to the high reproducibility material constants, the high current density behind the contact point (high accuracy thanks to photolithographic production) mainly de-

Fig. 7
a Design
b Characteristic curve
1 Contacts
2 Passivation (Nitride, oxide)
3 Si substrate
4 Counter-electrode without connection
R(T) Temperature-dependent resistor

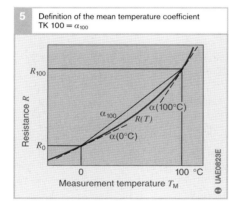

5 Definition of the mean temperature coefficient TK 100 = α_{100}

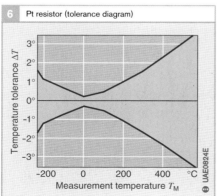

6 Pt resistor (tolerance diagram)

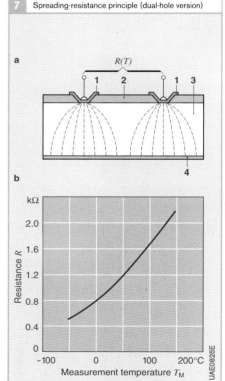

7 Spreading-resistance principle (dual-hole version)

fines the sensor's resistance. In order to make the sensor highly independent of polarity, these sensors are usually series-connected in pairs (dual-hole version, Fig. 7). The base electrode can be in the form of a temperature contact (no electrical function).

Measurement sensitivity is practically double that of a Pt resistor (TK = $7.73 \cdot 10^{-3}/K$). The progressive bend of the temperature curve is more pronounced than on a metallic sensor. The measuring range is limited to approx. +150 °C by the material's intrinsic conductivity (Fig. 7b). There are special versions (Fig. 8) available for operations up to 300 °C.

Thermocouples
Thermocouples are used in particular for measurement ranges >1,000 °C. They rely on the "Seebeck Effect" according to which there is a voltage between the ends of a metallic conductor when these are at different temperatures T_1 und T_2. This "thermo-voltage" U_{th} is solely a function of the temperature difference ΔT between the ends of the conductor (Fig. 9). The following applies:

$$U_{th} = c\,(T_2 - T_1) = c\,\Delta T, \qquad (9)$$

whereby the proportionality constant is material-specific and termed the "Seebeck-Effect".

Since the instrument leads used to measure this voltage across the metallic conductor must themselves be equipped with terminals (for instance made of copper), these are also subject to the same temperature difference, so that unfortunately only the difference between the metallic conductor and the instrument leads is measured. Thermoelectric voltages are always listed based on Platinum as the reference material (Table 3).

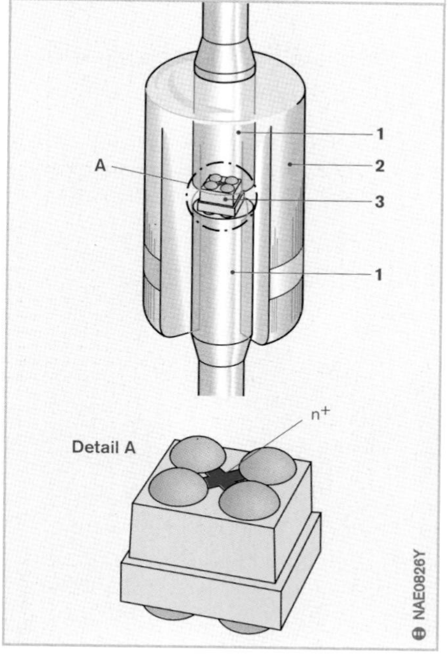

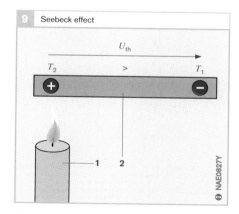

Fig. 8
1 Metal wire
2 Glass
3 Si crystal

Fig. 9
1 Heat source
2 Metallic conductor
+ High,
− Low thermal velocity
   of the electrons
T_2 High temperature
T_1 Low temperature
U_{th} Thermoelectric voltage

8 Spreading resistance sensor (unipolar version for temperatures up to 300 °C)

9 Seebeck effect

3 Thermoelectric voltage U_{th} of a number of metals

Material	Thermoelectric voltage U_{th} mV/100 °C
Constantan	−3.40
Nickel	−1.90
Paladium	−0.28
Platinum	0.00
Copper	+0.75
Manganin	+0.60
Iron	+1.88
Silicon	+44.80

Table 3

Fig. 10
1 Copper/Constantan
2 Iron/Constantan
3 Nickel-chromium/ Nickel
4 Platinum rhodium/ Platinum

Fig. 11
A/B Material pair (thermocouple legs)
1 Measurement point (electrically conductive junction)
2 Connection head
3 Equalising conductor
4 Reference point
5 Connection cable (Cu)
T_M Measurement temperature
T_R Reference temperature
U_{th} Thermoelectric voltage

Fig. 12
a Principle of the thermopile
b Example of application
1 Sensitive surface
2 "Hot" junctions at the measurement temperature T_M
3 "Cold" junctions at the reference temperature T_R
4 Thermopile

In order that the generated voltages are as high as possible, a number of material pairs have established themselves (Fig. 10, e.g. iron/constantan etc.). It is important that the "limbs" of such a thermocouple are joined at the end to which the heat is applied in such a manner that the joint is electrically conductive (by means of twisting, welding, soldering, etc., Fig. 11).

Thermocouples themselves are usually short, and the extensions up to the point where the signal is picked-off can be made with equalising conductors which use the same material pair as in the thermocouple itself. It is important that both free ends of the thermoelement configuration are at the same (reference) temperature, otherwise the temperature difference at the free ends will also be included in the measurement. Thermocouples, therefore, always measure only the temperature difference to a given reference point. If the measuring point's absolute temperature is to be measured, other devices (such as resistive sensors) must be used to also measure the temperature at the reference point.

The characteristic curve of the thermoelectric voltage against temperature is usually not as linear as that given in Equation (8). The sensor signals are usually small, and IC's are used for their amplification and for their linearisation. In order to increase the measurement voltage, it is common practice to connect a number of identical thermocouples in series. These have their "hot" junctions at the temperature to be measured, and their "cold" junctions at the reference temperature (Fig. 12, thermopile).

Although thermocouples are robust (for instance, high-level EMC due to low internal resistance), they are not particularly accurate as a measuring device. Their deviation can

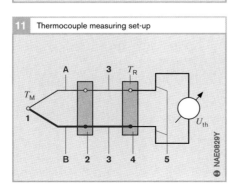

10 Common thermocouples (characteristic curves)

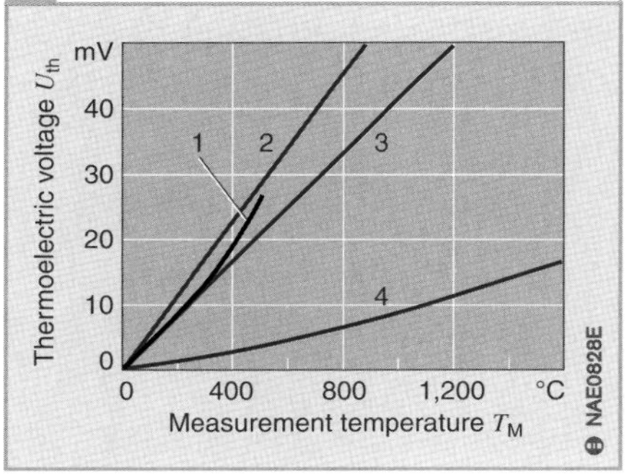

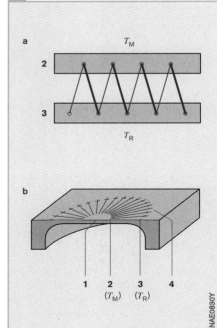

12 Thermocouples connected in series

11 Thermocouple measuring set-up

easily be in the 5°...15° range, and they are not outstanding regarding their resistance to ageing, which means that individual calibration does not result in a permanent improvement of their accuracy.

Of course, thermocouples can be manufactured using both thin-film and thick-film techniques. Metallic films stacked one on top of the other provide for excellent thermal contact, and extemely small thermocouples can be produced by applying microsystem technology. Thermocouples are particularly suitable for use in thermopiles comprised for instance from 50...100 individual thermocouples. They are used in non-contacting radiation thermometers (pyrometers).

Semiconductor barrier layers
Presuming a constant current, the forward voltage of semiconductor barrier layers (Fig. 13) such as those in diodes and in the basis-emitter path of a transistor, demonstrate very good linearity as a function of temperature:

coefficient (NTC). This current increases dramatically with rising temperature. Here, the silicon's intrinsic conductivity limits this sensor's application to temperature ranges <150 °C.

Sometimes, emitter-coupled transistor pairs are used in a similar manner for temperature measurement. With this form of temperature measurement, the ratios of the collector currents to each other represent a very good reproducible measure for the temperature. Usually, an integrated supplementary circuit is used for the "on-chip" conversion to an analog output voltage.

Zener diodes operated in the reverse direction can also be used as highly practical temperature sensors. Their voltage changes are highly dependent upon the Zener voltage. Here, the option exists of various levels of voltage reduction at Zener voltages <4.7 V, and voltage increases at Zener voltages >4.7 V.

Such sensors are often used for temperature compensation on the chip itself.

$$U_F(T) = \frac{k \cdot T}{q} \cdot \ln(\frac{I_F}{I_{sat}} + 1) \quad (10)$$

Where:
$I_{sat} = I_{sat}(T)$ and I_F = constant,
$q = 1.6 \cdot 10^{-19}$ C (elementary charge),
$k = 1.88 \cdot 10^{-23}$ JK^{-1} (Boltzmann's constant),
T Absolute temperature.

It is advantageous here when the sensor directly outputs a voltage which is a function of the temperature. The two-pole sensor is of course dependent upon polarity. Whereas, for all sensors, the forward voltage decreases by almost exactly 2 mV/°C, the absolute voltage at the barrier layer differs considerably from sensor to sensor, and may necessitate additional calibration elements in order for a precise measurement to be made. Above all, it is the temperature-dependent saturation current I_{sat} which is responsible for the negative temperature

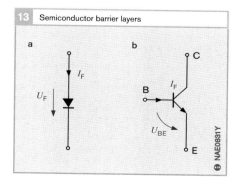

Fig. 13
a Diode
b Transistor
B Base
C Collector
E Emitter
I_F Conducting-state current
U_F Forward voltage
U_{BE} Voltage between base and emitter

Non-contacting temperature measurement, Pyrometry

The radiation emitted by a body is used for the non-contact (or proximity) measurement of its temperature. This radiation is for the most part in the infrared (IR) range (wavelength: 5...20 μm). Strictly speaking, it is the product of the body's radiated power and emission constant. The latter is a function of the material, but for materials which are technically of interest it is usually around 1, although for reflective (applies also to glass) and IR-permeable materials it is far less than 1.

The measuring point is projected onto a heat-sensitive element which, as a result, heats up slightly compared to its surroundings (typically by 0.01...0.001 °C). The element's temperature is a measure for the temperature of the body being measured. A given temperature difference at the object often corresponds to only 1/1,000th of this difference at the measuring point. Nevertheless, the object's temperature can be determined with an accuracy of ±0.5 °C.

Bolometer

The Bolometer is a highly sensitive resistance temperature sensor for measuring minute temperature increases (Fig. 14). A further sensor is needed for measuring the temperature of the sensor housing. If this device is to operate efficiently across a wide temperature range, it is necessary though for both these sensors to feature an extremely high degree of synchronism. The Bolometer housing is therefore usually thermostatically controlled (and well-insulated to the housing) so that the primary detecting element (sensor) always operates at the same temperature.

Thermopile sensor

When a very extensive temperature range is concerned, it is more practical for the temperature difference generated by the radiation from the object to be measured using thermocouples. In order to increase the measuring effect, a number of thermocouples are connected in series to form the so-called thermopile. Such a thermopile sensor (Fig. 15) is inexpensive to manufacture micromechanically. All its "hot" junctions are located on a thermally well insulated thin diaphragm, and all its "cold" junctions are in contact with the thicker chip rim (heat

Fig. 14
1. Lens housing with lens
2. Connections
3. Infrared-detector housing
4. Infrared window
5. Detector

Fig. 15
a Principle of the measuring element
1. Si chip
2. Thermocouples connected in series (i.e. Al/Poly-Si)
3. SiN diaphragm
4. Thermopile junctions
5. Absorber layer

b Sensor configuration
1. Thermocouple
2. "Cold" junction
3. Diaphragm
4. Absorber
5. Heat radiation
6. Electrical connection
7. Si_3N_4 layer
8. SiO_2 layer
9. Heat sink

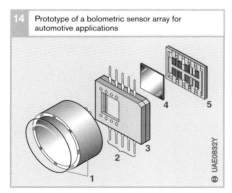

Prototype of a bolometric sensor array for automotive applications

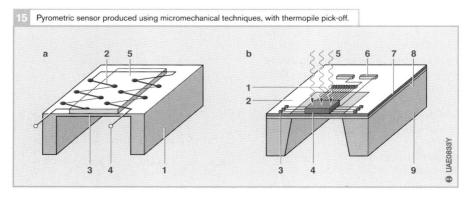

Pyrometric sensor produced using micromechanical techniques, with thermopile pick-off.

sink). Typically, the sensor's response time is approx. 20 ms. Using such a so-called "single-pixel sensor", it is an easy matter to determine the windshield's surface temperature so that measures can be taken to prevent misting-up should the dew point be dropped below.

Single-point sensors, image sensors
If a number of pixels are combined on a single chip (for example, 4x4) to form an array, this provide the basis for a rough form of image analysis (Fig. 16). The pixels must be thermally well insulated from each other, and there must not be too much insensitive surface between them. Due to the fact that each pixel can be electrically addressed, the chip has a large number of connections. On a TO5 casing for example, the ASIC for signal preamplification and series connection of the signal, must be located directly adjacent to the sensor chip. Usually, in the case of thermopile sensors, this ASIC also includes a reference-temperature sensor which measures the pixel's absolute temperature. This permits object temperatures to be measured with an accuracy of approx. ±0.5 K.

An IR imaging-optics system is required for the rough thermal display of an image on the sensor array. The very inexpensive curved mirror is usually ruled out due to it needing too much room. Glass lenses are impermeable for IR light, and plastic lenses are can only be used for operating temperatures of up to approx. 85 °C max. On the other hand, Si lenses are highly suitable for thermal radiation and up to diameters of approx. 4 mm micromechanical techniques can be used to inexpensively manufacture them in the form of a Fresnel or refraction lens. Fitted in the cover of a TO5 casing, these then also serve to protect the sensor against direct damage (Fig. 17). Even though filling the casing with an inert gas improves the crosstalk between the individual pixels, it also negatively affects their response time.

Examples of application
- Intake-air temperature
- Engine temperature
- Passenger-compartment temperature control
- Exhaust-gas high-temperature sensor
- Infrared image sensor

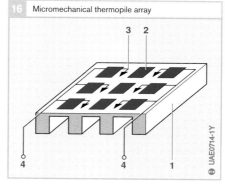

Fig. 16
1   Si chip
2   Pixel
3, 4 Pixel connections

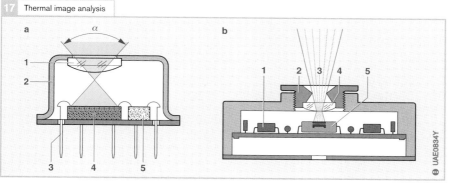

Fig. 17
a   IR image sensor
1   Si IR lens
2   TO5 casing
3   Terminal posts
4   Sensor chip
5   Evaluation ASIC
α   Viewing angle
b   Simple IR camera
1   Electronics
2   Lens system
3   Camera's field of view
4   Si IR lens
5   Sensor array

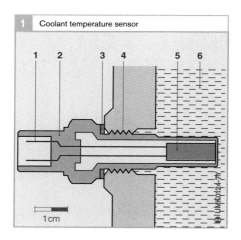

Fig. 1
1. Electrical connections
2. Housing
3. Gasket
4. Thread
5. Measuring resistor
6. Coolant

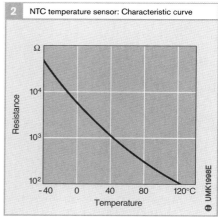

Temperature sensors

Applications

Engine-temperature sensor
This is installed in the coolant circuit (Fig. 1). The engine management uses its signal when calculating the engine temperature (measuring range – 40…+130 °C).

Air-temperature sensor
This sensor is installed in the air-intake tract. Together with the signal from the boost-pressure sensor, its signal is applied in calculating the intake-air mass. Apart from this, desired values for the various control loops (e.g. EGR, boost-pressure control) can be adapted to the air temperature (measuring range –40…+120 °C).

Engine-oil temperature sensor
The signal from this sensor is used in calculating the service interval (measuring range –40…+170 °C).

Fuel-temperature sensor
Is incorporated in the low-pressure stage of the diesel fuel circuit. The fuel temperature is used in calculating the precise injected fuel quantity (measuring range –40…+120 °C).

Exhaust-gas temperature sensor
This sensor is mounted on the exhaust system at points which are particularly critical regarding temperature. It is applied in the closed-loop control of the systems used for exhaust-gas treatment. A platinum measuring resistor is usually used (measuring range –40…+1,000 °C).

Design and operating concept

Depending upon the particular application, a wide variety of temperature sensor designs are available. A temperature-dependent semiconductor measuring resistor is fitted inside a housing. This resistor is usually of the NTC (Negative Temperature Coefficient, Fig. 2) type. Less often a PTC (Positive Temperature Coefficient) type is used. With NTC, there is a sharp drop in resistance when the temperature rises, and with PTC there is a sharp increase.

The measuring resistor is part of a voltage-divider circuit to which 5 V is applied. The voltage measured across the measuring resistor is therefore temperature-dependent. It is inputted through an analog to digital (A/D) converter and is a measure of the temperature at the sensor. A characteristic curve is stored in the engine-management ECU which allocates a specific temperature to every resistance or output-voltage.

The Bosch Boxberg Test Center

The practical tests which are performed at an early stage by the system supplier are an important factor in the development of modern automotive systems. But not all of these tests can be performed on public roads. The new Bosch Test Center in Boxberg between Heilbronn and Würzburg (southern Germany) has been in operation since mid-1998. Here, the most varied automotive systems and their components, no matter whether for driving, safety, comfort, or convenience, can be thoroughly tested on proving grounds covering an area of 92 ha (1 ha = 10^4 m^2). There are seven different track modules available for putting innovative systems through their paces under all possible driving situations and up to the limits of their physical capabilities. And all this, while still ensuring maximum-possible safety for the driver and for the vehicle.

The **rough-road tracks** (1) are designed for speeds up to max. 50 km/h (30 mph) or max. 100 km/h (60 mph). The following special types of track have been built:
- Pothole type
- Washboard type
- Shake and vibration type
- Belgian-block type and
- Tracks with varying degrees of unevenness

The asphalted **hill-climbing tracks** (2) for drive-off and acceleration tests on a hill with gradients of 5 %, 10 %, 15 %, and 20 %, also incorporate tile strips of varying widths that can be flooded with water.

Two **water-wading points** (3) with lengths of 100 m and 30 m (with depths of 0.3 and 1 m respectively) are also available.

Also at the disposal of the testers are **special, floodable tracks** (4) paved as follows:
- Chess-board-pattern pavement (with asphalt and tiles)
- Asphalt
- Tiles
- Blue basalt
- Concrete
- Aquaplaning tracks and
- Trapezoidal blue-basalt track

The 300 m diameter **vehicle-dynamic test-pad** (5) for special cornering tests is paved with asphalt, and can be partially flooded in order to simulate ice and aquaplaning conditions. The test pad is surrounded by a tire barrier as a precautionary measure for the safety of the drivers and vehicles.

The **high-speed oval** (6) is provided with three lanes and can be used by passenger cars as well as commercial vehicles. Design is such that tests can be performed at speeds of up to 200 km/h (120 mph.)

The **handling track** (9) is comprised of two different stretches of road: One for max. 50 km/h (30 mph) and the other for max. 80 km/h (50 mph). Both tracks feature curves with differing radiuses and inclinations. This track is mainly used for testing vehicle-dynamics systems (ESP).

View of the test-track section modules

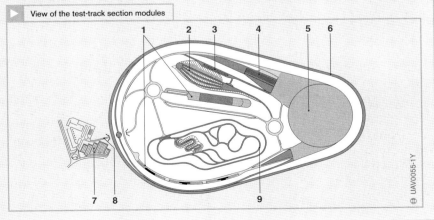

Fig. 1
1. Rough-road track
2. Hill-climbing track
3. Water-wading points
4. Special, floodable tracks
5. Vehicle-dynamics test pad
6. High-speed oval
7. Buildings
   - Workshops
   - Offices
   - Test benches
   - Laboratories
   - Filling station and
   - Welfare and medical facilities
8. Access road
9. Handling track

Prospects

Development trends

In future, it is to be expected that considerably more automotive sensors will be developed for the vehicle's immediate and more remote surroundings (Fig. 1) than for its drivetrain. These developments include the sensor technology which measures the vehicle's movement (kinematics), both as an entity on its own and as a moving component in the flow of traffic. Also included are sensors which directly register the contact of the vehicle's wheels with the road surface. Of decisive importance here are a number of extensive, new assignment areas:

- Guiding and steering the vehicle to a destination (navigation)
- Reliable, safe vehicle guidance and steering by means of electronic assistance systems (up to the limits of physical possibilities)
- Extended passenger protection systems with higher intelligence levels and preemptive effects, up to as far as the total prevention of collision as the optimum objective
- Safeguarding the vehicle by way of theft-deterrent systems using biometric sensors (e.g. passenger recognition by way of fingerprint)

Such sensors also form the basis for semi-autonomous (partially independent) vehicle driving, with full autonomy being the long-term objective.

Sensor examples

Image sensors (video)

In particular those sensors which generate images on the basis of visible light or infrared light will come to the forefront in ever increasing numbers. These will serve for passenger-compartment monitoring, and for observation outside the vehicle will be aligned to the vehicle's surroundings.

All of these sensors have one objective in view, and that is the simulation of the superior capabilities of the human eye and its mental recognition capabilities (of course, only to a very modest degree at first). It is certain that in the foreseeable future, the costs for image sensors, and the associated very-high-performance processors needed for the interpretation of a given scene, will become interesting from the automotive-applications viewpoint.

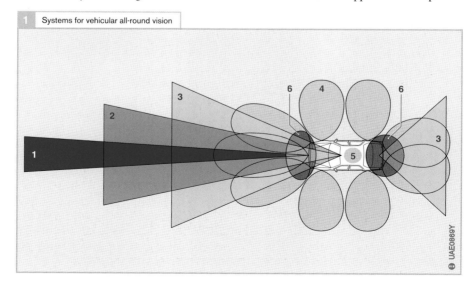

Fig. 1
1 Distant-zone radar 77 GHz, series production (distant zone: ≤120 m)
2 Distant-zone/ near-range infrared viewer (nightviewing)
3 Video coverage of the vehicle's immediate vicinity (nightviewing, mid-range ≤40 m)
4 Near-range radar 24 GHz (near range: ≤10 m)
5 Passenger-compartment video
6 Ultrasonic, series production (very near range ≤1.5 m)

1 Systems for vehicular all-round vision

In contrast to the human eye, common image sensors are sensitive in the near IR range (wavelength approx. 1 μm). With appropriate non-visible IR illumination, therefore, all imaginable applications in the vehicle become feasible, including nighttime operation.

In future, image sensors will be able to play a highly variegated role for the observation of the vehicle's interior (seating position, forward shift in case of a crash, size of seat occupants etc.), and of the vehicle's surroundings (vehicle tracking, collision prevention, parking and back-up aids, traffic-sign recognition).

Image sensors are a special case of "multi-sensor structures" formed from light-sensitive elements (pixels) which are arranged as matrix or line arrays and which receive their light through a conventional imaging-optics system. With the Si image sensors (**CCD** Charge-Coupled Devices, Figs. 2 and 3) available at present, the light entering through a transparent electrode generates charge carriers proportional to the light intensity and the exposure time. These are collected in a "potential layer" (Si-SiO$_2$ boundary layer). Further electrodes are used to transfer these charges into an opaque zone

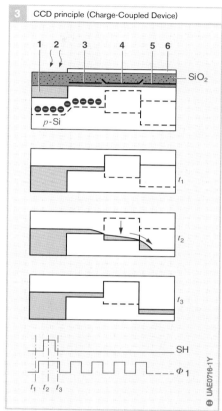

Fig. 3
1 Photodiode
2 Light
3 Storage elecrode
4 Shift gate
5 Transfer electrode
6 Optical masking

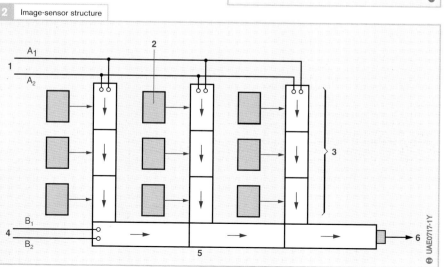

Fig. 2
1 Column clock pulse A_1/A_2
2 Photosensors
3 CCD array
4 Line clock pulse B_1/B_2
5 Output register
6 Video output

and by means of "analog" shift registers (bucket-brigade principle) are then transferred line by line into an output register which is then read out serially at a high clock-pulse rate.

Whereas, due to their limited dynamic light/dark response (50 dB), their read-out time, and their temperature range (<50 °C), CCD sensors are unsuitable for use in the automobile, innovative "smart" image sensors based on CMOS technology are apparently 100% suitable for such applications. Here, as well as having a dynamic response of 120 dB, the logarithmic light/signal curve which is possible corresponds to that of the human eye. This, for instance, not only makes an aperture control superfluous, but also provides for constant contrast resolution throughout the complete brightness range. These sensors permit random access to the individual pixels while at the same time permitting higher levels of sensitivity (higher readout rate). The first steps in pre-processing the signals on the image-sensor chip have already been implemented.

Optical sensors
Simple, optical sensors for contingency-triggered automatic cleaning of the vehicle's windshield or of the headlamp lenses are also aligned to the environment outside the vehicle.

Rain sensors
The rain sensor detects rain drops on the vehicle's windshield and triggers the operation of the windshield wipers. The relieves the driver of a number of operations that are needed with conventional wiper systems, and thus enables him/her to concentrate better on the road. Nevertheless, manual control is retained as an additional intervention. If the automatic control is required, the driver must activate it after starting the vehicle.

The rain sensor comprises an optical transmit/receive path (similar to the dirt sensor). An LED emits light which is coupled into the windshield at a given angle.

This light is reflected from the dry outside surface (total reflection) and reaches the receiver (photodiode) which is also aligned to the windshield at an angle. If there are water droplets on the windshield, a considerable portion of the light is refracted from them and is lost so that the signal received by the photodiode is correspondigly weaker. As from a certain level, the wiper also switches on automatically when there is dirt on the windshield. On newer sensor versions, infrared light is used instead of the visible light commonly employed.

The sensor controls the speed of the windshield wipers as a function of the amount of rain measured on the windshield. Together with the electronically controlled wiper drive, infinitely-variable wiper speeds are possible during interval operation. For instance, if the windshield is suddenly deluged by a gush of water when passing a truck, the system automatically switches on at top speed.

The rain sensor can also be used for closing the windows and the sunshine roof. Provided a second sensor is fitted, it can also control the vehicle headlights. When there is insufficient light, or when the vehicle enters a tunnel, it automatically switches on the headlamps without the driver having to do anything. It is even conceivable that the rain sensor's signals can be used to inform traffic telematics systems about the actual weather situation on a particular stretch of road.

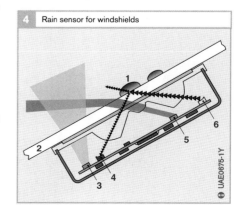

4 Rain sensor for windshields

Fig. 4
1 Raindrops
2 Windshield
3 Ambient-light sensor
4 Photodiode
5 Light sensor, aligned to far distance
6 LED

Dirt sensors

The dirt sensor (Fig. 5) detects the degree of dirt on the headlamp lenses and triggers an automatic cleaning process for them. The sensor's reflected-light barrier comprises a light source (LED) and an opto-receiver (phototransistor). It is located on the lens inner surface inside the cleaning area traversed by the headlamp-wiper blade, but not within the direct beam path of the light from the bulb. When the headlamp lens is clean or covered by raindrops, the measuring light (which is in the near IR range) passes through the headlamp lens practically unhindered, and only a negligible portion is reflected back to the opto-receiver. On the other hand, if the transmitted light hits dirt particles on the outside lens surface it is caused to scatter and reflects back to the opto-receiver. The degree of scatter is proportional to the degree of dirt, and above a certain level automatically triggers the headlamp wipers.

Near-range radar (24 GHz)

In case of broadside collisions, and frontal collisions on compact-class automobiles, the time available for triggering the safety and restraint systems is extremely short (≤5 ms), and there is very little space between the passengers and the intruding vehicle. And in contrast to frontal collisions on larger vehicles there are very few vehicle components available for distortion and for absorbing the crash energy. It would therefore be of considerable advantage to be able to reliably forecast the severity of the collision from the very first moment of impact. To this end, near-range radar sensors are being worked on which will be installed all round the vehicle to provide 100% coverage of its surroundings.

Further sensor systems

For the new assignments dealt with above, work is proceeding on the following sensor systems:
- Steering-torque sensing (electromotive power steering, "steer-by-wire" system)
- Drive-torque sensing (misfire detection, load signal)
- Braking-force sensing (electromotive braking systems, "brake-by-wire")
- Passenger protection (AOS **A**utomotive **O**ccupancy **S**ensing, **O**ut **o**f **P**osition sensing (OOP), passenger weight)
- Deformation sensors for broadside-collision sensing
- Pedestrian detection for triggering of engine-hood protective systems
- Registration of wheel forces (e.g. force-sensor technology integrated in the wheel bearing, and friction-coefficient potential)
- Liquid-measurement sensors (liquid levels, condition/quality of engine oil etc.)
- "Autonomous", that is 100% non-contacting sensors which can in some cases be scanned by radio, and which need no contacts for energy supply (plug-in contacts are still the most frequent causes of malfunction in the vehicle)

It is therefore obvious that the multiplicity of new electrical and electronic systems being introduced in the vehicle necessitates the development of a wide variety of new sensors. Of course, it still remains the objective that once they have completed an economically acceptable service life, existing sensors are replaced with new, more cost-efficient and better sensors produced using new technologies.

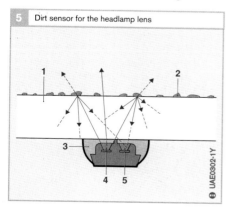

Fig. 5 Dirt sensor for the headlamp lens

Fig. 5
1. Lens
2. Dirt particles
3. Sensor housing
4. Transmitter
5. Receiver

Sensor-signal processing

Signal conditioning (Evaluation IC)

The sensor signals must be conditioned before they can be evaluated digitally (refer to the Section "Data Processing"). As far as required, this signal conditioning can include the following functions:
- Amplification (AC, DC)
- Rectification (also phase-synchronised)
- Threshold-value evaluation (also variable thresholds, pulse-shaping)
- Voltage/frequency conversion, pulse-duration modulation (pdm)
- Frequency filtering including interference-protection measures
- Analog/digital (AD) and digital/analog (DA) conversion
- Calibration of offset and amplification (characteristic curve in general), analog, digital (including E^2PROM)
- Linearisation
- Calibration of temperature compensation (analog, digital)
- Automatic zero reset, possibly also with calibration during operation
- Self-monitoring (On-Board-Diagnosis (OBD), diagnosis output) and test functions
- Control of the servo-controlled sensors (compensation method)
- Generation of AC voltage for carrier-frequency sensor systems
- Power-supply stabilisation
- Short-circuit-proof/overvoltage-proof output and driver stages
- Signal multiplexers, analog and/or digital serialisation of the signals, coding, including fault detection
- Bus interface (e.g. CAN) etc.

These functions are all available in the form of ASICs (Application-Specific Integrated Circuits). These circuits are tailor-made for the particular sensor application, and can either be installed locally (at the sensor) or at the ECU. In some cases, the functions are divided between both sides as far as this is expedient. Local integration of the circuit at the sensor (Fig. 1, integration stages 1 to 3) has the advantage that sensor and signal conditioning can be calibrated and compensated together. These then form an inseparable unit which is highly interference-proof and which must be replaced completely if one of the stages should fail.

Whereas previously, the functions described above were in some cases implemented as separate circuits (e.g. CMOS-IC for signal processing, bipolar IC as the interference-proof driver stage), present-day "mixed" technologies (e.g. BICMOS, BCD) also permit the integration on a single chip of the complete function including any digital, programmable memory-location cells which might be necessary (PROM). Basically, in practically all cases, monolithic integration of the sensor and the signal processing is possible (for instance Si manifold-pressure sensors and Hall-effect sensors). The euphoria which initially accompanied this integration has now given way to more sober considerations which take economic aspects more into account. At present, therefore, other state-of-the-art integration methods are in use which are more cost-effective (e.g. thick-film hybrids, combined "Lead frame" and combined chip housing). Such a concept, which in effect can be regarded as being modular, is also considerably more flexible since it can be more easily adapted to new assignments.

Considering the fact that the majority of sensors need an ASIC in order for them to operate correctly, and for their defined characteristics to apply, the wide variety of such signal-processing ASICs that have been created at Bosch represents a "treasure" of immense value. When sensors are produced not only for "in-house" use but also for sale outside, they should as far as possible only be marketed together with these signal-processing circuits.

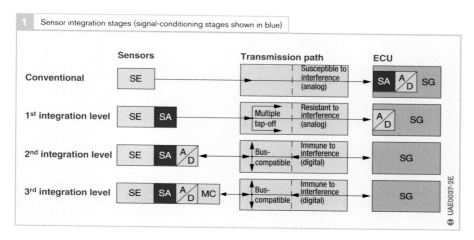

Fig. 1
SE Sensor(s)
SA Signal conditioning (analog)
A/D Analog/digital converter
SG ECU (digital)
MC Microcomputer

Examples of application

Originally, ASICs were conceived primarily for installation in the ECU. This meant though, that requirements which in part were based on special, and in some cases individual, sensor features could not be taken into account. As a rule, the ASICs are therefore now designed for direct installation on the sensor, and are able to store individual parameters for calibration and compensation, and use these to implement corrective measures in the sense of an "intelligent" sensor. Below, we deal with just a few examples of these ASICs.

ASIC CC212

The CC212 is an ASIC for the short-circuiting-ring sensor (for instance, the half-differential version) used for the measurement of displacement, travel, and angle. Due to their considerable measuring effect and their moderate operating frequency (5...50 kHz) short-circuiting-ring sensors do not necessarily need local electronic circuitry. The ASIC CC212 is therefore installed in the ECU where it combines all electronic functions as needed in the ECU, for instance, for the triggering and evaluation of the sensors for Electronic Diesel Control (EDC).

Since this ASIC is already installed in the ECU, the advantages of individual curve and temperature compensation aligned to the special sensor version were dispensed with on purpose. This electronic circuitry could have been installed directly on the sensor and could have simplified the sensor as a result (refer to ASIC CC400). This automatically detects short circuits and cable breaks in and at the sensor, and suppresses any interference peaks which might be present on the sensor output lead.

The ASIC drives a half-differential short-circuiting-ring sensor as an AC voltage divider (10 kHz), whereby the end of the divider is fed with a constant amplitude. Phase-opposed voltage is applied to the other end of the divider, the amplitude of which is regulated (closed-loop controlled) until the output at the divider's pick-off reduces to zero. The closed-loop-controlled output voltage is at the same time the analog output signal (Fig. 2, next page).

Sensor-signal processing Examples of application

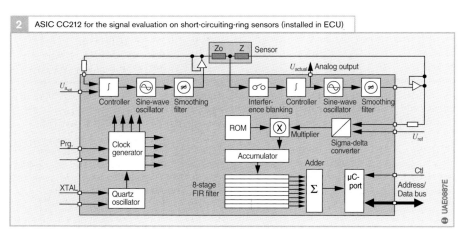

2 ASIC CC212 for the signal evaluation on short-circuiting-ring sensors (installed in ECU)

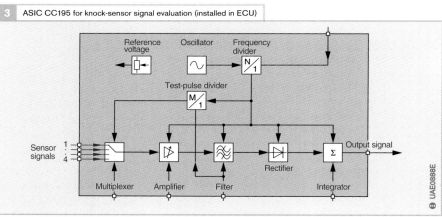

3 ASIC CC195 for knock-sensor signal evaluation (installed in ECU)

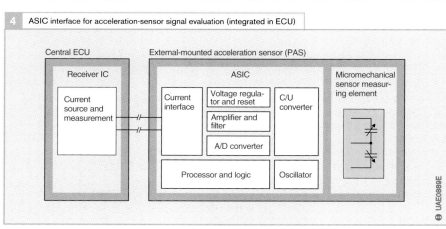

4 ASIC interface for acceleration-sensor signal evaluation (integrated in ECU)

ASIC CC195

The CC195 is a knock-sensor ASIC. Knock sensors are mounted directly on the engine where they detect acceleration signals in the form of structure-borne noise. The combustion-knock signals are typically in the 5...15 kHz range, and must be filtered out. A time-window control suppresses precisely that phase of the working cycle during which the signal can theoretically occur and allocates this to a given cylinder. The signal occurring in the critical frequency range is rectified and averaged and then evaluated by the ECU. As a result, the ECU shifts the ignition point until knock stops. The ASIC which is responsible for the above functions is inside the ECU and is able to evaluate the signals from up to 4 knock sensors (Fig. 3).

Interface ASIC

In case of a *frontal* collision, acceleration sensors trigger the vehicle's restraint systems in order to protect the passengers. They are located directly in the airbag ECU which is usually installed in the vehicle's console.

The triggering of the protection system for *side-on* collisions must be much faster, and it is therefore necessary to locate the respective acceleration sensors at the vehicle's periphery (for instance at the chassis cross member). From here, the sensors transmit their signals digitally to the central ECU through a two-wire connection.

As its name implies, the PAS (**P**eripheral **A**cceleration **S**ensor) is located at the periphery of the vehicle, and in a two-chip concept contains the capacitive acceleration sensor itself as well as its triggering circuitry and evaluation electronics (Fig. 4). The ASIC used here, incorporates not only the sensor-triggering circuitry and sensor-signal evaluation, but also triggering for the output interface and sensor self-monitoring.

ASIC CC340

The CC340 is a universal, digitally controllable signal amplifier (Fig. 5, three-chip concept). This module applies CMOS technology and is in fact an analog DC difference amplifier. Depending upon a temperature signal, it can simultaneously control the offset and the amplification by means of a correction circuit.

Using this ASIC therefore, and provided that the bridge operating temperature t is precsiely monitored, it is possible to precisely amplify the output voltage of the pressure-sensor DMS bridge. Here, on the one side the advantages of simple broad-band, no-delay analog amplification are retained, and on the other, temperature correction takes place in a fully digitised circuit stage which need not be subjected to any demands

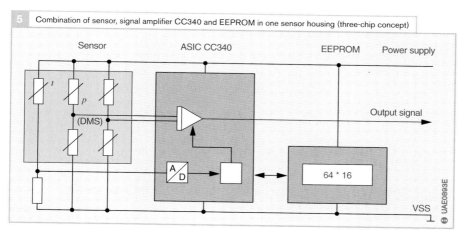

5 Combination of sensor, signal amplifier CC340 and EEPROM in one sensor housing (three-chip concept)

Fig. 5
p Pressure
t Temperature

at all regarding high working speeds and high resolution.

Using 6 bits (64 stages), the temperature signal is roughly digitised. With this digital word, an offset factor and an amplification factor are read out of an EEPROM. These each comprise 8 bits, and can be applied to the amplifier so that extremely non-linear temperature responses can be corrected across a wide range. A selectable basic amplification and a basic offset are also stored in the EEPROM.

When the design of this ASIC is updated, both the EEPROM and the bipolar protective circuit could be integrated in a single chip (Fig. 6). In the first versions of this ASIC, these had to be separated in the three-chip concept in line with the state-of-the-art at that time.

ASIC CC400

Using the CC400, it is possible to digitally evaluate inductive sensors using calibration and correction functions that have been specifically aligned to the special sensor design (Fig. 7). With this ASIC, it is possible to vastly improve the characteristics not only of micromechanical sensors but also of macro-mechanical versions (for instance, inductive or capacitive sensors). This takes place by integrating the electronics, a step which at the same time leads to sensor simplification.

In the measuring systems, using a simple self-oscillating circuit, the inductance L of a travel or angle sensor, and its operating-temperature as registered by an NTC temperature sensor, are converted into an easily digitised period of oscillation. By means of these two values, the relevant practically faultless measurement values are then read out of a two-dimensional "look-up table".

Only a few values are stored on the ASIC in order to get by with very little memory space. When necessary, the ASIC performs a linear interpolation between these values. The values in the "look-up table" are calculated in a once-only calibration process and stored in the (EE)PROM of the ASIC. Total measure and calculation time is less than 0.5 ms.

Thanks to the measured-value correction as described above, on short-circuiting-ring sensors for instance, the linearisation contour for the laminated iron core can be dispensed with as can a second, fixed-adjusted reference system. This serves to simplify the sensor somewhat. Nonetheless, the CC400

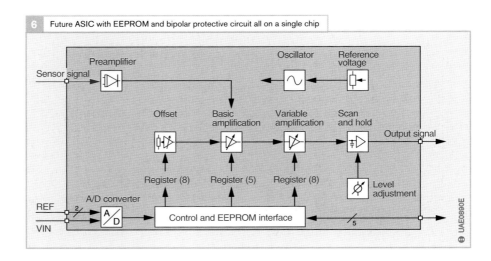

6 Future ASIC with EEPROM and bipolar protective circuit all on a single chip permits a considerable increase in sensor accuracy compared with conventional evaluation. Across the complete temperature range, this applies up to 0.1% of the measuring range. The only principle limitation here is the ageing stability of the travel/angle sensors and of the temperature sensor used.

Fig. 8 shows an example of a capacitive acceleration sensor with evaluation circuit.

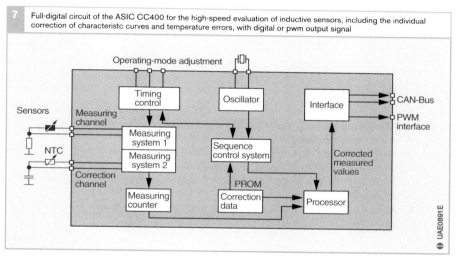

7 Full-digital circuit of the ASIC CC400 for the high-speed evaluation of inductive sensors, including the individual correction of characteristc curves and temperature errors, with digital or pwm output signal

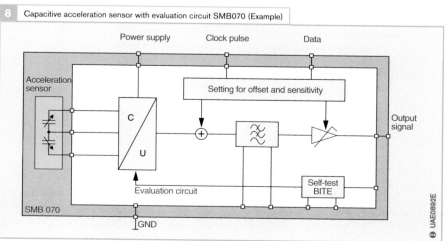

8 Capacitive acceleration sensor with evaluation circuit SMB070 (Example)

Data processing in the vehicle

Requirements

Highly sophisticated state-of-the-art open-loop and closed-loop control concepts are essential for meeting the demands for function, safety, environmental compatibility and comfort associated with the wide range of automotive subsystems installed in modern-day vehicles. Sensors monitor the reference and controlled variables, which an electronic control unit (ECU) then converts into the signals required to adjust the final controlling elements/actuators. The input signals can be analog (e.g. voltage characteristic at pressure sensor), digital (e.g. switch position) or pulse-shaped (i.e. information content as a function of time; e.g. engine-speed signal). These signals are processed after being conditioned (filtering, amplification, pulse shaping) and converted (analog/digital); digital signal-processing methods are preferred.

Thanks to modern semiconductor technology, powerful computer units, with their accompanying program and data memories, and special peripheral circuitry, designed specifically for real-time applications, can all be integrated on only a few chips.

Modern vehicles are equipped with numerous digital control units (ECUs), e.g. for engine management, ABS, and transmission-shift control. Improved performance and additional functions are obtained by synchronizing the processes controlled by the individual control units, and by adapting (in real time) their respective parameters to each other. An example of this type of function is a traction control system (TCS) which reduces the driving torque when the drive wheels spin.

Up to now, data between the control units (in the example cited above, ABS/TCS and engine management) has been exchanged mostly through separate lines. However, this type of point-to-point connection is only suitable for a limited number of signals. The data-transmission potential between the individual ECUs can be enhanced by using a simple network topology designed specifically for serial data transmission in automotive applications.

Microcomputer

The microcomputer comprises both the central processing unit (CPU) for processing arithmetic operations and logical relationships, and special function modules to monitor external signals and to generate the control signals for external servo elements. These peripheral modules are largely capable of assuming complete control of real-time operations. The program-controlled CPU could only discharge these at the price of both additional complication and curtailment in the number of functions (e.g. determining the moment at which an event occurred).

Computing power

Apart from the architecture (e.g. accumulator, register machine) and the word length (4 ... 32 bits), the product of the internal clock frequency and the average number of clock pulses required per instruction determines the CPU's power:
- Clock frequency: 1 ... 40 MHz (typical)
- Clock pulses per instruction:
  1 ... 32 pulses (typical), depending on the CPU's architecture and the instruction (e.g. 6 pulses for addition, 32 pulses for multiplication)

Electronic control unit (ECU)

Digital input signals
Register a switch position or digital sensor signals (e.g. rotational-speed pulses from a Hall-effect sensor).
Voltage range: 0 V to battery voltage.

Analog input signals
Signals from analog sensors (lambda sensor, pressure sensor, potentiometer).
Voltage range: Several mV up to 5 V.

Pulse-shaped input signals
Signals from inductive rpm sensors. After signal conditioning, they are further processed as digital signals.
Voltage range: 0.5 V to 100 V.

Data processing in the vehicle Electronic control unit (ECU) 487

1 Microcomputer

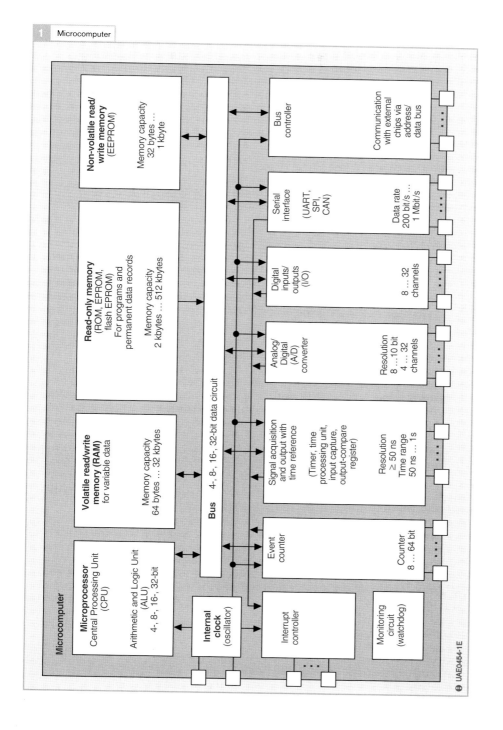

Signal conditioning

Protective circuits (passive: R and RC circuits; active: special surge-proof semiconductor elements) are used to limit the voltage of the input signals to acceptable levels (microcomputer operating voltage). Filters remove most of the superimposed noise from the useful signals, which are then amplified to the microprocessor's input voltage. Voltage range: 0 V to 5 V.

Signal processing

ECUs usually process signals in digital form. Rapid, periodic, real-time signals are processed in hardware modules specifically designed for the particular function. Results, e.g. a counter reading or the time of an event, are transmitted in registers to the CPU for further processing. This procedure substantially reduces the CPU's interrupt-response requirements (μs range).

The amount of time available for calculations is determined by the open-loop or closed-loop control system (ms range).

The software contains the actual control algorithms. Depending on the data, an almost unlimited number of logic operations can be established and data records stored and processed in the form of parameters, characteristic curves and multidimensional program maps.

Output signals

Power switches and power-gain circuits amplify the microprocessor's output signals (0 V...5 V, several mA) to the levels required by the various final-controlling elements/actuators (battery voltage, several A).

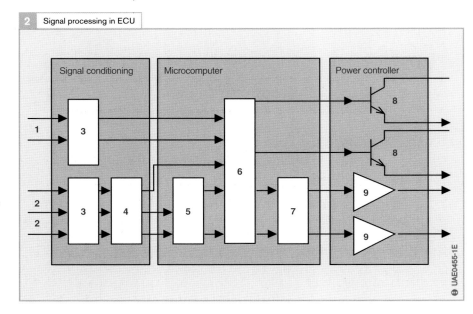

Fig. 2
1. Digital input signals
2. Analog input signals
3. Protective circuit
4. Amplifier, filter
5. A/D converter
6. Digital signal processing
7. D/A converter
8. Circuit-breaker
9. Power amplifier

Complete system

Logistical concept (CARTRONIC)
This concept divides the vehicle's complete electrical system into conveniently dimensioned subsystems. Units with closely-coupled functions (that is, units with high rates of mutual data exchange) are combined in a sub-network. Although this logistical concept results in sub-networks with varying requirements on transmission capacity, demands on data exchange do not vary.

Topology
At the logical level, all the known communications systems developed for automotive applications are based on a single serial connection of the ECUs. The physical layout employs one-wire or differential two-wire interfaces in bus form to interconnect the control units.

Protocol
The protocol consists of a number of a specific collection execution statements which are used to control data communications between the individual control units. Procedures have been laid down for bus access, message structure, bit and data coding, error recognition and response, and the identification of faulty bus users (CAN).

Transmission speed
Multiplex bus: 10 kbit/s...125 kbit/s,
Drivetrain bus: 125 kbit/s...1 Mbit/s,
Telecommunications bus:
10 kbit/s...125 kbit/s.

Latency time
Latency time is defined as the time that elapses between the transmitter's send request and the target station's receipt of the error-free message.
Multiplex bus: 5 ms...100 ms,
Drivetrain bus: 0.5 ms...10 ms,
Telecommunications bus: 5 ms...100 ms.

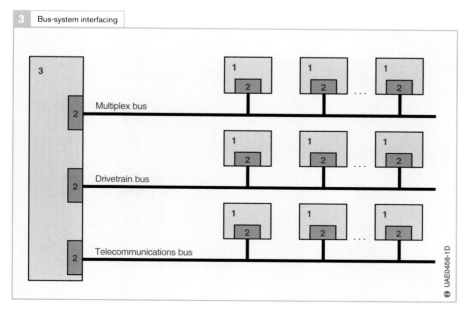

Fig. 3
1 ECU
2 Bus controller
3 Gateway

Data transfer between automotive electronic systems

Today's vehicles are being equipped with a constantly increasing number of electronic systems. Along with their need for extensive exchange of data and information in order to operate efficiently, the data quantities and speeds concerned are also increasing continuously.

For instance, in order to guarantee perfect driving stability, the Electronic Stability Program (ESP) must exchange data with the engine management system and the transmission-shift control.

System overview

Increasingly widespread application of electronic communications systems, and electronic open and closed-loop control systems, for automotive functions such as

- Electronic engine-management (EDC and Motronic)
- Electronic transmission-shift control (GS)
- Antilock braking system (ABS)
- Traction control system (TCS)
- Electronic Stability Program (ESP)
- Adaptive Cruise Control (ACC) and
- Mobile multimedia systems together with their display instrumentation

has made it vital to interconnect the individual ECUs by means of networks.

The conventional point-to-point exchange of data through individual data lines has reached its practical limits (Fig. 1), and the complexity of current wiring harnesses and the sizes of the associated plugs are already very difficult to manage. The limited number of pins in the plug-in connectors has also slowed down ECU development work.

To underline this point:
Apart from being about 1 mile long, the wiring harness of an average middle-class vehicle already includes about 300 plugs and sockets with a total of 2,000 plug pins. The only solution to this predicament lies in the application of specific vehicle-compatible Bus systems. Here, CAN has established itself as the standard.

Serial data transfer (CAN)

Although CAN (Controller Area Network) is a linear bus system (Fig. 2) specifically designed for automotive applications, it has already been introduced in other sectors (for instance, in building installation engineering).

Data is relayed in serial form, that is, one after another on a common bus line. All CAN stations have access to this bus, and via a CAN interface in the ECUs they can receive and transmit data through the CAN bus line. Since a considerable amount of data can be exchanged and repeatedly accessed on a single bus line, this networking results in far fewer lines being needed.

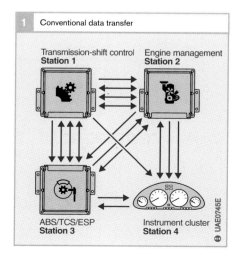

1 Conventional data transfer

Transmission-shift control Station 1
Engine management Station 2
ABS/TCS/ESP Station 3
Instrument cluster Station 4

Applications in the vehicle
For CAN in the vehicle there are four areas of application each of which has different requirements. These are as follows:

Multiplex applications
Multiplex is suitable for use with applications controlling the open and closed-loop control of components in the sectors of body electronics, and comfort and convenience. These include climate control, central locking, and seat adjustment. Transfer rates are typically between 10 kbaud and 125 kbaud (1 kbaud = 1 kbit/s) (low-speed CAN).

Mobile communications applications
In the area of mobile communications, CAN networks such components as navigation system, telephone, and audio installations with the vehicle's central display and operating units. Networking here is aimed at standardizing operational sequences as far as possible, and at concentrating status information at one point so that driver distraction is reduced to a minimum. With this application, large quantities of data are transmitted, and data transfer rates are in the 125 kbaud range. It is impossible to directly transmit audio or video data here.

Diagnosis applications
The diagnosis applications using CAN are aimed at applying the already existing network for the diagnosis of the connected ECUs. The presently common form of diagnosis using the special K line (ISO 9141) then becomes invalid. Large quantities of data are also transferred in diagnostic applications, and data transfer rates of 250 kbaud and 500 kbaud are planned.

Real-time applications
Real-time applications serve for the open and closed-loop control of the vehicle's movements. Here, such electronic systems as engine management, transmission-shift control, and electronic stability program (ESP) are networked with each other. Commonly, data transfer rates of between 125 kbaud and 1 Mbaud (high-speed CAN) are needed to guarantee the required real-time response.

Bus configuration
Configuration is understood to be the layout and interaction between the components in a given system. The CAN bus has a linear bus topology (Fig. 2) which in comparison with other logical structures (ring bus and/or star bus) features a lower failure probability. If one of the stations fails, the bus still remains fully accessible to all the other stations. The stations connected to the bus can be either ECUs, display devices, sensors, or actuators. They operate using the Multi-Master principle, whereby the stations concerned all have equal priority regarding their access to the bus. It is not necessary to have a higher-order administration.

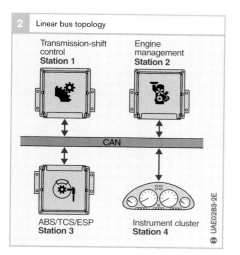

2 Linear bus topology

Index of technical terms

Technical Terms

0-Ampere speed (n0), 155
100% maintenance-free battery, 102
3-phase AC, 122

A

Absolute rotating-speed measurement, 395f
AC bridge circuit, 123
ACC Adaptive Cruise Control, 385
ACC sensor ECU, 373
Acceleration sensor, 317
Acceleration sensors, 414, 419, 422
Acceleration sensors, 412ff
Accelerator-pedal sensors, 378f
Accumulator architecture, microcontrollers, 322
Acid values (diluted sulfuric acid), starter batteries, 210
Acoustic sensors (ultrasonic), 372
Active material, starter batteries, 93
Active wheel-speed sensors, 403
Actuators, 288
Adjustment mechanisms, headlight levelling control, 257
Aerotwin wiper blade, 280
Aiming conditions, headlamps (Europe), 252
Aiming conditions, headlamps (North America), 252
Aiming information, headlamps (Europe), 253
Air-quality sensors, 457
Air-temperature sensor, 474
Airflow measurement, 442
Alternator, 155
Alternator circuitry, 156f
Alternator current output, 12
Alternator design, 130
Alternator drive, 160
Alternator installation and drive, 159
Alternator operation in the vehicle, 158ff
Alternator output power, 118
Alternator versions, 132ff
Alternators, 116-120
American sealed-beam headlamp system, 246
AMLCD, 270f
Analog circuits, ICs, 305
Analog input signals, ECU, 486
Anechoic chamber, EMC, 74
anemometers, 442
Angle-measuring (torsion-measuring) sensors, 438

Angle-of-rotation sensors in the range above 180°, 366
Angle-of-rotation sensors in the range up to 180°, 365
Application, electrically powered vehicles, 110
Applications in the vehicle, sensors, 350
Approval codes and symbols, 223f
ASIC CC195, 483
ASIC CC212, 481
ASIC CC340, 483f
ASIC CC400, 484
Asymmetrical projection patterns, head lamps, 227
Asynchronous drive, 112
Atmospheric-pressure sensor, 428
Automatic headlight-levelling control, 256
Automotive lamps, 233ff
Automotive lighting technology, 222ff
Automotive microelectronics, 288ff
Automotive sensors, 350
auxiliary diodes, 124
Auxiliary diodes at the star (neutral) point, 156
Auxiliary driving lamps, 259
Auxiliary loads, electrical, 83
Axle sensors, 382

B

Backup (reversing) lamps, 262
Backup mode, battery charger, 217
Barber Pole, sensors, 369
Basic physical principles, 121ff
Batteries, conventional, 92
Batteries, maintenance-free, 94
Battery care, starter batteries, 211
Battery case, starter batteries, 92
Battery changeover 12/24 V, 106
Battery changeover relay, 108
Battery charger, service technology, starter batteries, 213
Battery charging relay, 108
Battery charging voltage, vehicle electrical systems, 14
Battery construction, starter batteries, 92
Battery covers, 101
Battery cutoff relay, 108
Battery design, starter batteries, 80
Battery disposal, 105
Battery electrolyte, starter batteries, 210
Battery for extended current output, 104

Battery history, 109
Battery master switch, 107
Battery plates, starter batteries, maintenance-free, 94
Battery relay, 107
Battery selection, 104
Battery storage, 212
Battery systems, drive/traction batteries, 113
Battery types, 99
Battery versions, 102
Battery-cable terminals, 101
Battery-charge equation, vehicle electrical systems, 15
BCI method, 76
Belt drive, 160
Benz, 164
Bernoulli's equation, 444
Bi-Litronic "Projection", 252
Bi-Litronic "Reflection", 251f
Bipolar transistors, 301
Block diagram, 35
Bolometer, 472
Boost charging, starter batteries, 214
Bosch part number, TTNR, starter batteries, 99
Bosch Service, 83, 105, 208
Brake-fluid pressure sensor, 431
Breakdown voltage, p-n junction, 295
Bridge circuit for the rectification of the 3-phase AC, 123
Broad-band Lambda sensors, 462
Bulk current injection method, EMC, 76
Bulk micromechanics, 334
Bulk silicon acceleration sensors, 419
Bus architectures, 329
Bus configuration, 491
Bus, ICs, 308

C

CAD systems, 204
Capacitive coupling, EMC, 66
Capacitive pick-off, pressure sensor, 425
Capacitors, 296
Capacity, starter batteries, 89
CCD principle, 477
CCD sensors, 478
Cell connectors, starter batteries, 94
Cell voltage, starter batteries, 87
Centrifugal pumps, 286
Ceramic multilayer substrates, 340
Characteristic curve of power input, 155
Characteristic curves, 154f

Index of technical terms

Characteristics (summary), 120
Characteristics of compound motor, 179
Characteristics of permanent-magnet DC motor, 176
Characteristics of permanent-magnet motor with flux concentrators, 178
Characteristics of series-wound motor, 177
Characteristics, maintenance-free starter batteries, 96
Charge-balance calculation, 118
Charge-coupled device (CCD), 315
Charge-indicator lamp, 126, 162
Charge-voltage setting, battery charger, 216
Charging curves, service technology, starter batteries, 215
Charging methods, service technology, starter batteries, 214
Charging, lead storage cell, 85
Charging, service technology, starter batteries, 213
Circuit diagram for passenger cars, 47
Circuit diagrams, 34
Circuit diagrams and symbols, 26
Classification, sensors, 351f
Claw-pole alternators with collector rings, 133
claw-pole chamfer, 152
Clearance lamps, 261, 263
"Clear" outer lenses, headlamps, 239
Climatic loading, starter batteries, 81
Clock-pulse generator (CPG), 308
CMOS transistors, 303
CO sensor, 457
Codes, lighting technology, 223
cold room, 206
Cold-discharge test current, starter batteries, 91
Cold-starting tests, 206
Color filters, lamps, 241
Combined lens sectors, headlamps, 239
Commercial-vehicle circuits, batteries, 106
Compact alternators, 136
Compact-diode-assembly alternators (LIT), 133
Compass sensors (earth's-field sensors), 371
Components, starter batteries, 107
Compound motor, 179
Computer simulation, 204
Computing power, Microcomputer, 486
Concentration sensors, 454ff

Conduction of electricity, 292
Conductivity, 294
Conductor cross-sectional areas, 20
Conductors (Metals), 292
Connection, 123
Consequential-damage protection device, 148
Consumer power demand, 14
Control units (ECUs), 288
Controls, 265f
Conventional batteries, 92
Conventional headlamps, 247
Conventional printed-circuit boards, 335ff
Conventional wiper blades, 278f
Coolant temperature sensor, 474
Cooling, 150
Cooling and noise, 150f
Cooling without fresh-air intake, 150
Copper losses, 153
Copper wires, dimensioning of wires, 20
Coriolis accelerations, 395
Cover lens, clear, 240
CPU, 320
CPU (Central Processing Unit), 321
Cumulative frequency, vehicle electrical systems, 13
Current characteristic curve (I), 155
Curve types, sensors, 352
Customer search number, KSN, starter batteries, 100

D

Data addressing, Microcontrollers, 325
Data processing in the vehicle, 486ff
Data transfer between automotive electronic systems, 490f
Daytime running lamps, 261
DC motors, 174
Decoding complexity, Microcontrollers, 323
Deep-cycle-resistant battery, 103
Defects in the battery, 212
Defects in the vehicle electrical system, 212
Defects, starter batteries, 212
Degree of integration, ICs, 304
"Delta" connection, 122
Design configurations, head lamps (North America), 246
Design configurations, headlamps (Europe), 245
Design criteria, 132
Design factors, 119
Designations for electrical devices, 42

Designs, battery, 100
Designs, lamps, 241
Detection of child's safety seat, 440
Development, 204
Development of starting systems, 164
Development and production of alternators and starter motors, 203ff
Development of lighting technology, 224ff
Development trends, sensors, 476
Device classification, circuit diagrams, 57
Diagnosis applications, CAN, 491
Diaphragm-type sensors, 425f
Diesel rail-pressure sensor, 431
Differential Hall-effect rod sensors, 400
Differential Hall-effect sensors, 364
Differential magnetoresistive sensors, 367f
Digital circuits, ICs, 306
Digital input signals, ECU, 486
Dimensioning of starting systems, 172
Dimensioning of wires, 20
Diode cooling, 151
Diodes, 298
Direct charging in the vehicle, starter batteries, 216
Direct pressure measurement, 424
Direct-contact sensors, 465ff
Direct-current drive, separately-excited, 111
Direct-drive, 179
Direct-drive starter motors for cars, 190
Dirt sensor, 283
Dirt sensors, 479
Discharge processes, overview, 87
Discharge, lead-acid battery, 86
Discrete semiconductor components, 297
Displacement or travel-measuring systems, sensors, 412ff
Display elements, 266f
Display types, 270ff
DMA (Direct Memory Access) I/O, Microcontrollers, 327
DMS1) pick-off, pressure sensors, 425
Doping, 293
Doping by diffusion, 331
Doping by ion implantation, 331
Doping during crystal growth, 330
Doping processes, 330ff
Doppler-effect radar systems, 396
Double-sided, interconnected circuit board, 335

Index of technical terms

Double-sided, non-interconnected circuit board, 335
Drive motors, wiper systems, 284ff
Drive/traction batteries, 110
Drives, electric, 111
Drives, electrically powered, 110
Drivetrain bus, 489
Drivetrains, electric vehicles, 111
DSTN LCD, 270
Dual micromechanical sensors, 421
Dual-headlamp systems, 245
dual-heat-sink system, 151
Dual-voltage electrical system, 17
Dynamic RAM (DRAM), 312
Dynamic system characteristics, vehicle electrical systems, 14

E
Eddy-current sensors, 360, 439
EEPROM, 313
Effects on mobile radio reception of high-frequency signal feedback in the electrical system, EMC, 69ff
Efficiency, 153
Electric drives, 111
Electric motors and other electromechanical components, EMC, 77
Electric washer-system pump, 286
Electrical conductivity, 292ff
Electrical connection, starter batteries, 100
Electrical copper wire for motor vehicles, 20f
Electrical data and sizes, 132
Electrical energy management, EMM, 18
Electrical loads, 116
Electrical loads, vehicle electrical system, 80
Electrical power generation using DC generators, 120
Electrical power generation using alternators, 118
Electrical power supply in conventional vehicle electrical systems, 12
Electrical power supply in conventional vehicles, 12
Electrical system design, 15
Electrical system simulation, 15
Electrical systems, 42-V, 17
Electrically powered vehicles, 110
Electrochemical processes in lead storage cell, starter batteries, 84
Electrodynamic principle, 121
Electrolyte, 87

Electrolyte specific gravity, starter batteries, 210
Electromagnetic compatibility (EMC), 64ff
Electromagnetic sensors (radar), 372
Electromagnetic voltage regulators, 142
Electronic battery charger, 214
Electronic components, 296ff
Electronic control unit (ECU), 486ff
Electronic diagnosis, 289
Electronic Service Information ESI[tronic], 209
Electronic systems, demands, 290
Electronic voltage regulators, 143
Electronically controlled wiper drive, 285
Electronics in Motor Vehicles, 289
Elements of digital circuits, ICs, 307
Elements, starter batteries, 93
EMC anechoic chamber, 73
EMC between the vehicle and its surroundings, 72ff
EMC between various vehicular systems, 65ff
EMC in the electronic control unit (ECU), 77
EMC ranges, 64
EMC standards for motor vehicles, 78f
EMC, standards, 79
Emergency identification lamps, 272
Emitted interference, EMC, 72f
endurance test, 205
Energy balance in the vehicle, 158
Energy supply, electric vehicles, 110
Engine control unit, 17
Engine fuel consumption, vehicle electrical systems, 14
engine idle speed, 126
Engine-oil temperature sensor, 474
Engine-temperature sensor, 474
Envelope-type separator, maintenance-free starter batteries, 95
Epitaxy, 331
EPROM, 313
Erasable ROM, 313
European type number, ETN, starter batteries, 99
Excitation circuit, 127
Excitation-current rectification, 126
exciter diodes, 124
Exhaust-gas temperature sensor, 474
External excitation, 122
External loading, starter batteries, 81

F
Field-effect transistors (FETs), 302
Filling battery electrolyte, 210
Film and hybrid circuits, 340f
Film circuits, 340
Finite-element calculation, 205
Fitting the wiper blade, 279
Flash EEPROM, 313
Flasher signals, 259
Floating operation, starter batteries, 215
Floodlamps, 272f
Flow measurement, 442
Flow meters, 442ff
Flow profiles, 443
FLRY/FLY wires, dimensioning of wires, 21
Fluorescent lamps and fluorescent foils, 267
Flywheel starter motor, 165
Fog lamps, 257
Fog warning lamps, 263f
Force sensor, 426
Force sensors, 433
Force sensors, 432ff
Forward bias, p-n junction, 295
Free-form technology, fog lamps, 257
Free-wheeling diode, 149
Freezing point, electrolyte, 211
fresh-air intake, 151
Front and side turn-signal lamps, 260f
Front fog lamps, 257f
Front lighting system, components, 242ff
Front parking lamps, 261
Front turn signals, lamps, 261
Front-end lighting, lighting technology, 222ff
Fuel-flow measurement, 442ff
Fuel-level sensor, 377
Fuel-temperature sensor, 474
Full-wave bridge rectifier, 124
Further sensor systems, 479
Future electrical systems, 16

G
Galvanic coupling, EMC, 65
Galvanomagnetic sensors, 363
Gas measurement, 454
Gas sensors, 454
Gas-analysis processes, gas sensors, 454
Gaseous-discharge lamps, 232f
Gasoline rail-pressure sensor, 431
Gates, ICs, 307
Gauge factor, 426

Index of technical terms

Gauge factors, 426
Gearbox-rpm sensors, 404
General power-supply battery, 17
Generation of electrical energy in the motor vehicle, 116
Generator circuit, 128
Giant magnetoresistive (GMR) elements, 394f
Glare, 225
Glossary and tables for automotive microelectronics, 342ff
Glove-compartment lighting, 265
GMR sensors, 371
Gradient sensors, 393f
Graphics displays for instrument clusters, 270

H

Half-differential short-circuiting-ring sensors, 376
Hall-effect acceleration sensors, 418
Hall-effect phase sensors, 400f
Hall-effect rod sensors, 400
Hall-effect sensor for detecting magnetic fields, 316
Hall-effect sensors for transistorized ignition, 407
Hall-effect steering-wheel-angle sensor, 380
Hall-effect switch, 363
Hall-effect vane switches, 392
Halogen bulbs, 231
Harvard architecture, busses, 329
Hazard-warning flashers, 259
Head up Display (HUD), 271
Headlamp adjustment, 252ff
Headlamp alignment equipment, 254
Headlamp cleaning systems, 282ff
Headlamp dirt sensor, 283
Headlamp elements, 236ff
Headlamp systems (Europe), 245
Headlamp systems (North America), 246
Headlamp test screen, 253
Headlamp testing, 254
Headlamp versions ff, 247
Headlamps with complex-surface reflectors, 248
Headlamps with stepless reflectors, 247
Headlamps with stepped reflectors, 247
Headlamps without focal lenses (clear outer lens), 247
Headlamps, PES, 248
Headlight levelling control, 255ff

Heated tube-type (finger) sensor LSH24, 459
Heavy-duty-battery, 104
High beam (North America), 245f
High beam, headlamps (Europe), 242
High beams: Illumination technology (Europe), 244
High beams: Installation (Europe), 244
High-pressure sensors, 431
High-pressure washer system, 282
High-speed CAN, 491
High-voltage ignition, EMC, 77
History of vehicle electrical systems, 10
HNS reflectors, 237
Home battery chargers, 214
Homofocal reflectors, headlamps, 247
Hot-film air-mass meter HFM2, 451
Hot-film air-mass meter HFM5, 452f
Hot-wire air-mass meter, 446
Hot-wire air-mass meter HLM, 450
Hot-wire/Hot-film anemometers, 445ff
Human eye, 229
Humidity sensor, 456f
Hybrid batteries, 94
Hybrid circuits, 340f
Hybrid regulators, 144

I

IC classification, 305
Illumination, 228
Illumination for controls and comfort and convenience equipment, 265
Illumination for display instruments, 265
Image sensors, 473, 477
Image sensors (video), 476
Image sensors based on CMOS technology, 478
Image-sensor structure, 477
Immunity to interference, 77
Impact-pressure airflow sensor, 448
Impact-pressure flow meter, 444f
Implementation location, Microcontrollers, 325
Incandescent (vacuum) bulb, 231
Incandescent bulbs, 267
Incident radiation via antenna, EMC, 77
Incident radiation, EMC, 73ff
Incorrect-polarity protection, battery charger, 217
Inductances, 297
Induction-type sensors for transistorized ignition, 406
Inductive coupling, EMC, 66
Inductive engine-speed sensors, 398

Inductive-type sensors, 389f
Input/output (I/O), 308
Installation, 159
Installation point, starter batteries, 81
Installation, substitute battery, 105
Instruction execution, Microcontrollers, 324
Instruction set, Microcontrollers, 323
Instruction-set architectures, Microcontrollers, 323
Instrument clusters, 268f
Instrument-panel illumination, 265
Insulators, 292
Integrated crankshaft starter alternator, 19
Interface ASIC, 483
Interference from electrostatic charges, EMC, 72
Interference pulses, 67
Interference receptors, 67
Interference sources, 67
Interference suppression, 78
Interference suppression, 64f, 77
Interference-signal spectrum, EMC, 68
Interference-suppression classes, EMC, 71
Interference-suppression measures, 157
Interior lamps, 265
Interior lighting, 265
Interior lighting system: Components, 265ff
Internal resistance, starter batteries, 88
Interrupt-driven I/O, Microcontrollers, 327
Intrinsic conductivity, 294
Inverters, ICs, 308
Iron losses, 153
Isolated peripherals (isolated I/O), Microcontrollers, 326

J

Janus principle, 397
Jointless wiper blade, 280

K

K factor, 435
Knock sensor, 423

L

Laboratory measurements, EMC, 74
Lambda oxygen sensor, 454, 458
Lamps combining fresnel and reflector optics, 240
Lamps with fresnel optics, 240
Lamps with reflector optics, 240

Index of technical terms

Laser diode, 315
Latency time,
   data communications, 489
Lead storage cell, 84
Lead-acid battery,
   drive/traction batteries, 113
Lead-antimony alloy,
   starter batteries, 93
Lead-calcium alloy, maintenance-free
   starter batteries, 94
LED, 267
Legal framework, lighting technology, 223f
Legal regulations, main headlamps
   (North America), 246
Legends, circuit diagram, 39
Length of instructions,
   Microcontrollers, 323
Lens materials, headlamps, 239
Lenses, headlamps, 238f
License-plate lamps, 264
Light generation, 232
Light sources, 230ff
Light sources, lighting technology, 267
Light-distribution pattern and range, 250
Light-distribution patterns, 227
Light-emitting diode, 314
Light-emitting-diode display, 266
Lighting equipment, 222
Lighting technology, 222-228
Lighting, instrument clusters, 269
Lights and lamps, 239ff
Linear temperature coefficient (TC), 467
Liquid cooling, 151
Liquid-cooled, windingless-rotor
   compact alternator (LIF), 140
Liquid-crystal display, 267
Lithium-based batteries, 115
Lithium-ion battery, 115
Lithium-polymer battery, 115
Litronic, 248ff
Litronic 2 System, 251
Litronic PES projection headlamps, 250
Litronic quad system, 251
Loads, electrical, 80
Logistical concept (CARTRONIC), 489
Longitudinal gauge factor, 426
Loss distribution in an alternator, 153
Low beam (North America), 245
Low beam, headlamps (Europe), 242
Low beams: Illumination technology
   (Europe), 244

Low beams: Installation (Europe), 244
Low-speed CAN, 491
Low/high beam:
   Switching (Europe), 244
Luminous flux, 233
Luminous intensity, 230
Luminous intensity and range, 242

M

Magnetically inductive sensors, 360ff
Magneto-elastic effect, 434
Magneto-elastic force sensors, 434
Magneto-elastic principle,
   force sensors, 433ff
Magnetoresistive NiFe
   thin-film sensors, 368ff
Magnetoresistive steering-wheel-angle
   sensor LWS, 381
Magnetoresistors, 300
Magnetostatic sensors, 362ff, 392
Main headlamps (North America), 245
Main headlamps (Europe), 242ff
Main operations,
   Microcontroller components, 321
Main requirements, sensors, 352
(Main) stop lamps, 263
Maintenance, battery, 210
Maintenance-free batteries, 94, 102
Manifold or boost-pressure sensor, 429f
Manifold-pressure or boost-pressure
   sensor, 428
Manual headlight-levelling control, 257
Manufacture of semiconductor
   components and circuits, 330ff
Mass flow rate, 443
Material, active, starter batteries, 93
Measuring incident interference,
   EMC, 70
Measuring instruments, 268
Measuring specific gravity, 211
mechanical losses, 153
Mechanical wiper mechanism, 276f
Memory capacity, Electronics, 291
Memory, Microcontroller components, 320
Memory-mapped peripherals
   (memory-mapped I/O),
   microcontrollers, 326
Memory-memory architecture,
   Microcontrollers, 322
Memory-register architecture,
   Microcontrollers, 322
Method of operation,
   starter batteries, 84

Micro-contacts,
   plug-in connections, 24
Microcomputer, 486
Microcomputers, 308ff
Microcontroller busses, 328f
Microcontroller components, 320f
Microcontroller memories, 325f
Microcontroller peripheral modules, 326f
Microcontrollers, 310
Microcontrollers, 318ff
Microelectronics, 288
Micromechanical acceleration sensors, 420
Micromechanical bulk silicon
   acceleration sensors, 419
Micromechanical hot-film air-mass
   meter, 447
Micromechanical pressure sensors, 428ff
Micromechanical sensors, 298f, 316
Micromechanical yaw-rate sensor
   MM1, 410
Micromechanical yaw-rate sensor
   MM2, 411
Micromechanical yaw-rate sensors, 410f
Micromechanics, 334
Microprocessor, 310
Mileages and maintenance intervals, 162
Miniaturisation, Electronics, 291
Miniaturization,
   micromechanical sensor, 357
Minimum starting temperature, 170
Minimum starting temperature,
   vehicle electrical systems, 12
Mixed-signal ICs, 307
Mixing battery electrolyte, 210
Mixing the electrolyte,
   starter batteries, 210
Mobile communications applications,
   CAN, 491
Moisture measurement, 454f
Monocrystalline silicon semiconductor
   resistors (PTC), 468f
Monolithic integrated circuits, 297
Monolithic integrated circuits (ICs), 304ff
Monolithic integration, 304
Monolithic regulators, 145
Motor-vehicle bulbs, specifications, 234f
Mounting, starter batteries, 101
Multi-Master principle, 491
Multifocal reflector, headlamps, 247

Index of technical terms

Multifunctional voltage regulators, 145
Multilayer circuit board, 335
Multiplate overrunning clutch, 183ff
Multiplex applications, CAN, 491
Multiplex bus, 489

N
n-type doping, 293
nA Cutting-in speed, 155
Narrow-band interference, 69
Near-range radar (24 GHz), 479
Needle-motion sensor, 405
Negative electrode,
  starter batteries, 85
Networks, 288f
Nickel-based batteries, 114
NMOS transistors, 303
Noise, 151
Nominal capacity, starter batteries, 90
Nominal voltage, starter batteries, 87
Non-conductors (Insulators), 292
Non-contacting temperature
  measurement, 472f
Normal charging, starter batteries, 214
Notes on operation, 161
NOX sensor, 457
NTC resistors (thermistors), 300
NTC temperature sensor, 474

O
Occupant classification (OC), 440f
Off-load voltage, starter batteries, 87
Ohmic resistors, 296
Oil and fuel-pressure sensor, 428
Oil-pressure sensors, 428
One-piece cover, starter batteries, 92
"Open-flank" belt, 160
Operand addresses,
  Microcontrollers, 323
Operand memories,
  Microcontrollers, 322f
Operating conditions,
  starter batteries, 80
Operating sequence of the starter
  motor, 166
Operating temperature,
  starter batteries, 210
Operation of alternators in parallel, 157
Optical element, headlamps, 239
Optical lens elements, 239
Optical sensors, 478f
Optical-imaging, 248
Opto-electronic components, 314f
Optoelectronic components, 298
Orthogonal printed resistors,
  force sensors, 436f

Oscillation gyroscope, 395f
Outer lens with light-dispersion optics, 238
Output signal, sensors, 352
Output signals, ECU, 488
Outside temperatures,
  starter batteries, 83
Overrunning clutch, 183
Overrunning clutches, 183
overvoltage, 146
Overvoltage in vehicle electrical
  system, 146
Overvoltage protection, 146
Overvoltage-protection devices
  (only for 28 V alternators), 147
Overvoltage-protection devices,
  non-automatic, 147
Overvoltage-protection devices,
  automatic, 147

P
p-n junction, 294f
p-n junction with external voltage, 295
p-n junction without external voltage, 294
p-type doping, 293
Paraboloid, fog lamps, 257
Parallel and series circuits,
  batteries, 106
Parallelogram wiper arm, 278
Parameters, starter batteries, 87
Passive (inductive) wheel-speed
  sensors, 402
Passive components, Electronics, 296f
Peripheral illumination,
  lighting technology, 225
Peripheral modules: Addressing, Microcontrollers, 326
Peripheral modules: Operating mode,
  Microcontrollers, 327
Peripherals, Microcontroller
  components, 320
Permanent-magnet motor, 175
Permanent-magnet motor with flux
  concentrators, 178
Permanently-excited synchronous
  drive, 113
PES fog lamps, 258
PES headlamps, 248
PES PLUS concept, headlamps, 248
Photodiode, 314
Photolithography, 332
Photoresistor, 314
Phototransistor, 315
Photovoltaic cell, 314

Physical principles,
  lighting technology, 229ff
Piezoelectric "oscillating drum"
  yaw-rate sensors, 409
Piezoelectric "tuning-fork"
  yaw-rate sensor, 408
Piezoelectric acceleration sensors, 422
Piezoelectric bimorphous plates, 417
Piezoelectric charge coefficient d, 416
Piezoelectric effect, 415
Piezoelectric knock sensors, 423
Piezomodule K, 416
Pinion engagement, 166
Pinion-engaging mechanism, 186
Planar broad-band Lambda oxygen
  sensors, 462f
Planar Lambda oxygen sensors, 460
Planar process, 332
Plate strap, 94
Plug-in connections, 25
PMOS, transistors, 303
Polled I/O, Microcontrollers, 327
Poly-V belt, 161
Position sensors (travel/angle), 358ff
Positive electrode, starter batteries, 87
Potentiometer-type sensors, 359
Power circuit, 201
Power demand, 82, 83
power diodes, 156
Power input, maintenance-free starter
  batteries, 97
Power losses, 153
Power requirements, 117
pre-control relay, 194
Pre-excitation circuit, 126
Pre-excitation on alternators with multi-
  functional voltage regulator, 127
Preconditions for starting, 168ff
Pressure sensors, 317, 424ff
Principle of operation of the alternator, 122
Printed-circuit board, 335
Prismatic lens sector, headlamps, 239
Production (starter motors), 206
Production process, plating, 336
Programmed I/O,
  Microcontrollers, 327
Programming model,
  Microcontrollers, 321f
Programming, Microcontrollers, 310
Protocol, data communications, 489
Pseudohall, sensors, 369
Pt resistor, 469
PTC resistors (thermistors), 300

Index of technical terms

PTC thin-film/thick-film metallic resistors, 467
Pulsation in the vehicle electrical system, EMC, 67f
Pulse-shaped input signals, ECU, 486
Pyrometers, 471
Pyrometric sensor, 472
Pyrometry, 472f

Q
Quad headlamp systems, 245
Quality management, 203
quality standards, 206
Quick-Clip, wipers, 278

R
Radar sensors, 396f
Radial-tooth overrunning clutch, 185
Radio-interference suppression, standards, 78
Rain sensor, 281, 283
Rain sensors, 478
Random-access memory (RAM), 311
Range, head lamps, 227
Ranging radar, 384f
Rapid-start battery charger, service technology, starter batteries, 214
Ratings for fuses, 22
Read-only memory, 313
Real-time applications, CAN, 491
Rear lighting system: Components, 262ff
Rear parking lamps, 263
Rear turn-signal lamps, 262
Rear-end lighting, lighting technology, 223
Rear-window cleaning, 281
Recharging, starter batteries, 211
Rectification of the AC voltage, 123
Rectifier circuits, 124
rectifier diode, 123, 299
Rectifier diodes, 125
rectifier losses, 153
Rectifiers for high reverse voltages, 299
reduction of alternator noise, 151
reduction-gear starter motors, 179
Reduction-gear starter motors for cars, 190
Reflection headlamps, Litronic, 250f
Reflection, headlamps, 236, 238
Reflector focal length, headlamps, 237
Reflector materials, headlamps, 237
Reflectors, headlamps, 236
Refraction, headlamps, 238
Regeneration of lead batteries, 213

Register-register architecture, Microcontrollers, 323
Regulation of excitation current, 129
Regulations, driving lamps, 259
Regulations, headlight levelling control, 255
Regulations, main headlamps (Europe), 244
Reinstallation, starter batteries, 212
Relative rpm and speed measurement, 389ff
Removal and installation, substitute battery, 105
Removal, battery, 105
Removal, starter batteries, 211
Replaceable Bulb Headlamp (RBH), 246
Representation, circuit diagram, 34
Representation, symbols, 28
Representations, schematic diagram, 36
Requirements to be met by automotive generators, 120
Residual, 122
Resistance to exhaustive discharge, maintenance-free starter batteries, 98
Resistance to incident radiation, EMC, 75
Resistance to overcharge, maintenance-free starter batteries, 98
Resistance, internal, 88
Resistive sensors, 465
Retrofit equipment, starter batteries, 83
Reverse bias, p-n junction, 295
Reverse-current block, 124
Reynolds number Re, 443
Ribbed-V belt, 160
Road-surface illumination, 225
Roller-type overrunning clutch, 183
Rotational-speed (rpm) sensors and incremental angle-of-rotation sensors, 399
Rotors, rpm sensors, 388
Rubber wiper element, 280f

S
Safety requirements, battery charging, 213
Safety, lighting technology, 229
Schematic diagram, 36, 60
Schmitt trigger, 363
Schottky diodes, 299
Screw-type terminals, starter batteries, 100

Sealed beam, head lamps (North America), 246
Seat profile of the human body, 441
Seat-occupation, 440
Section designations, circuit diagrams, 57
Section identification, schematic diagram, 39
Seebeck Effect, 469
Selected circuit symbols, 29
Self-discharge, maintenance-free starter batteries, 96
self-excitation, 122
Semiconductor barrier layers, 471
Semiconductor components, 330
Semiconductor components, 297ff
Semiconductor memories, 311ff
Semiconductor resistors, 300
Semiconductor technology, basic principles, 292ff
Semiconductors, 292ff
Sensitivity, human eye, 229
Sensor contact, plug-in connections, 25
Sensor examples, development trends, 476ff
Sensor integration stages, 481
Sensor plates, 444
Sensor, version with reference vacuum in special chamber, 429
Sensor, version with the reference vacuum on the component side, 428
Sensor-flap (impact-pressure) air-flow sensor LMM, 448f
Sensor-plate potentiometer, 374
Sensor-signal processing, 480ff
Sensors, 288, 350ff
Sensors for measuring angular difference, 438
Separately-excited direct-current drive, 111
Separators, starter batteries, 93
Serial data transfer (CAN), 490f
Series D78 direct-drive starter motor, 190
Series-wound direct-current drive, 111
Series-wound motor, 176
Service AWN, 208
Service technology, 208
Setpoint generators, 288
Short-circuiting-ring sensors, 360
Si image sensors, 477
Side turn signals, lamps, 261
Side-marker and clearance lamps, 261
Side-marker lamps, 261

Index of technical terms

Signal conditioning (Evaluation IC), 480
Signal conditioning, ECU, 488
Signal identification range, lighting technology, 229
Signal image, lighting technology, 228
Signal processing, ECU, 488
Signaling of changes in direction, lighting technology, 228
Simple Hall-effect rod sensors, 393
single-element double-contact regulator, 142
single-element, single-contact regulator, 142
Single-point sensors, 473
Single-sided circuit board, 335
Six-headlamp systems, 245
Smart Bags, 440
Solder-type terminals, 100
Solenoid switch, 181
Solenoid-plunger sensors, 362
Sources of power loss, 153
Special cases, starter batteries, 106
Special-purpose lamps, 272f
Specific gravity, starter batteries, 210
Spectrum, EMC, 69
Speed and rpm sensors, 386ff
Spinning-current principle, 363
Spot lamps, 273
Spreading resistance sensor, 469
Spreading-resistance principle, 468
Standard-range compact-diode-assembly alternators G1, K1, and N1, 134
"Star" connection, 122
Starin-gauge force sensors, 435
Start-assist using battery charger, 217
Start-assist with jumper cables, 217
Starter alternator, 19
Starter batteries, 80
Starter battery, vehicle electrical systems, 17
Starter motors, 164ff
Starter motors for commercial vehicles, 191
Starter motors with pre-engaged starter pinion engagement mechanism incorporating motor-assisted pinion rotation, 197
Starter motors with pre-engaged starter pinion engagement mechanism incorporating mechanical pinion rotation, 196
Starter-battery checks, 212
Starter-motor batteries, 202
Starter-motor control, 200

Starter-motor design, 174
Starter-motor design variations, 188
Starter-motor main circuit, 174
Starter-motor power cables, 201
Starter-motor type designations, 188
Starter-motor types, 188
Starting and overrunning, 168
Starting power, maintenance-free starter batteries, 97
Starting systems, 80
Starting temperature, starter batteries, 81
Starting the internal-combustion engine, 166
State of charge, starter batteries, 210
Static RAM (SRAM), 312
Steady-state voltage, starter batteries, 88
Steering-wheel-angle sensors, 380f
STN LCD, 270
Stop (brake) lamps, 263
Strain-gauge (DMS) principle (piezoresistive), force sensor, 435f
Strain-measuring sensors, 438
Stripline method, EMC, 75
Substitute batteries, 99, 104
Sulfation, starter batteries, 212
Supplementary high-mount stop lamps, 263
suppressor diode, 149
Surface micromechanical acceleration sensors, 420f
Surface micromechanics, 334
Surface-mounting method, 337
Susceptible devices, 67
Switch illumination, 266
Switches, 265f
Symbols, 26
Symbols, lighting technology, 223
Synchronous drive, permanently-excited, 113
System characteristics, vehicle electrical systems, 15
Systems for measuring mechanical strain and stress, sensors, 415ff

T

Tables for automotive microelectronics, 348
Tail lamps, 262
Tail lamps and clearance lamps, 262f
Tandem-pattern wiper system, 277
Tangential sensors, 394
Technical demands, lighting technology, 222

Technology of electrical starting systems, 200
Telecommunications bus, 489
TEM cell, 76
Temperature Coefficient, 467
Temperature coefficient (NTC), 471
Temperature coefficient (TK), 466
Temperature sensor, 430
Temperature sensors, 474
Temperature sensors , 464ff
Temperatures in the vehicle, 465
Term "electronics", 349
Terminal "W", 157
Terminal designations, 47
Terminal diagram, 40
Terminal posts, starter batteries, 94, 101
Terminal voltage, starter batteries, 88
Terms and definitions, sensors, 350
Test techniques, EMC, 70
The alternator's circuits, 126
Thermal image analysis, 473
Thermal radiators, lighting technology, 230ff
Thermocouple measuring, 470
Thermocouples, 469f
Thermoelectric voltage, 469
Thermopile array, 473
Thermopile sensor, 472
Thick-film circuits, 340
Thick-film pressure sensors, 427
Thick-film resistors (PTC/NTC), 468
Thin-film circuits, 340
third harmonic, 157
Throttle-valve sensor, 375
Through-fitting method, 337
Thyristors, 303
TN-LCD, 270
Topology, ECUs, 489
Torque sensors, 432ff
Torque sensorsff, 437
transistor regulator using hybrid technology, 144
Transistors, 301ff
Transmission speed, data communications, 489
Transport, batteries, 105
Transputer, 310
Transverse electromagnetic (TEM), 76
Transverse gauge factor, 426
Travel/angular position, 358
Troubleshooting, starter batteries, 213
Trunk lighting, 265
Tube-type (finger) sensors, 458
"Tuning-fork" piezo yaw-rate sensor, 408

Index of technical terms

Tunnel diode, 299
Turn signals, 259
Turn signals and flashers, 259f
Turn-signal and flasher systems for vehicles with/without trailers, 260
Turn-signal and flasher systems for vehicles without trailers, 260
Turn-signal lamps, 260
Turning the engine, 166
Two-battery systems, 16
Two-step Lambda oxygen sensors, 458f
Type B (LIC-B) Compact alternators, 137
type designation, 155
Type designation, starter batteries, 99
Type DT1 compact-diode-assembly alternators, 136
Type E and P (LI-E and LI-P) compact alternators, 137
Type EL hybrid regulator, 145
Type HEF109-M starter motor for commercial vehicles, 195
Type HEF95-L starter motor for commercial vehicles, 192
Type LIC compact alternators, 136
Type N3 compact-diode-assembly alternators, 139
Type RE86 and HE(F)95 starter motors for commercial vehicles, 192
Type T1 compact-diode-assembly alternators, 136
Type TB/TF pre-engaged starter motor, 199
Type U2 salient-pole collector-ring alternators, 141
Type X (LI-X) compact alternators, 138
Types of protection, 146

U
Ultrasonic sensors, 383
Unheated finger sensor LS21, 458f
Universal adapter, wipers, 278
Useful life, maintenance-free starter batteries, 98

V
V-belt, 160
Variable orifice plates, 444
Variable-capacitance diode, 299
Varistors, 300
Vehicle electrical system, 118
Vehicle electrical systems, 10
Vehicle Headlamp Aiming Device (VHAD), 246
Vehicle interior, lighting technology, 223

Vehicle measurements, EMC, 73
Vehicle operation, vehicle electrical systems, 13
Vehicle's lighting equipment, 222
Vehicular all-round vision, 476
Vibration sensors, 412ff
Vibration-proof battery, 103
Vision, lighting technology, 230
Vision, optical, 229
Visual acuity evaluating, 229
Visual perception, 229
Visual range, lighting technology, 229
Voltage at commencement of gassing, starter batteries, 89, 215
Voltage drop, 23
Voltage regulation, 129
Voltage-regulator characteristic, 130
Voltage-regulator versions, 142ff
Volume flow rate, 443
Von-Neumann architecture, busses, 329

W
Wafer bonding, 334
Washer systems, 276
Water consumption, maintenance-free starter batteries, 95
Water reservoirs, 286
Wave-propagation sensors, 372f
Wheel-speed sensors, 402f
Wide and narrow-band limits, 72
Wide-band interference, 69
Windingless rotor, 139
Windingless-rotor alternators without collector rings, 138
Windshield cleaning, 274ff
Windshield rain sensor, 283
Windshield wiper systems, 275
Wipe/wash system, 282
Wipe/wash systems, 276
Wiper and washer systems, 274ff
Wiper arm with four-bar-linkage elongation control, 278
Wiper arm with wiper-blade control, 278
Wiper arms, 277f
Wiper blade, jointless, 280
Wiper blades, conventional, 278
Wiper drive, electronically controlled, 285
Wiper motor, 284
Wiper rubber, 280
Wiper systems, 274f
Wiper-system linkage mechanism, 277
Wiper-type potentiometer, 359
Wire cross-sections, 21

Z
Zener diodes, 125, 299

Abbreviations

A
A/D: **A**nalog/**D**igital converter
ABS: **A**ntilock **B**raking **S**ystem
AC: **A**lternating **C**urrent
ACC: **A**daptive **C**ruise **C**ontrol
ADC: **A**nalog-**D**igital **C**onverter
EGR: **E**xhaust-**G**as **R**ecirculation
ALE: **A**ddress **L**atch **E**nable
ALU: **A**rithmetic **L**ogic **U**nit
AMLCD: **A**ctive-**M**atrix **L**iquid-**C**rystal **D**isplay
AMR: **A**nisotropic **M**agneto **R**esistive
AOS: **A**utomatic **O**ccupancy **S**ensing
ARD: **A**ctive surge dumping (German: **A**ktive **R**uckel**d**ämpfung)
ARS: **A**ngle of **R**otation **S**ensor
ASIC: **A**pplication **S**pecific **I**ntegrated **C**ircuit
ASSP: **A**pplication-**S**pecific **S**tandard **P**roduct
AST: **A**utomatic **S**hift **T**ransmission
AT: **A**utomatic **T**ransmissions
ATF: **A**utomatic **T**ransmission **F**luid
AWN: **A**sanetwork **W**orkshop **N**etwork
AZG: **A**daptive cylinder equalization (German: **A**daptive-**Z**ylinder**g**leich-stellung)

B
BCI: **B**ulk **C**urrent **I**njection
BIP: **B**egin of **I**njection **P**eriod
Bit: **B**inary dig**it**
BITE: **B**uilt-**I**n **T**est (self-test)
BDC: **B**ottom **D**ead **C**enter (piston/crankshaft)

C
CA: **C**amshaft
CAD: **C**omputer-**A**ided **D**esign
CAE: **C**omputer-**A**ided **e**ngineering
CAL: **C**omputer **A**ided **L**ighting
CAM (1): **C**omputer-**A**ided **M**anufacturing
CAM (2): **C**ontent **A**ddressable **M**emory, or Associative Memory
CAN: **C**ontroller **A**rea **N**etwork
CCD: **C**harge-**C**oupled **D**evice
CCFL: **C**old-**C**athode **F**luorescent **L**amp
CHMSL: **C**enter **H**igh-**M**ounted **S**top **L**amp
CISC: **C**omplex **I**nstruction-**S**et **C**omputer
CMOS: **C**omplementary **M**etal-**O**xide **S**emiconductor transistor
CPG: **C**lock-**p**ulse **g**enerator
CPU: **C**entral **P**rocessing **U**nit
CR: **C**ommon **R**ail
CSWS: **C**omposite **S**eal **W**ith **S**ensor
CUR: **C**onstruction and **U**se **R**egulations
CVT: **C**ontinuously **V**ariable **T**ransmissions

D
DAC: **D**igital-**A**nalog **C**onverter
DC: **D**irect **C**urrent
DI: **D**irect **I**njection
DMA: **D**irect-**M**emory **A**ccess
DMS: Strain gauge/Strain-gauge measuring resistor (German: **D**ehn**m**ess**s**treifen)
DRAM: **D**ynamic **RAM**
DRO: **D**ielectric **R**esonance **O**scillator
DRS-MM: Yaw-rate sensor, microme-chanical (German: **D**reh**r**ate**s**ensor, **m**ikro**m**echanisch)
DSTN-LCD: **D**ouble **S**uper **T**wisted **N**ematic **L**iquid-**C**rystal **D**isplay
DWS: Angle of rotation sensor (German: **D**reh**w**inkel**s**ensor)
DZG: Speed sensor (German: **D**reh**z**ahl**g**eber)

E
EAB: Solenoid-operated shutoff valve (German: **E**lektrisches **Ab**stellventil)
EC: **E**uropean **C**ommunity
ECE: **E**conomic **C**ommission for **E**urope
ECU: **E**lectronic **C**ontrol **U**nit
EDC: **E**lectronic **D**iesel **C**ontrol
EEC: **E**uropean **E**conomic **C**ommission
EEM: **E**lectrical **E**nergy **M**anagement
EEPROM: **E**lectrically **E**rasable **P**rogrammable **R**ead-**O**nly **M**emory
ELAB: Solenoid-operated shutoff valve (German: **E**lektrisches **Ab**stellventil)
EMC: **E**lectromagnetic **c**ompatibility
EMV: **E**lectromagnetic compatibility (EMC)
EOBD: **E**uropean **O**n-**B**oard-**D**iagnosis

Index of technical terms Abbreviations

EOL: End-of-Line programming
EPROM: Erasable Programmable Read-Only Memory
ESD: Electrostatic discharge
ESI[tronic]: Electronic Service Information
ESP: Electronic Stability Program
ETN: European Type Number
EU: European Union

F
FEM: Finite Elements Method
FET: Field-effect transistor
FGB: Vehicle-speed limitation (German: Fahrgeschwindigkeitsbegrenzung)
FGR: Vehicle-speed controler (Cruise control) (German: Fahrgeschwindigkeitsregelung)
FIFO: First in, first out
FIR-F: Finite Impulse Response filter
FLL: Frequency-Locked Loop
FMCW: Frequency Modulated Continuous Wave
FMEA: Failure Modes and Effects Analysis
FMVSS: Federal Motor Vehicle Safety Standard
FSR: Force-Sensitive Resistance

G
GMR: Giant Magneto Resistive
GP: Rotated Parabel (German: Gedrehte Parabel)
GSK: Glow plug (German: Glühstiftkerze)
GZS: Glow plug control unit (German: Glühzeitsteuergerät)

H
HD: Heavy Duty
HDK: Sensor with semidifferential short-circuiting ring (German: Halb-Differenzial-Kurzschlussringsensor)
HDL: Hardware Description Language
HFM: Hot-film air-mass meter
HGB: Maximum-speed limiter (German: Höchstgeschwindigkeitsbegrenzung)
HLM: Hot-wire air-mass meter (German: Hitzdraht-Luftmassenmesser)
HNS: Homogeneous Numerically calculated Surface
HUD: Head up Display

I
IC: Integrated Circuit
ICT: In-circuit test
IDI: Indirect Injection
ISO: International Organization for Standardization
I/O: Input/output
IWZ: Incremental angle-time signal (German: Inkrementales-Winkel-Zeit-Signal)

J
JFET: Junction-gate Field-Effect Transistor

K
KS: Knock Sensor
KSN number: Customer search number (German: Kundensuchnummer)

L
LCD: Liquid Crystal Display
LED: Light-Emitting Diode
LIFO: Last in, first out
LKS: Position Check System (German: Lagekontrollsystem)
LMM: Air-flow sensor (German: Luftmengenmesser)
LS: Lambda Oxygen Sensor
LSI: Large-scale integration
LSU: Broadband oxygen sensor Type LSU
LTCC: Low-temperature cofired ceramic
LWS: Steering-wheel-angle Sensor (German: Lenkwinkelsensor)

M
MAB: Fuel cutout (German: Mengenabschaltung)
MAR: Control of injected-fuel-quantity compensation (German: Mengenausgleichsregelung)
MC: Microcomputer
MEMS: Micro Electro Mechanical System
MI: Main Injection
MIL: Malfuction Indicator Lamp (diagnosis lamp)
MM: Micromechanics
MMA: Fuel-quantity mean-value adaptation (German: Mengenmittelwertadaption)
MOS: Metal-oxide semiconductor
MSG: Engine ECU (German: Motorsteuergerät)

Automotive Handbook

The Automotive Handbook is a comprehensive 960-page reference work. The 5th edition has again been completely revised and updated by experts from Bosch and from the automotive industry.

The state-of-the-art in automotive technology is presented comprehensibly, in a well laid out and concise form. A wide variety of system schematics, illustrations, and tables provide an insight into a fascinating technological world.

Added subjects:
- Fuel cells
- Cartronic
- EHB for passenger cars
- Natural gas as engine fuel
- Adaptive cruise control (ACC)
- Display instruments
- Fuel filters
- MED Motronic
- Plug-in connections

Completely revised subjects:
- Engine management for spark-ignition engines
- Engine management for diesel engines
- Emissions-control engineering
- Safety systems
- Lighting technology
- Electronics
    - Materials science
    - Vibration and oscillation
    - and many more subjects.

The Bosch "Blue Book" has in the meantime advanced to become a bestseller in the area of technical literature.

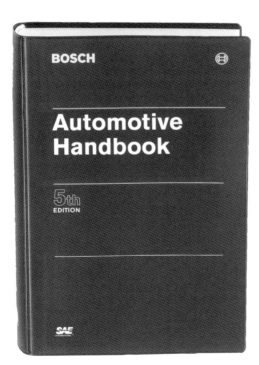

Hardcover,
12 x 18 cm format,
5th Edition, completely revised and extended,
962 pages,
softcover,
with numerous illustrations.

ISBN
0-7680-0669-4